S. Hrg. 114–31

"THE STATE OF TECHNOLOGICAL INNOVATION RELATED TO THE ELECTRIC GRID"

HEARING

BEFORE THE

COMMITTEE ON ENERGY AND NATURAL RESOURCES UNITED STATES SENATE

ONE HUNDRED FOURTEENTH CONGRESS

FIRST SESSION

ON

"THE STATE OF TECHNOLOGICAL INNOVATION
RELATED TO THE ELECTRIC GRID"

MARCH 17, 2015

Printed for the use of the
Committee on Energy and Natural Resources

"THE STATE OF TECHNOLOGICAL INNOVATION RELATED TO THE ELECTRIC GRID"

S. Hrg. 114–31

"THE STATE OF TECHNOLOGICAL INNOVATION RELATED TO THE ELECTRIC GRID"

HEARING

BEFORE THE

COMMITTEE ON
ENERGY AND NATURAL RESOURCES
UNITED STATES SENATE

ONE HUNDRED FOURTEENTH CONGRESS

FIRST SESSION

ON

"THE STATE OF TECHNOLOGICAL INNOVATION
RELATED TO THE ELECTRIC GRID"

MARCH 17, 2015

Printed for the use of the
Committee on Energy and Natural Resources

U.S. GOVERNMENT PUBLISHING OFFICE

94–050   WASHINGTON : 2015

For sale by the Superintendent of Documents, U.S. Government Publishing Office
Internet: bookstore.gpo.gov   Phone: toll free (866) 512–1800; DC area (202) 512–1800
Fax: (202) 512–2104   Mail: Stop IDCC, Washington, DC 20402–0001

CONTENTS

OPENING STATEMENTS

WITNESSES

ALPHABETICAL LISTING AND APPENDIX MATERIAL SUBMITTED

"THE STATE OF TECHNOLOGICAL INNOVATION RELATED TO THE ELECTRIC GRID"

TUESDAY, MARCH 17, 2015

U.S. SENATE,
COMMITTEE ON ENERGY AND NATURAL RESOURCES,
Washington, DC.

The Committee met, pursuant to notice, at 10:07 a.m. in room SD–366, Dirksen Senate Office Building, Hon. Lisa Murkowski, Chairman of the Committee, presiding.

OPENING STATEMENT OF HON. LISA MURKOWSKI, U.S. SENATOR FROM ALASKA

The CHAIRMAN. Good morning.

We're calling to order the Energy Committee. We are here for a Grid 2.0 hearing, and it should excite us all.

This is a good subject. It is an opportunity for us this morning to hear testimony from a panel of experts to evaluate what's actually happening with technological innovation for the electric grid. No other electricity network on Earth provides as much power to as many people as reliably and affordably as our American grid. It really is a modern marvel, and I think we should be proud of what we have accomplished.

Keeping the lights on, as all segments of the industry have for the past 100 years, is not just about flicking the switch or changing a bulb. It is a highly complex and technical undertaking that requires scores of talented individuals.

We have also reached a very good point in time to be looking at this subject this morning. The grid was built incrementally, but the rate at which it changes, or is compelled to change, appears to be accelerating. A combination of market forces, technological innovation, and policy directives at both the federal and the state levels could well result in an unprecedented transformation of the electricity sector.

Today's developments have tremendous potential, but they also present a number of challenges, like smoothing out the intermittency of variable, weather dependent generation. With the rise of distributed generation and smart grid technologies, Americans are gaining more control over how they use and consume electricity, but the grid must be even more closely integrated as a result.

Innovation and new technologies, such as commercially viable storage, are clearly necessary to assist in this transformation. We are talking a lot in my state about microgrids. Many people don't think of Alaska as being a pioneer in some of these areas, but what

(1)

we are doing with microgrids—which used to be called ''isolated islanded grids''—is really making a difference in a state where sometimes it is tough to keep warm or keep the lights on.

It is always best when public policy responds to proven technology advancements. Today, however, we must hope technology can deliver solutions to meet political mandates while satisfying our expectations for reliability and affordability. Two major questions that I have, and I will be proposing these to the panelists this morning, are over what time period can solutions be developed, and at what cost? Credible estimates hold that new grid technologies alone will require a cumulative investment of between $300 and $500 billion over the next 20 years.

This Administration has rightly classified the electric grid as uniquely critical infrastructure because so many things—communications, roadways, hospitals—depend upon a functioning grid. The reliability of our nation's grid is therefore paramount, and the impact of policy directives must be seriously considered—not dismissed as somehow anti-environment or anti-future.

I focused our Committee on drafting broad energy legislation and expect that electric issues will be a key part of it. We have an impressive group of panelists with us this morning whose testimony will assist us in this effort.

We have Dr. Michael Howard, who is President and CEO of the Electric Power Research Institute. He is going to start us off with an overview of the changing integrated grid. He had hoped to use a computer for his presentation, but we are still using 19th century technology here in the Senate, so we are stuck with posters, but thank you for that.

We have also Dr. Peter Littlewood, who is the Director of Argonne National Laboratory. He is joining us because research is really the foundation of the federal role, and his lab is a key part of our efforts on grid modernization and energy storage.

We also have Dr. Jeff Taft, who is the Chief Architect for Electric Grid Transformations at the Pacific Northwest National Lab. He's here to tell us more about PNNL's work with smart grid technologies and grid modernization.

We have Lisa Barton, who is the Executive Vice President for Transmission for American Electric Power. She is here to share the perspective of a major utility that operates the nation's largest electricity transmission system with 40,000 miles of transmission lines to deliver electricity in 11 states.

To round out the panel this morning we have The Honorable Lisa Edgar, a Commissioner at the Florida Public Service Commission and President of NARUC, the National Association of Regulatory Utility Commissioners. She will present the state's perspective which, of course, is invaluable because so much of this transition is happening at the retail level.

So I welcome all of our panelists, and I look forward to your presentations.

We do have a couple of votes coming up a little after 11 o'clock this morning, so we may be a little bit disjointed here, but we certainly intend to spend the full time this morning getting the most that we can from our panelists.

With that, I would like to turn to my Ranking Member, Senator Cantwell.

STATEMENT OF HON. MARIA CANTWELL, U.S. SENATOR FROM WASHINGTON

Senator CANTWELL. Thank you, Madam Chairman and thank you for holding this important hearing today. I also join you in thanking the witnesses for being here to discuss these issues.

As our economy has evolved and information technology has evolved, we have seen it has disrupted many industries and business models, everything from the telecom industry to the music industry. So I think today we'll have a little bit of discussion about what that disruption is also going to mean for the electricity industry and how the advanced electrical grid is part of efficiency that will drive savings to consumers and businesses.

The grid of tomorrow should offer new opportunities for consumers, savings on their electricity bills, and lower costs for businesses through new technology. These aren't obscure academic or regulatory debates. These things are hitting the streets today.

States as diverse as Idaho, Georgia, New Jersey, and California have sped ahead with distributed generation, smart meters, and net metering. Some places around the world, like South Africa and Uganda, have skipped the capital-intensive steps of developing centralized grids and just used pre-pay options so consumers can benefit from cheap electricity using American technology. Cities in the United States like Spokane, Monterey, Salt Lake, and McAllen, Texas, have installed or are considering electric bus designs that include wireless charging stations embedded in roadways.

My point is that we can't predict where the technology will take us, but we can invest in efficient electricity grids that make these innovations possible and give consumers more options.

Our hearing today is about the savings, no matter what the source of generation, and putting that to use in a smarter way. It doesn't matter if your state relies on hydro power, like my State of Washington, or nuclear or fossil fuels. The cost of wind and solar has come down, and the cost of natural gas has come down along with the integration of smart appliances. The grid is being used in new ways to drive double digit savings.

The challenges and opportunities we face in upgrading our electricity grid and thinking about the global markets is what the debate of this panel is today.

According to a 2011 Electric Power Research Institute report, investments in the grid will require $300 to $500 billion of new investment over the next 20 years.

Bloomberg New Energy Finance calculated that global smart grid investments alone reached $15 billion in 2013, and Pike Research estimates that global spending on this will reach $34 billion by 2020.

I say those numbers because there's an opportunity here for the United States to continue to perfect technology that will then become a global platform.

The first job of utilities, power producers, and technology vendors in each of our states is to sell and deliver reliable electricity. The federal government is uniquely situated to take the broadest and longest view of that electrical grid as a platform for economic development and diversification. This broad view enables smart people like the national labs, that are here today, and a program like ARPA–E, to explore solutions that are creative and promising, but offer no short term return and will challenge us on how to implement these over the long run.

The grid's efficiency, enhancing its resiliency, security, and new technologies are all part of the decisions that we're going to hear from actual regulators today and how they're implementing those. Obviously some of these solutions are already being pushed in the marketplace and can deliver new efficiencies.

As Chairman Murkowski and I discuss with our colleagues our broader energy policy for this Congress, I hope we can find some common ground on continuing the federal investment in grid technologies. It does pay off for consumers and our economy.

The Bonneville Power Administration helped lead the way 15 years ago towards a responsive grid by installing the first network of sensors to take wide-area measurements of transmission systems.

We'll hear from Dr. Taft today from the Pacific Northwest National Lab, headquartered in Richland, Washington. The lab has a long history of working hand-in-hand with industry on pioneering new methods of controlling our rapidly changing electrical grid, with all sorts of new energy storage, new tools for predicting, and integrating the output of variable generation such as wind and solar. It was also an integral part of the largest smart grid demonstration project in the country.

Grid technology companies like Itron, Schweitzer, and Alstom all have roots in Washington State and employ thousands of people. And as our economy grows, it continues to find new sources of distributed generation.

So I want to applaud Secretary Moniz and the Department of Energy for convening the Grid Modernization Laboratory Consortium. This will help spread the wealth of the creativity of our national labs through our state, private, and academic partners.

As I've said, I will continue to work with the Chair as we think about energy policy on how to support the investments in a smarter grid. Thank you.

The CHAIRMAN. Thank you, Senator Cantwell. With that, we will begin with our panel. We'll start with you, Dr. Howard and just go straight on down the line.

I ask that you to keep your comments to five minutes. Your full testimony will be included as part of the record.

Welcome and good morning.

STATEMENT OF DR. MICHAEL HOWARD, PRESIDENT AND CEO, ELECTRIC POWER RESEARCH INSTITUTE

Dr. HOWARD. Thank you very much, Chairman Murkowski, Ranking Member Cantwell, and members of the Committee. I'm Mike Howard, President and CEO of the Electric Power Research Institute.

This is EPRI's 43rd year. We were started by the electric utility industry to focus on advancing the science and technologies needed to ensure that society continues to have reliable, affordable, and environmentally responsible electricity. That's been our focus for 43 years.

We are a global organization. We work with utilities across the country and in 30 different countries globally, so we have a global perspective of some of the challenges that we're facing.

Our research at EPRI focuses on the generation of electricity to the delivery of electricity to the end use of electricity including energy efficiency. Last year we published a report. The report is entitled, ''The Integrated Grid, Realizing the Full Value of Central and Distributed Energy Resources,'' and my remarks today are going to focus on that report and some of the key insights from that report.

I'm going to use this chart, and you should have a handout as well. It's a simplified version of the electric utility industry, but I'm going to refer to that.

Today's power system is extremely complex. It's an interconnected machine that includes everything from generation on the left-hand side, to delivery in the middle, to consumers on the right-hand side. Traditionally, the system was designed to flow electricity from the left-hand side all the way to the right-hand side.

The system has to make sure that at every given second there is enough generation to supply consumer's demand for electricity, and that is a paramount criteria for the safe operation of the electrical power system. But the entire power system is changing, and it's changing at a very, very fast pace. Faster than I've seen in my 35 year career in this industry, and that's an exciting time to observe and be part of the technologies that are evolving.

As an example of these changes, let's look at the customer side generation, which is on the right-hand side. It's causing power, in some cases, to flow from the right to the left, and this is introducing a two way power flow. It's just one example of some of the changes that are occurring.

Most of these changes occurring in the power system are occurring on the right-hand side of this chart, which is at the edge of the distribution system, and they're related to technologies referred to as distributed energy resources. These important technologies include energy storage, demand response, and smart thermostats, which is another example of some of the technologies that are occurring.

Our research suggests that the successful integration of these distributed energy resources must begin with the existing power system. However, the existing power system, specifically the distribution system, was not designed to accommodate this high penetration of distributed energy resources. To fully realize the value of these distributed energy resources and to serve all customers at the established standards of quality and reliability, we must integrate these technologies into the planning and operations of the entire power system.

Most grid connected distributed energy resources benefit from the electrical support and the flexibility and reliability of the entire power system but are not integrated into the power system. For example, customers with distributed generation may not consume any net energy from the grid, yet they need the power at times when their own generation, such as rooftop PV, does not provide enough immediate electricity. So consumers need the ability or capacity to tap into the grid and even use grid power, though at different times of the day they may return actual energy to the grid. So with increased distributed energy resources capacity related costs will become an increased portion of the overall cost of electricity.

One of the key points that I want to emphasize today is the importance of understanding capacity and energy and the impact that it has on the system.

In my written testimony I've outlined in more detail the important items policy makers should consider to enable the integrated grid. I appreciate the opportunity to be here this morning, and I look forward to your questions. Thank you.

[The prepared statement of Dr. Howard follows:]

Written Testimony

Hearing of the Energy and Natural Resources Committee

United States Senate

Dr. Michael Howard
President and Chief Executive Officer
Electric Power Research Institute

March 17, 2015

EPRI is an independent, non-profit research organization with close to $400 million in annual research funding principally from electric utility companies in more than 30 countries. EPRI was started 43 years ago with a mission to advance safe, reliable, affordable and environmentally responsible electricity for society through global collaboration, thought leadership and science and technology innovation. This remains the EPRI mission today.

EPRI brings together electric utility companies, scientists and engineers, along with experts from academia, industry and other centers of research to:

- Collaborate in solving challenges in electricity generation, delivery and use;
- Provide technical, scientific, and economic analyses to drive long-range research and development planning;
- Support multi-disciplinary, objective research in emerging technologies; and
- Help accelerate the commercial deployment of advanced electricity technologies for the benefit of the public.

Early last year EPRI issued an in-depth technical look at the changing power system. This report, *The Integrated Grid: Realizing the Full Value of Central and Distributed Energy Resources*, was the first phase in a larger EPRI project to chart the transformation of the power system. Remarks today will focus on the Integrated Grid concept.

The power system is a complex machine that includes everything from how consumers use and interact with electricity to how electricity is delivered over a vast network of distribution and transmission wires and ultimately the generation of electricity. The end-to-end power system is an *interconnected* machine that is critically important to the economic well-being of this country. An interconnected machine or *power system* means that all devices are electrically connected together. With the tremendous advancements that are occurring in how electricity is generated, delivered and now personally managed, changes in regulations and management of the power system will also change.

Please take a look at the diagram below. This is a simplified drawing but a useful, high-level view of the power system.

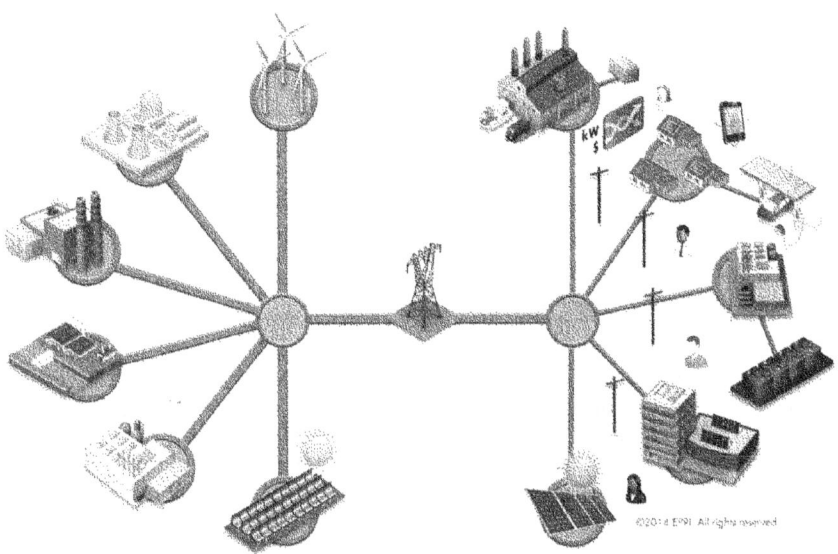

Many of the innovations are occurring at the edge of the distribution system, the far right-hand side, where customers connect to the power system. Because the power system is interconnected, changes that occur at the edge can impact the entire power system, including the rest of the distribution system, the transmission system and central generation on the far left-hand side.

The power system was originally designed to connect large generation plants with customers ranging in size from small residential to major industrial manufacturers. The U.S. power system is anchored on the left-hand side by approximately 1,000 gigawatts (GW) of central generation and on the other end by customers who expect reliable and affordable electricity.

Electricity flows in one direction from power plants on the left to substations and then to customers on the right. The amount of generation is matched with the customer's needs on a second by second basis. This has been the cornerstone of reliable grid operation for more than 100 years.

However, the entire power system is changing at a fast pace, driven by technology and customer expectations. For example, customer-sited generation is causing power to flow in some cases from the load, which is the right-hand side, to the substation further to the left-hand side, thereby introducing two-way power flows. Technologies like smart thermostats are resulting in unique ways customers can manage their energy. Variable generation that depends on wind and sun cannot be dispatched to meet customer demands, but is available when the resources are available.

Most of the changes occurring at the edge of the distribution system are related to a class of technologies called Distributed Energy Resources or DER. Customers, energy suppliers, and developers are increasingly adopting these DER technologies with the aim to supplement or supplant grid-provided electricity. These important resources include distributed generation resources as well as storage, demand response resources and a range of technologies that allow remote control of electricity use.

EPRI research suggests that **if** the various components of the power system are also *integrated* together, the entire power system can realize the full benefit of all the pieces including the significant innovations that are occurring with Distributed Energy Resources and central generation. This is outlined in much more detail in the EPRI study referenced earlier, *The Integrated Grid: Realizing the Full Value of Central and Distributed Energy Resources,* and provided to this Committee as part of written testimony.

EPRI research further suggests that the successful integration of DER begins with the existing power system. The existing power system, especially the distribution system, was not designed to accommodate a high penetration of DER while sustaining high levels of electric quality and reliability. The technical characteristics of certain types of DER, such as variability and intermittency, are quite different from central power stations.

To fully realize the value of distributed resources and to serve all consumers at established standards of quality and reliability, DER must be integrated into the planning and operation of the power system. Again, this is what is referred to as the *Integrated Grid*.

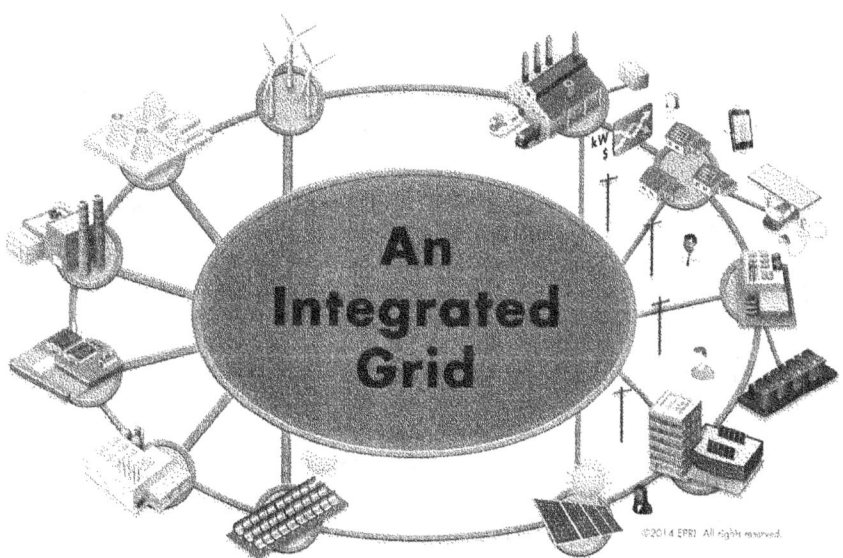

An Integrated Grid should not favor any particular energy technology, power system configuration or power market structure. Instead, it should make it possible for stakeholders to identify optimal architectures and the most promising configurations – recognizing that the best solutions vary with local or regional circumstances, goals, and interconnections.

Most grid-connected DER sources benefit from the electrical support, flexibility, and reliability of the grid but they are not integrated into the grid's operation. Consequently, the full value of DER is not realized in providing support for grid reliability, voltage, frequency, and reactive power - all essential for an integrated grid system.

Customers with distributed generation may not consume any net energy (KWH) from the grid, yet they need the grid at times when their generation does not provide them enough immediate electricity. Consumers need the ability – or **capacity** – to tap into the grid and even use grid power, though at different times of day they might return actual **energy** into the grid.

A consumer's intermittent need for electricity and the constant ability, or capacity, to deliver it by a utility has financial implications. For residential customers, the costs for generation, and transmission and distribution (T&D) components can be broken down into two parts:

1. Costs of the actual **energy** used by the customer, and
2. Costs to provide the **capacity** to deliver the energy and grid-related services.

Based on the U.S. Department of Energy's *Annual Energy Outlook 2012*, an average customer consumes 982 kWh per month, paying an average bill of $110 per month of which approximately $59 per month is the cost of actual energy and $51 per month is the cost to provide generation capacity and other services such as load following, voltage support (known as ancillary services) and transmission and distribution capacity.

Bottom line: A consumer with DER is not necessarily a self-sufficient energy consumer.

With the growing penetration of variable generation, capacity-and ancillary service-related costs will become an increasing portion of the overall cost of electricity. However, with an Integrated Grid, DER could more efficiently contribute to the capacity and ancillary services needed to operate the grid.

Policy and regulatory frameworks are needed to encourage the effective and efficient introduction of new technologies, and also provide equitable allocation and recovery of costs incurred to transform to an Integrated Grid. New market frameworks will have to evolve to assess potential contributions of distributed and central resources, allocate costs, and quickly integrate new interconnection and communication technologies to system capacity and energy costs.

EPRI views the following four items as important considerations for policymakers to enable an Integrated Grid:

1. Interconnection rules and communication technologies and standards;
2. Assessment and deployment of advanced distribution and reliability technologies;
3. Strategies for integrating DER with grid planning and operation; and
4. Enabling policy and regulation.

In addition, EPRI research results suggest that the policy and technology transformation of the power system has three important and concurrent paths.

First, while moving toward an Integrated Grid, ensure that the current power system assets continue to operate safely and with always improving performance. The need to Perform while changing is essential.

Second, the existing and emerging technologies must <u>Adapt</u> to a future state to make the most of current power system investments. This will include enabling central generation to perform flexibly as well as variable generation to contribute toward system capacity and ancillary services.

Third, while ensuring that the electricity system Performs well and Adapts, new technologies <u>Create</u> new ways to deliver safe, affordable, reliable and environmentally responsible electricity.

At EPRI, there is a pride in our objective and collaborative electric research, done across companies, service territories and the technology applications. The industry has an important job ahead to ensure the public has a supply of electricity that is clean, affordable, safe and secure. The utility industry, together with other stakeholders, is up to that challenge.

EPRI looks forward to offering continued technical support to the electricity sector, public policy-makers and other stakeholders to achieve the vision of an Integrated Grid.

The Integrated Grid

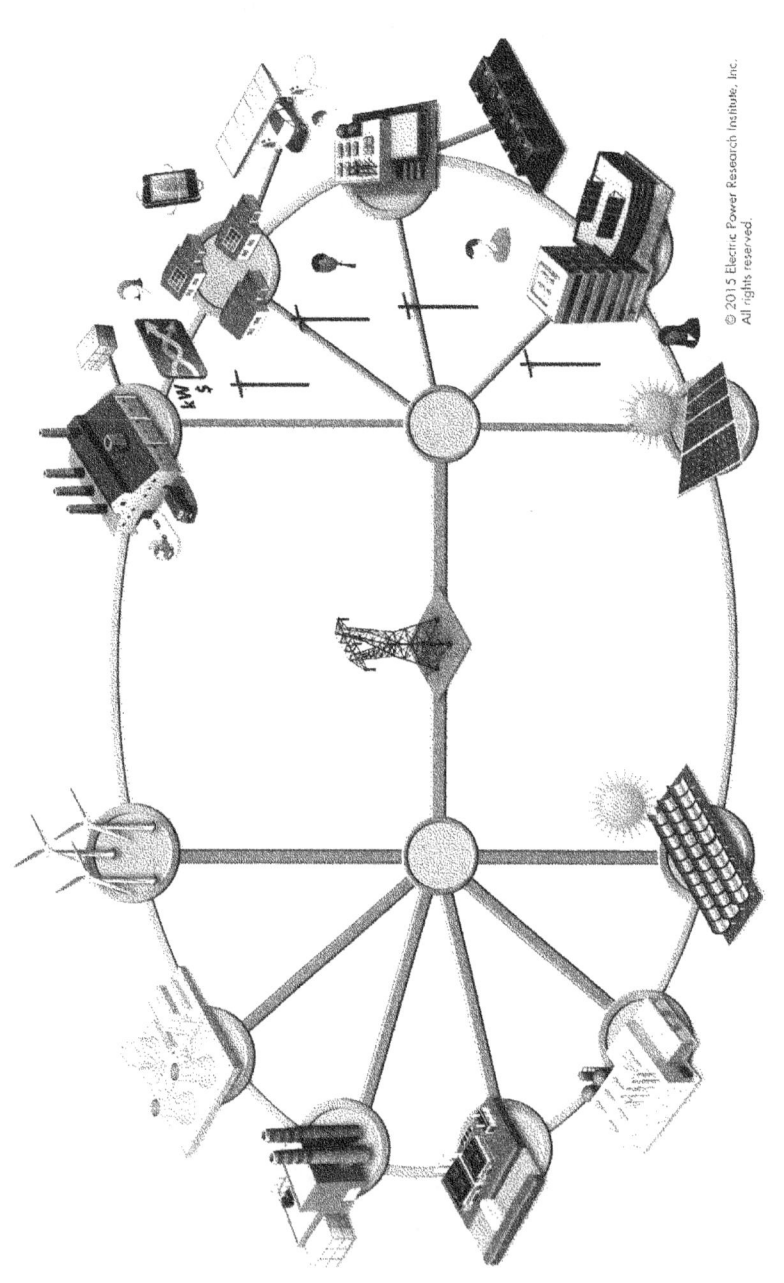

EPRI | ELECTRIC POWER RESEARCH INSTITUTE

13

EPRI Mission

Advancing safe, reliable, affordable and environmentally responsible electricity for society through global collaboration, thought leadership and science & technology innovation.

EPRI | ELECTRIC POWER
RESEARCH INSTITUTE

THE
INTEGRATED GRID

REALIZING THE FULL VALUE OF CENTRAL
AND DISTRIBUTED ENERGY RESOURCES

Table of Contents

© 2014 Electric Power Research Institute (EPRI), Inc.
The Integrated Grid

16

Executive Summary

The electric power system has evolved through large, central power plants interconnected via grids of transmission lines and distribution networks that feed power to customers. The system is beginning to change – rapidly in some areas – with the rise of distributed energy resources (DER) such as small natural gas-fueled generators, combined heat and power plants, electricity storage, and solar photovoltaics (PV) on rooftops and in larger arrays connected to the distribution system. In many settings DER already have an impact on the operation of the electric power grid. Through a combination of technological improvements, policy incentives, and consumer choices in technology and service, the role of DER is likely to become more important in the future.

The successful integration of DER depends on the existing electric power grid. That grid, especially its distribution systems, was not designed to accommodate a high penetration of DER while sustaining high levels of electric quality and reliability. The technical characteristics of certain types of DER, such as variability and intermittency, are quite different from central power stations. To realize fully the value of distributed resources and to serve all consumers at established standards of quality and reliability, the need has arisen to integrate DER in the planning and operation of the electricity grid and to expand its scope to include DER operation – what EPRI is calling *the Integrated Grid*.

The grid is expected to change in different, perhaps fundamental ways, requiring careful assessment of the costs and opportunities of different technological and policy pathways. It also requires attention to the reality that the value of the grid may accrue to new stakeholders, including DER suppliers and customers.

This paper is the first phase in a larger Electric Power Research Institute (EPRI) project aimed at charting the transformation to the Integrated Grid. Also under consideration will be new business practices based on technologies, systems, and the potential for customers to become more active participants in the power system. Such information can support prudent, cost-effective investment in grid modernization, and the integration of DER to enable energy efficiency, more responsive demand, and the management of variable generation such as wind and solar.[1]

Along with reinforcing and modernizing the grid it will be essential to update interconnection rules and wholesale market and retail rate structures so that they adequately value both capacity and energy. Secure communications systems will be needed to connect DER and system operators. As distributed resources penetrate the power system more fully, a failure to plan for these needs could lead to higher costs and lower reliability.

Analysis of the integrated grid, as outlined here, should not favor any particular energy technology, power system configuration or power market structure. Instead, it should make it possible for stakeholders to identify optimal architectures and the most promising configurations -- recognizing that the best solutions vary with local circumstances, goals, and interconnections.

Because local circumstances vary, this paper illustrates how the issues that are central to the integrated grid are playing out in different power systems. For example, Germany's experience illustrates consequences for price, power quality and reliability when the drive to achieve a high penetration of distributed wind and PV results in outcomes that were not fully anticipated. As a result, German policymakers and utilities now are changing interconnection rules, grid expansion plans, DER connectivity requirements, wind and PV incentives, and operations to integrate distributed resources.

In the United States, Hawaii has experienced a rapid deployment of distributed PV technology that is challenging the power system's reliability. In these and other jurisdictions policymakers are considering how best to recover the costs of an integrated grid from all consumers that benefit from its value.

---

[1] This paper is about DER, but the analysis is mindful of the ways that DER and grid integration could affect energy efficiency and demand response as those could have large effects as well on the affordability, reliability, and environmental cleanliness of the grid.

Action Plan

The current and projected expansion of DER may significantly change the technical, operational, environmental, and financial character of the electricity sector. An integrated grid that optimizes the power system while providing safe, reliable, affordable, and environmentally responsible electricity will require global collaboration in the following four key areas:

1. Interconnection Rules and Communications Technologies and Standards

    - *Interconnection rules* that preserve voltage support and grid management
    - *Situational awareness* in operations and long term planning, including rules-of-the-road for installing and operating distributed generation and storage devices
    - Robust *information and communication technologies*, including high-speed data processing, to allow for seamless interconnection while assuring high levels of cyber security
    - A *standard language and a common information model* to enable interoperability among DER of different types, from different manufacturers, and using different energy management systems

2. Assessment and Deployment of Advanced Distribution and Reliability Technologies

    - *Smart inverters* that enable distributed energy resources to provide voltage and frequency support and to communicate with energy management systems [1]
    - *Distribution management systems and ubiquitous sensors* through which operators can reliably integrate distributed generation, storage and end-use devices while also interconnecting those systems with transmission resources in real time [2]
    - *Distributed energy storage and demand response*, integrated with the energy management system [3]

3. Strategies for Integrating Distributed Energy Resources with Grid Planning and Operation

    - *Distribution planning and operational processes* that incorporate DER
    - *Frameworks for data exchange and coordination* among DER owners, distribution system operators (DSOs) and organizations responsible for transmission planning and operations
    - Flexibility to *redefine roles and responsibilities* of DSOs and independent system operators (ISOs)

4. Enabling Policy and Regulation

    - *Capacity-related costs* must become a distinct element of the cost of grid-supplied electricity to ensure long-term system reliability
    - *Power market rules* that ensure long-term adequacy of both energy and capacity
    - *Policy and regulatory framework* to ensure costs incurred to transform to an integrated grid are allocated and recovered responsibly, efficiently, and equitably
    - *New market frameworks* using economics and engineering to equip investors and other stakeholders in assessing potential contributions of distributed resources to system capacity and energy costs

Next Steps for EPRI and Industry

EPRI has begun work on a three-phase initiative to provide stakeholders with information and tools that will be integral to the four areas of collaboration outlined above.

* **Phase I** – A concept paper (this document) to align stakeholders on the main issues while outlining real examples to support open fact-based discussion. Input and review were provided by various stakeholders from the energy sector including utilities, regulatory agencies, equipment suppliers, non-governmental organizations (NGOs) and other interested parties.
* **Phase II** – This six-month project will develop a framework for assessing the costs and benefits of the combinations of technology that lead to a more integrated grid. This includes recommended guidelines, analytical tools and procedures for demonstrating technologies and assessing their unique costs and benefits. Such a framework is required to ensure consistency in the comparison of options and to build a comprehensive set of data and information that will inform the Phase III demonstration program. Phase II output will also support policy and regulatory discussions that may enable Integrated Grid solutions.
* **Phase III** – Conduct global demonstrations and modeling using the analytics and procedures developed in Phase II to provide comprehensive data and information that stakeholders will need for the system-wide implementation of integrated grid technologies in the most cost effective manner.

Taken together, Phases II and III will help identify the technology combinations that will lead to cost-effective and prudent investment to modernize the grid while supporting the technical basis for DER interconnection requirements. Additionally, interface requirements that help define the technical basis for the relationship between DER owners, DSOs, and transmission system operators (TSOs) or independent system operators (ISOs) will be developed. Finally, the information developed, aggregated, and analyzed in Phases II and III will help identify planning and operational requirements for DER in the power system while supporting the robust evaluation of the capacity and energy contribution from both central and distributed resources.

The development of a consistent framework supported by data from a global technology demonstration and modeling program will support cost effective, prudent investments to modernize the grid and the effective, large-scale integration of DER into the power system. The development of a large collaborative of stakeholders will help the industry move in a consistent direction to achieve an Integrated Grid.

© 2014 Electric Power Research Institute (EPRI), Inc.

20

Key Points – The Integrated Grid

Several requirements are recognized when defining an integrated grid. It must enhance electrical infrastructure, must be universally applicable and should remain robust under a range of foreseeable conditions.

* Consumers and investors of all sizes are installing DER with technical and economic attributes that differ radically from the central energy resources that have traditionally dominated the power system.
* So far, rapidly expanding deployments of DER are *connected* to the grid but not *integrated* into grid operations, which is a pattern that is unlikely to be sustainable.
* Electricity consumers and producers, even those that rely heavily on distributed energy resources, derive significant value from their grid connection. Indeed, in nearly all settings the full value of DER requires grid connection to provide reliability, virtual storage and access to upstream markets.
* DER and the grid are not competitors but complements, provided that grid technologies and practices develop with the expansion of distributed energy resources.
* We estimate that the cost of providing grid services for customers with distributed energy systems is about $51/month on average in the typical current configuration of the grid in the United States; in residential PV systems, for example, providing that same service completely independent of the grid would be four to eight times more expensive.
* Increased adoption of distributed resources requires interconnection rules, communications technologies and standards, advanced distribution and reliability technologies, integration with grid planning, and enabling policy and regulation.
* Experience in Germany provides a useful case study regarding the potential consequences of adding extensive amounts of DER without appropriate collaboration, planning and strategic development.
* While this report focuses on DER, a coherent strategy for building an integrated grid could address other challenges such as managing the intermittent and variable supply of power from utility-scale wind and solar generators.

The Integrated Grid © 2014 Electric Power Research Institute (EPRI), Inc. 7

Today's Power System

Today's power system was designed to connect a relatively small number of large generation plants with a large number of consumers. The U.S. power system, for example, is anchored by ~1,000 gigawatts (GW) of central generation on one end, and on the other end consumers that generally do not produce or store energy [4] [5]. Interconnecting those is a backbone of high voltage transmission and a medium- and low-voltage distribution system that reaches each consumer. Electricity flows in one direction, from power plants to substations to consumers, as shown in Figure 1. Even with increasing penetration, U.S. distributed resources account for a small percent of power production and consumption and have not yet fundamentally affected that one-way flow of power.

Energy, measured in kilowatt-hours (kWh), is delivered to consumers to meet the electricity consumption of their lighting, equipment, appliances and other devices, often called load. *Capacity* is the maximum capability to supply and deliver a given level of energy at any point in time. *Supply capacity* comprises networks of generators designed to serve load as it varies from minimum to maximum values over minutes,

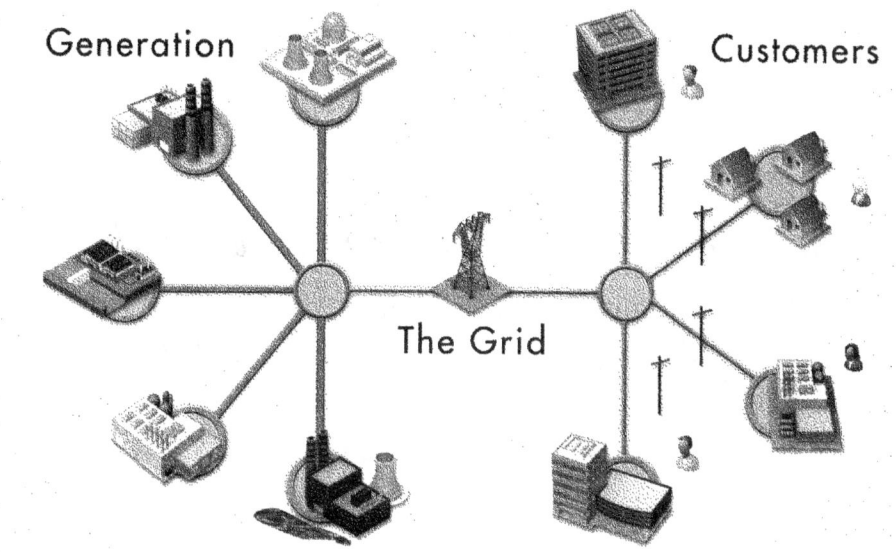

Figure 1: Today's Power System Characterized by Central Generation of Electricity, Transmission, and Distribution to End-Use Consumers

hours, days, seasons, etc. *Delivery capacity* is determined by the design and operation of the power transmission and distribution systems that deliver the electricity to consumers. The system's supply and delivery capacity plan is designed to serve the expected instantaneous maximum demand over a long-term planning horizon.

Because the whole grid operates as a single system in real time and the lead times for building new resources are long, planning is essential to ensuring the grid's adequacy. Resource adequacy planning determines the installed capacity required to meet expected load, with a prescribed reserve margin that considers potential planned and unplanned unavailability of given generators. In addition to providing sufficient megawatts to meet peak demand, the available generation (along with other system resources) must provide specific operating capabilities to ensure that

the system operates securely at all times. These ancillary services include frequency regulation, voltage support, and load following/ramping. As a practical matter, the reliability of grid systems is highly sensitive to conditions of peak demand when all of these systems must operate in tandem and when reserve margins are smallest.

Today's power system has served society well, with average annual reliability of 99.97%, in terms of electricity availability [6]. The National Academy of Engineering designated electrification enabled by the grid as the top engineering achievement of the twentieth century. Reliable electrification has been the backbone of innovation and growth of modern economies. It has a central role in many technologies considered pivotal for the future, such as the internet and advanced communications.

Today's power system has served society with average annual system reliability of 99.97% in the U.S., in terms of electricity availability.

The Growth in Deployment of Distributed Energy Resources

The classic vision of electric power grids with one-way flow may now be changing. Consumers, energy suppliers, and developers increasingly are adopting DER to supplement or supplant grid-provided electricity. This is particularly notable with respect to distributed PV power generation—for example, solar panels on homes and stores—which has increased from approximately 4 GW of global installed capacity in 2003 to nearly 128 GW in 2013 [7]. In Germany, the present capacity of solar generation is approximately 36 GW while the daily system peak demand ranges from about 40 to 80 GW. By the end of 2012, Germany's PV capacity was spread across approximately 1.3 million residences, businesses and industries, and exceeded the capacity of any other single power generation technology in the country [8]. This

rapid spread of DER reflects a variety of public and political pressures along with important changes in technology. This paper focuses on system operation impacts as DER reaches large scales.

By the end of 2013, U.S. PV installations had grown to nearly 10 GW. Although parts of the U.S. have higher regional penetration of PV, this 10 GW represents less than 2% of total installed U.S. generation capacity [9], which matches German PV penetration in 2003 (Figure 2). With PV growth projected to increase in scale and pace over the next decade, now is the time to consider lessons from Germany and other areas with high penetration of distributed resources.

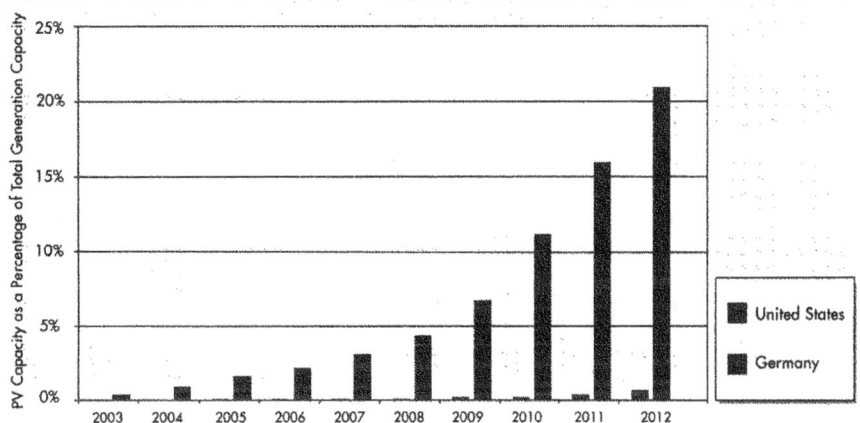

Figure 2: U.S. PV Capacity as a Percentage of Total Capacity Compared with Germany at the Beginning of Its "Energy Transformation"

The Integrated Grid

In addition to Germany, high penetration of distributed PV is evident in California, Arizona, and Hawaii and in countries such as Italy, Spain, Japan and Australia [7]. Beyond PV, other distributed resources are expanding and include such diverse technologies as batteries for energy storage, gas-fired micro-generators, and combined heat and power (CHP) installations – often referred to as *cogeneration*. In the United States natural gas prices and the cost and efficiency of gas-fired technologies have made these options effectively competitive with retail electricity service in some regions, for some consumers [10]. In jurisdictions where power prices are high, even more costly DER such as solar PV can be competitive with grid-supplied power.

In most cases, grid-connected DER benefit from the electrical support, flexibility and reliability that the grid provides, but they are not integrated with the grid's operation. Consequently the full value of DER is not realized with respect to providing support for grid reliability, voltage, frequency and reactive power.

Distributed PV power generation has increased from approximately 4 GW of global installed capacity in 2003 to nearly 128 GW in 2013.

Germany's Experience: More Distributed but Not Integrated

The circumstances surrounding Germany's extensive deployment of distributed solar PV and wind offers important lessons about the value of planning for integration of DER, both economic and technical. Germany's experience is unique for these reasons:

- Germany represents a large interconnected grid with extensive ties with other grids, which is similar to the U.S. and other countries.
- The penetration of DER over the past decade is substantial (~68 GW of installed capacity of distributed PV and wind generation over 80 GW of peak load). The observed results, in terms of reliability, quality, and affordability of electricity, are not based on a hypothetical case or on modeling and simulations.
- This growth in penetration of DER occurred without considering the integration of these resources with the existing power system.
- Germany has learned from this experience, and the plan for continuing to increase the deployment of solar PV and wind generation hinges on many of the same integrated grid ideas as outlined in this paper.

German deployment was driven by policies for renewable generation that have commanded widespread political support. PV and wind generation are backed by the German Renewable Energy Sources Act (EEG), which stipulates feed-in tariffs[2] (FIT) for solar power installations. This incentive, which began in 2000 at €0.50/kWh ($0.70/kWh) for a period of 20 years, has stimulated major deployment of distributed renewable generation.

In the meantime, electricity rates have increased in Germany, for various reasons, to an average residential rate in 2012 of €0.30/kWh ($0.40/kWh), more than doubling residential rates since 2000 [8]. These higher electricity rates and lower costs for DER, due to technology advancements and production volume, have turned the tables in Germany. Today the large FIT incentives are no longer needed, or offered, to promote new renewable installations.[3]

Notably, the desire to simultaneously contain rising electricity rates while promoting deployment of renewable energy resources has led to an evolution in German incentive policy for distributed renewable generation. For residential PV the FIT has dropped from ~ €0.50/kWh in 2000 to ~ €0.18/kWh today. An electricity price greater than the FIT has resulted in a trend of self-consumption of local generation. To ensure that all customers are paying for the subsidy for PV, the German cabinet in January, 2014 approved a new charge on self-consumed solar power. Those using their own solar generated electricity will be required to pay a €0.044kWh ($0.060/kWh) charge. Spain is considering similar rate structures to ensure all customers equitably share the cost. Still to be resolved is how grid operating and infrastructure costs will be recovered from all customers who utilize the grid with increasing customer self-generation.

Technical repercussions have resulted from DER's much larger share of the power system. Loss of flexibility in the

---

[2] Feed-in tariffs are a long-term guaranteed incentive to resource owners based on energy production (in kWh), which is separately metered from the customer's load.

[3] PV installations commissioned in July 2013, receive €0.104 to €0.151/kWh ($0.144 to $0.208/kWh) for a period of 20 years.

generation fleet prompted the operation of coal plants on a "reliability-must-run" basis. Distributed PV was deployed with little time to plan for effective integration. Until the last few years and the advent of grid codes, PV generators were not required to respond to grid operating requirements or to be equipped to provide grid support functions, such as reactive power management or frequency control. Resources were located without attention to the grid's design and power flow limitations. The lack of coordination in planning and deploying DER increases the cost of infrastructure upgrades for all customers and does not provide the full value of DER to power system operation. Rapid deployments have led to several technical challenges:

1. Local over-voltage or loading issues on distribution feeders. Most PV installations in Germany (~80%) are connected to low-voltage circuits, where it is not uncommon for the PV capacity to exceed the peak load by 3-4 times on feeders not designed to accommodate PV. This can create voltage control problems and potential overloading of circuit components [11].

2. Risk of mass disconnection of anticipated PV generation in the event of a frequency variation, stemming from improper interconnection rules.[4] This could result in system instability and load-shedding events [12]. The same risk also exists from both a physical or cyber security attack.

3. Resource variability and uncertainty have disrupted normal system planning, causing a notable increase in generation re-dispatch[5,6] events in 2011 and 2012 [13].

4. Lack of the stabilizing inertia from large rotating machines that are typical of central power stations[7] has raised general concern for maintaining the regulated frequency and voltage expected from consumers, as inverter-based generation does not provide the same inertial qualities [14].

---

[4] Distributed PV in Germany initially was installed with inverters that are designed to disconnect the generation from the circuit in the event of frequency variations that exceed 50.2 Hz in their 50 Hz system. Retrofits necessary to mitigate this issue are ongoing, and estimated to cost approximately $300 million [12].

[5] German transmission system operator Tennett experienced a significant increase in generation re-dispatch events in 2011 and 2012 relative to previous years. Generation scheduling changes are required to alleviate power flow conditions on the grid or resource issues that arise on short notice rather than in the schedule for the day.

[6] While the primary driver to re-dispatch issues has been a reduced utilization of large nuclear generators, the increase in wind generation and PV in Germany is expected to continue changing power flow patterns.

[7] Many distributed energy resources connect to the grid using inverters, rather than the traditional synchronous generators. Increasing the relative amount of distributed and bulk system inverter-based generation that displaces conventional generation will negatively impact system frequency performance, voltage control and dynamic behavior, if the new resources do not provide compensation of the system voltage and frequency support.

Smart inverters capable of responding to local conditions or requests from the system operator can help avoid distribution voltage issues and mass disconnection risk of DER. This type of inverter was not required by previous standards in Germany, although interconnection rules are changing to require deployment of smart inverters. (see highlight box for further information)

The rate impacts and technical repercussions observed in Germany provide a useful case study of the high risks and unintended consequences resulting from driving too quickly to greater DER expansion without the required collaboration, planning, and strategies set forth in the Action Plan. The actions in Phases II and III should be undertaken as soon as it is feasible to ensure that systems in the United States and internationally are not subjected to similar unintended consequences that may negatively impact affordability, environmental sustainability, power quality, reliability and resiliency in the electric power sector.

Smart Inverters and Controls

With the current design emphasis on distribution feeders supporting one-way power flow, the introduction of two-way power flow from distributed resources could adversely impact the distribution system. One concern is over-voltage, due to electrical characteristics of the grid near a distributed generator. This could limit generation on a distribution circuit, often referred to as *hosting capacity*. Advanced inverters, capable of responding to voltage issues as they arise, can increase hosting capacity with significantly reduced infrastructure costs [15], [16].

German Grid Codes

In Germany, grid support requirements are being updated so that distributed resources can be more effectively integrated with grid operation [17], [18]. These requirements, called *grid codes*, are developed in tandem with European interconnection requirements recommended by the European Network of Transmission System Operators (ENTSO-E) [19], [20]:

1. Frequency control is required of all generators, regardless of size. Instead of disconnecting when the frequency reaches 50.2 Hz, generator controls will be required to gradually reduce the generators' active power output in proportion to the frequency increase (Figure 3). Other important functions, such as low-voltage ride-through, are also required at medium voltage.

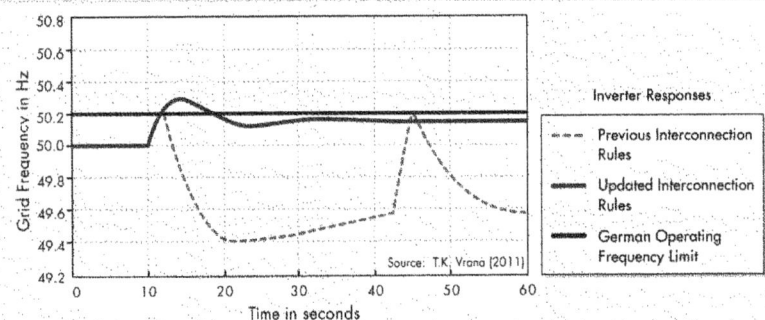

Figure 3: Example of Improved Performance with Inverter Controls That Implement a Droop Function for Over-Frequency Conditions Rather Than Tripping.

2. Voltage control functions are required from inverters, depending on the requirements of the DSO. Control methods include fixed power-factor operation, variable power factor as a function of active power, or reactive power management to provide voltage control.

3. Communication and energy management functions are now required of distributed resources, receiving commands from the system operator for active and reactive power management. As of 2012, this capability is required for all installations greater than 30 kVA. Systems less than 30 kVA without this capability are limited to 70% of rated output.

Germany is requiring that all existing inverters with a capacity greater than 3.68 kVA be retrofitted to include the droop function rather than instantly tripping with over-frequency. The cost of the retrofit associated with this function is estimated to be $300 million.

While necessary, these steps are probably still not enough to allow full integration of DER into the grid. Significant investment in the grid itself will be needed, including development of demand response resources (for example, electric transportation charging stations with time of use tariffs), and various energy storage systems. Also needed are markets and tariffs that value capacity and replacement of fossil-fueled heating plants with electric heating to take advantage of excess PV and wind capacity. German energy agency DENA determined that German distribution grids will require investment of €27.5 billion to €42.5 billion ($38.0 billion to $58.7 billion) by 2030. This includes expanding distribution circuits between 135,000 km and 193,000 km [21]. Extensive research is under way to develop and evaluate technologies to improve grid flexibility and efficiency with even more renewable capacity.

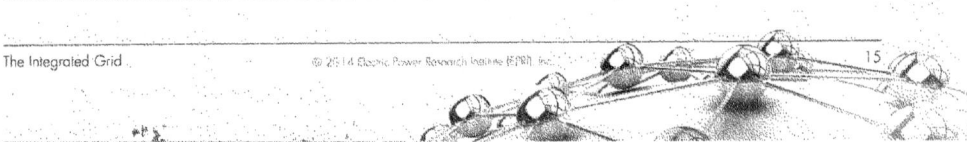

Assessing the Cost and Value of Grid Services

An electric grid connection, in ways different from a telephone line, provides unique and valuable services. Thirty percent of landline telephone consumers have canceled this service, relying solely on cellular service [22]. In contrast, virtually all consumers that install distributed generation remain connected to the grid. The difference is that the cellular telephone network provides functionality approximately equal to landline service, while a consumer with distributed generation will still need the grid to retain the same level of service. Unlike a cellphone user, operating without interconnection to this grid will require significant investment for on-site control, storage, and redundant generation capabilities.

This section characterizes the value of grid service to consumers with DER, along with calculations illustrating costs and benefits of grid connection. Subsequent sections focus on the value that DER can provide to the grid. In the context of value it is important to distinguish the difference between value and cost. Value reflects the investments that provide services to consumers. It guides planning and investment decisions so that benefits equal or exceed costs. The costs that result are recovered through rates that in a regulated environment are set to recover costs, not to capture the full value delivered.

Value of Grid Service: Five Primary Benefits

Often, the full value of a grid connection is not fully understood. Grid-provided energy (kWh) offers clearly recognized value, but grid connectivity serves roles that are important beyond providing energy. Absent redundancy provided by the grid connection, the reliability and capability of the consumer's power system is diminished. Grid capacity provides needed power for overload capacity, may absorb energy during over-generation and supports stable voltage and frequency. The primary benefits of grid connectivity to consumers with distributed generation are shown in Figure 4 and are described below.

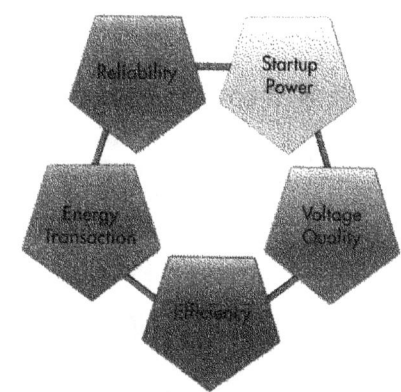

Figure 4: Primary Benefits of Grid Connectivity to Consumers with Distributed Generation

30

1. **Reliability** – The grid serves as a reliable source of high-quality power in the event of disruptions to DER. This includes compensating for the variable output of PV and wind generation. In the case of PV, the variability is not only diurnal but as shown in Figure 5, overcast conditions or fast-moving clouds can cause fluctuation of PV-produced electricity. The grid serves as a crucial balancing resource available for whatever period—from seconds to hours to days and seasons—to offset variable and uncertain output from distributed resources. Through instantaneously balancing supply and demand, the grid provides electricity at a consistent frequency. This balancing extends beyond real power, as the grid also ensures that the amount of reactive power in the system balances load requirements and ensures proper system operation.[8]

The need for reliability is fundamental to all DER, not just variable and intermittent renewable sources. For example, a customer depending solely on a gas-fired generator, which has an estimated reliability of 97%, is projected to experience 260 hours of power outage [23] compared with the 140 minutes of power outage that U.S. grid consumers experience on average (excluding major events such as hurricanes) [6]. Improvements in reliability are generally achieved through redundancy. With the grid, redundant capacity can be pooled among multiple consumers, rather than each customer having to provide its own backup resources. This reduces the overall cost of reliability for each customer [23].

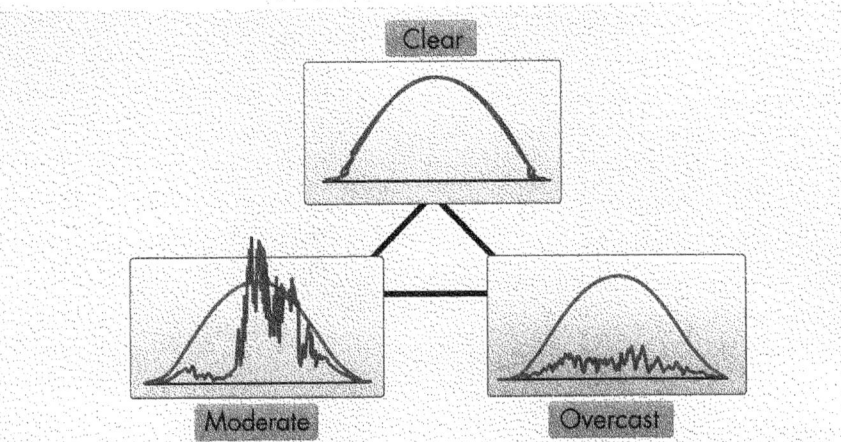

Figure 5: The Output of PV is Highly Variable and Dependent on Local Weather

---

[8] Consumer loads typically require two different kinds of power, both real and reactive. Real power is a function of the load's energy consumption, and is used to accomplish various tasks. Reactive power is transferred to the load during part of the cycle, and returned during the other part, doing no work. Balancing both real and reactive power flow is a necessary function of a reliable electric grid.

2. Startup Power – The grid provides instantaneous power for appliances and devices such as compressors, air conditioners, transformers and welders that require a strong flow of current ("in-rush" current) when starting up. This enables them to start reliably without severe voltage fluctuation. Without grid connectivity or other supporting technologies[9], a conventional central air conditioning compressor relying only on a PV system may not start at all unless the PV system is oversized to handle the in-rush current. A system's ability to provide this current is directly proportional to the fault contribution level[10]. Even if a reciprocating engine distributed generator is used as support, its fault level is generally five times less than the grid's [23]. The sustained fault current from inverter-based distributed resources is limited to the inverter's maximum current and is an order of magnitude lower than the fault level of the grid.

Figure 6 illustrates the instantaneous power required to start a residential air conditioner. The peak current measured during this interval is six to eight times the standard operating current [24]. While the customer's PV array could satisfy the real power requirements of the heating, ventilating and air conditioning (HVAC) unit during normal operation, the customer's grid connection supplies the majority of the required starting power.

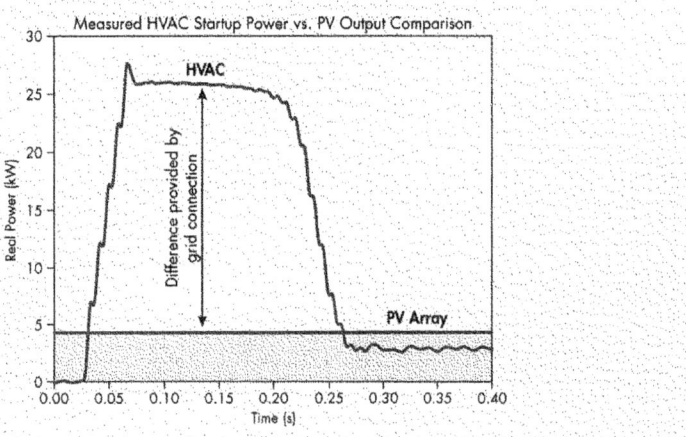

Figure 6: The Grid Provides In-Rush Current Support for Starting Large Motors Which May Be Difficult to Replicate with a Distributed Generator

---

[9] Supporting technologies include variable-frequency drive (VFD) systems, which are able to start motors without the in-rush current common in "across-the-line" starting [24].

[10] "Fault level" is a measure of the current that would flow at a location in the event of short circuit. Typically used as a measure of electrical strength, locations with a high fault level are typically characterized by improved voltage regulation, in-rush current support, and reduced harmonic impact. Locations with a low fault level are more susceptible to voltage distortion and transients induced by harmonic-producing loads.

3. Voltage Quality – The grid's high fault current level also results in higher quality voltage by limiting harmonic distortion[11] and regulating frequency in a very tight band, which is required for the operation of sensitive equipment. Similarly, the inherent inertia of a large, connected system minimizes the impact of disturbances, such as the loss of a large generators or transmission lines, on the system frequency. As shown in Figure 7, grid-connected consumers on average will experience voltage that closely approximates a sinusoidal waveform with very little harmonic distortion.

In contrast, voltage from a distributed system that is not connected to the grid will generally have a higher voltage harmonic distortion, which can result in malfunction of sensitive consumer end-use devices. Harmonics cause heating in many components, affecting dielectric strength and reducing the life of equipment, such as appliances,[12] motors, or air conditioners [25]. Harmonics also contribute to losses that reduce system efficiency. In addition, a disturbance occurring inside the unconnected system will create larger deviations in frequency than if the system maintained its connection to the larger grid.

4. Efficiency – Grid connectivity enables rotating-engine-based generators to operate at optimum efficiency. Rotating-engine-based distributed resources, such as micro-turbines or CHP systems are most efficient when operating steadily near full output [26]. This type of efficiency curve is common for any rotating machine, just as automobiles achieve the best gasoline mileage when running at a steady, optimal speed. With grid

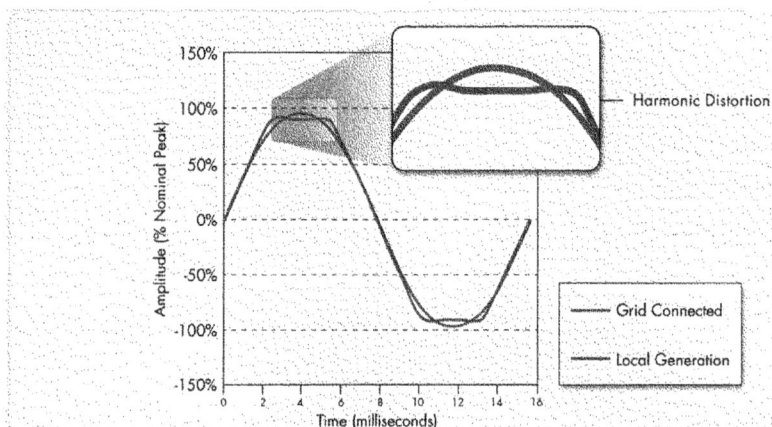

Figure 7: The Grid Delivers High Quality Power with Minimal Harmonic Distortion

---

[11] Harmonics are voltages or currents that are on the grid, but do not oscillate with the main system frequency (60Hz in the United States). The magnitude of the harmonics, when compared to the magnitude of the 60Hz component, is referred to as the harmonic distortion.

[12] Technological improvements are available, such as uninterruptible power supplies (UPS) that reduce the sensitivity of loads to poor power quality, but at an additional cost.

connectivity, a distributed energy resource can always run at its optimum level without having to adjust its output based on local load variation. Without grid connectivity, the output of a distributed energy resource will have to be designed to match the inherent variation of load demand. This fluctuating output could reduce system efficiency as much as 10%–20% [26].

5. **Energy Transaction** – Perhaps the most important value that grid connectivity provides consumers, especially those with distributed generation, is the ability to install any size DER that can be connected to the grid. A utility connection enables consumers to transact energy with the utility grid, getting energy when the customer needs it and sending energy back to the grid when the customer is producing

more than is needed. This benefit, in effect, shifts risks with respect to the size of the energy resource from the individual user to the party responsible for the resources and operation of the grid. Simulated system results for such transactions are provided in Figure 8.

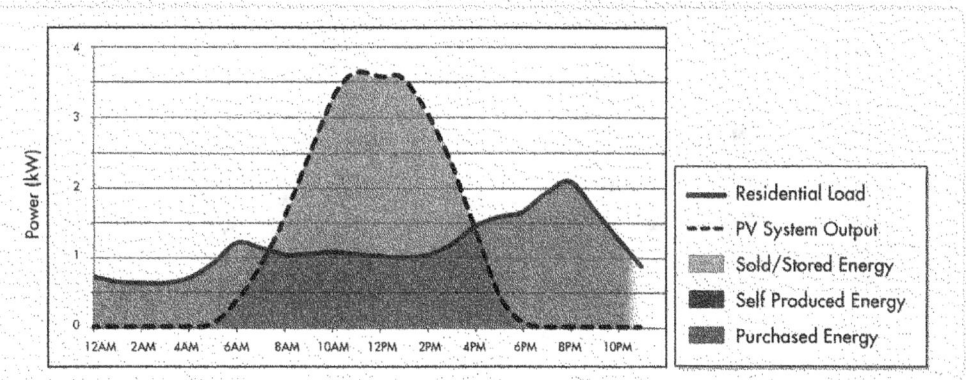

Figure 8: Because Residential Load and PV System Output Do Not Match, Owners of Distributed Generation Need the Grid for Purchasing or Selling Energy Most of the Time

Cost of Grid Service: Energy and Capacity Costs

For residential customers, the cost for generation and transmission and distribution components can be broken down as cost related to serve the customer with *energy* (kWh) and cost related to serve the customer with *capacity* that delivers the energy and grid-related services. The five main benefits of grid connectivity discussed in the previous section span both capacity and energy services. Figure 9 shows that, based on the U.S. Department of Energy's *Annual Energy Outlook 2012*, an average customer consumes 982 kWh per month, paying an average bill of $110 per month, with the average cost of $70 for generation of electricity. That leaves $30 for the distribution system and $10 for the transmission system [27] – known together as "T&D". These are average values, and costs vary among and within utilities and across different types of customers. (See *Appendix A* for explanation of calculations in this section.)

The next step in the analysis is to allocate these costs (generation and T&D) into fractions that are relevant for analyzing how the grid works with DER. In this analysis we focus on capacity and grid-related services because they are what enable robust service even for customers with DER. Indeed, consumers with distributed generation may not consume any net energy (kWh) from the grid yet they benefit from the same grid services as consumers without distributed generation.

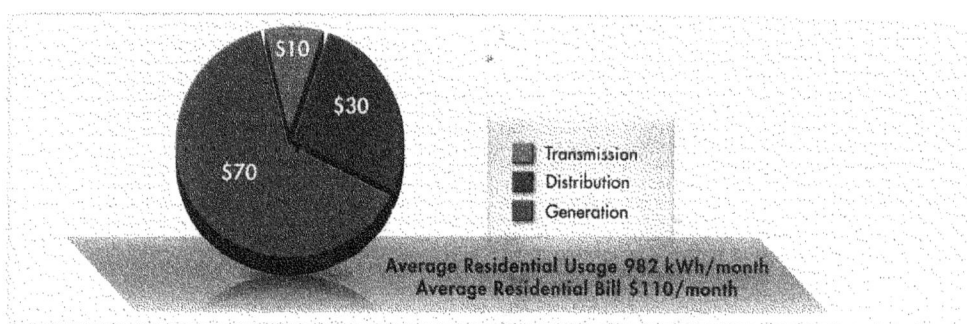

Figure 9: Cost of Service Breakdown for Today's Grid-Connected Residential Customer [27]

Calculating the total cost of capacity follows the analysis summarized in Figure 10. These values are based on the assumption that most costs associated with T&D are related to capacity (except for a small fraction representing system losses – estimated to be $3 per month per customer from recent studies in California) [28]. Working with recent data from PJM [29] regarding the cost of energy, capacity, and ancillary services it is possible to estimate that 80% of the cost of generation is energy related, leaving the rest for capacity and grid services. This 80-20 split will depend on the market and in the case of a vertically integrated utility will depend on the characteristics of the generation assets and load profile, but it is a useful average figure with which some illustrative calculations follow.

As illustrated, the combination of transmission, distribution and the portion of generation that provides grid support averages $51/month while energy costs average $59/month. These costs vary widely across the United States and among consumers, and also will vary with changes in generation profile and the deployment of new technologies such as energy storage, demand response-supplied capacity, and central generation. The values are shown to illustrate that capacity and energy are both important elements of cost and should be recovered from all customers who use capacity and energy resources. Customers with distributed generation may offset the energy cost by producing their own energy, but as illustrated in previous sections they still utilize the non-energy services that grid connectivity provides.

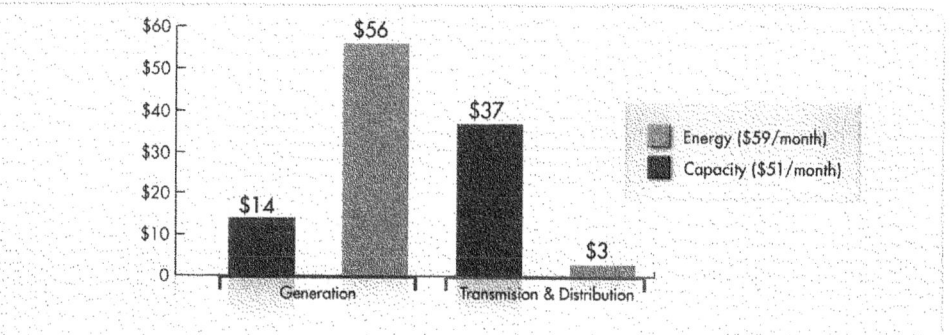

Figure 10: In Considering the Value of the Integrated Grid, Costs of Generation, Transmission, and Distribution Can Be Further Determined for Energy and Capacity

Cost of Service Without Grid Connectivity

Technologies are available that enable consumers to self-generate and disconnect from the grid. To estimate the capacity-related cost for such investments, a simplified analysis examined a residential PV system. The analysis was based on estimating the additional costs of providing the five services that grids offer—as outlined earlier in this section. For illustration, consider a residential PV system that is completely disconnected from the grid, amortized over 20 years and presented as a monthly cost. Reinforcing the system for an off-grid application required the following upgrades:

- Additional PV modules beyond the requirements for offsetting annual energy consumption in order to survive periods of poor weather;
- Multi-day battery storage with dedicated inverter capable of operating in an off-grid capacity;
- Backup generator on the premises designed to operate for 100 hours per year; and
- Additional operating costs including inverter replacement and generator maintenance.

In simulation, the cost to recreate grid-level service without a grid connection ranges from $275–$430 per month *above* that of the original array. Expected decreases in the cost of battery and PV module technology could reduce this to $165–$262 within a decade. Further information on this analysis is provided in Appendix A. Costs for systems based on other technologies, or larger deployments such as campus-scale microgrids, could be relatively lower, based on economies of scale. However, even if amortized capital costs are comparable to grid services, such isolated grids will result in deteriorating standards of reliability and quality of electricity service and could require extensive use of backup generators whose emissions negatively impact local air quality.

Such isolated grids will result in deteriorating standards of reliability and quality of electricity service and could require extensive use of backup generators whose emissions negatively impact local air quality.

Enabling Policy and Regulation

A policy and regulatory framework will be needed to encourage the effective, efficient, and equitable allocation and recovery of costs incurred to transform to an integrated grid. New market frameworks will have to evolve in assessing potential contributions of distributed and central resources to system capacity and energy costs. Such innovations will need to be anchored in principles of equitable cost allocation, cost-effective and socially beneficial investment, and service that provides universal access and avoidance of bypass.

As discussed, the cost of supply and delivery capacity can account for almost 50% of the overall cost of electricity for an average residential customer. Traditionally, residential rate structures are based on metered energy usage. With no separate charge for capacity costs, the energy charge has traditionally been set to recover both costs. This mixing of fixed and variable cost recovery is feasible when electricity is generated from central stations, delivered through a conventional T&D system, and with an electromechanical meter that measures energy use only by a single entity [30] [31].

Most residential (and some commercial) rate designs follow this philosophy, but the philosophy has not been crisply articulated nor reliably implemented for DER. Consequently, consumers that use distributed resources to reduce their grid-provided energy consumption significantly, but remain connected to the grid, may pay significantly less than the costs incurred by the utility to provide capacity and grid connectivity. In effect, the burden of paying for that capacity can potentially shift to consumers without DER [32].

A logical extension of the analysis provided here, as well as many other studies that look at DER under different circumstances, is that as DER deploy more widely, policy makers will need to look closely at clearly separating how customers pay for actual energy and how they pay for capacity and related grid services.

A policy and regulatory framework will be needed to encourage the effective, efficient, and equitable allocation and recovery of costs incurred to transform to an integrated grid.

Realizing the Value of DER through Integration

The analysis of capacity-related costs (including cost of ancillary services) in the previous sections is based on today's snapshot of the components that make up the grid and also based on minimum contribution from DER to reduce the capacity cost. With increasing penetration of variable generation (distributed and central) it is expected that capacity- and ancillary service-related costs will become an increasing portion of the overall cost of electricity [33].

However, with an integrated grid there is an opportunity for DER to contribute to capacity and ancillary services that will be needed to operate the grid. The following considerations will affect whether and how DER contribute to system capacity needs:

* **Delivery Capacity** – The extent to which DER reduce system delivery capacity depends on the expected output during peak loading of the local distribution feeder, which typically varies from the aggregate system peak. If feeder peak demand occurs after sunset, as is the case with many residential feeders, local PV output can do nothing to reduce feeder capacity requirements. However, when coupled with energy storage resources dedicated to smoothing the intermittent nature of the resources, such resources could significantly reduce

capacity need. Similarly, a smart inverter, integrated with a distribution management system may be able to provide distributed reactive power services to maintain voltage quality.

* **Supply Capacity** – The extent to which DER reduce system supply capacity depends on the output expected during high-risk periods when the margin between available supply from other resources and system demand is relatively small. If local PV production reduces high system loads during summer months but drops significantly in late evening prior to the system peak, it may do little to reduce system capacity requirements. Conversely, even if PV production drops prior to evening system peaks, it may still reduce supply capacity requirements if it contributes significantly during other high-risk periods such as shoulder months when large blocks of conventional generation are unavailable due to maintenance. Determining the contribution of DER to system supply capacity requires detailed analysis of local energy resources relative to system load and conventional generation availability across all periods of the year and all years of the planning horizon.

With an integrated grid there is an opportunity for DER to contribute to capacity and ancillary services.

- **System Flexibility** – As distributed, variable generation is connected to the grid it may also impact the nature of the system supply capacity required. Capacity requirements are defined by the character of the demand they serve. Distributed resources such as PV alter electricity demand, changing the distributed load profile. PV is subject to a predictable diurnal pattern that reduces the net load to be served by the remaining system. At high levels, PV can alter the net load shape, creating additional periods when central generation must "ramp" up and down to serve load. Examples are early in the day when the sun rises and PV production increases and later, as the sun sets, when PV output drops, increasing net load. The net load shape also becomes characterized by abrupt changes during the day, as when cloud conditions change significantly.

- **Integration of DER Deployment in Grid Planning** – Adequacy of delivery and supply capacity are ensured through detailed system planning studies to understand system needs for meeting projected loads. In order for DER to contribute to meeting those capacity needs in the future, DER deployment must be included in the associated planning models. Also, because DER are located in the distribution system, certain aspects of distribution, transmission, and system reliability planning have to be more integrated. (Read more in the section, *Importance of Integrated Transmission and Distribution Planning and Operation for DER.*)

- **DER Availability and Sustainability over Planning Horizon** – For either delivery or supply capacity, the extent to which DER can be relied upon to provide capacity service and reduce the need for new T&D and central generation infrastructure depends on planners' confidence that the resource will be available when needed across the planning horizon. To the extent that DER may be compensated for providing capacity and be unable or unwilling to perform when called upon, penalties may apply for non-performance.

In addition to altering the system daily load curve, wind and solar generation's unscheduled, variable output will require more flexible generation dispatch. For example, lower cost and generally large and less operationally flexible plants today typically carry load during the day. These resources may have to be augmented by smaller and more flexible assets to manage variability; however, this flexibility to handle fast ramping conditions comes with a cost. [34] [35] The potential for utilizing demand response or storage should not be overlooked, as rapid activation (on the order of seconds or minutes) could provide additional tools for system operators. Improving generator scheduling and consolidating balancing areas could improve access and utilization of ramping resources, preventing the unnecessary addition of less-efficient peaking units [36].

In addition to altering the system daily load curve, wind and solar generation's unscheduled, variable output will require more flexibile generation dispatch.

Figure 11 illustrates the importance of understanding the system to determine the value of DER. The graph shows the German power system's load profile and the substantial impact of PV power generation at higher penetration [37]. In this case, the PV resource's peak production does not coincide with the system peak, and therefore does not contribute to an overall reduction in system peak. From the single average plots in Figure 11 it is unclear to what extent PV might contribute to system capacity needs during critical supply hours outside of absolute system peak. During system peak, which for Germany is winter nights, the ~36 GW of installed PV does not contribute to reducing that peak. This is based on the requirements of "reliably available capacity" [38] which is defined as the percentage of installed capacity that is 99% likely to be available.

The ~33 GW of wind is also credited to a minor extent towards meeting the winter peak demand. Hydro power provides the bulk of the 12 GW of renewable resource that is considered as reliable available capacity to meet the 80 GW of winter peak load. However in the United States, where the PV peak coincides more with the system peak (depending on the facility's orientation, shading, and other factors), the results could be different. In general, however, PV without storage to achieve coincidence with system peak will be relatively ineffective in reducing capacity costs due to its variable, intermittent nature.

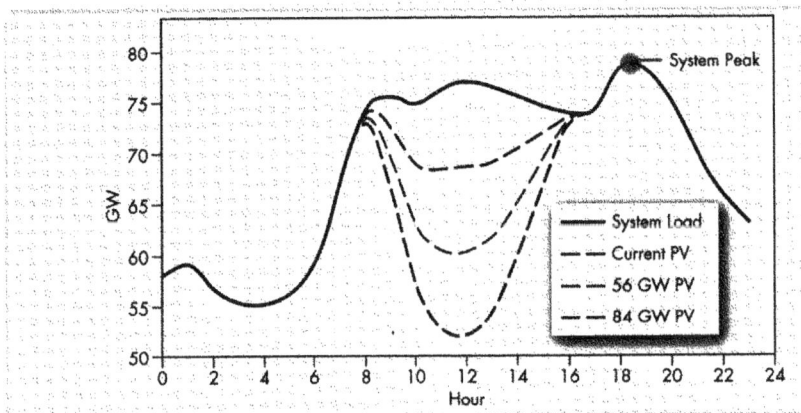

Figure 11: Peak Load Reduction and Ramp Rate Impacts Resulting from High Penetration of PV [39]

Importance of Integrated Transmission and Distribution Planning and Operation for DER

To realize their full value while ensuring power quality and reliability for all customers, DER must be included in distribution planning and operation, just as central generation resources are included in transmission planning and operation. As DER penetration increases and becomes concentrated in specific areas, their impact can extend beyond the distribution feeders to which they are interconnected, potentially affecting the sub-transmission and transmission systems. The aggregated impact of DER must be visible and controllable by transmission operators and must be included in transmission planning to ensure that the transmission system can be operated reliability and efficiently. Additionally, the T&D system operators must coordinate to expose DER owners to reliability needs and associated price signals. This will require significantly expanded coordination among transmission and distribution system planners and operators, as well as the development and implementation of new analysis tools, visualization capabilities, and communications and control methods.

Integrated transmission and distribution planning methods that include DER are not yet formalized, even in regions with high DER penetration levels such as Germany, Arizona, California, and Hawaii. Without a framework for integration into both transmission and distribution system operations, the cost of integration will increase significantly and the potential value of DER will not be fully realized. For example, DER installations in sub-optimal locations, such as the end of long feeders, may require significant feeder upgrades to avoid impacts to voltage quality. When strategically located, however, DER may require little or no upgrade of the feeder while delivering multiple benefits.

Examples of Integration of DER in Distribution Planning and Operations

The Hawaiian Electric Company (HECO) system on the island of Oahu had more than 150 MW of installed distributed PV in mid-2013. At this level of penetration, HECO has found it necessary to develop PV fleet forecasting methods, which it uses to provide operators with geographic information on expected PV output and potential impact on local feeder operations, as well as aggregate impact on system balancing and frequency performance. Additionally, HECO has developed detailed distribution feeder models that incorporate existing and expected future PV deployments for considering PV in planning. Although still in development, HECO is taking these steps to ensure reliability by integrating distributed PV into their operational and planning processes.

To realize their full value while ensuring power quality and reliability for all customers, DER must be included in distribution planning and operation, just as central generation resources are included in transmission planning and operation.

Realizing the importance of planning in DER procurement and operation, regulatory commissions in some cases have decided that distributed resource needs are best served by utility ownership or at least utility procurement of required distributed resources [40], [41]. Competitive procurement often reduces the asset cost while proper planning reduces integration costs and often maximizes the opportunity for capitalizing on multiple potential DER value streams. A recent ruling from the California Public Utilities Commission (CPUC) highlighted this consideration by requiring utilities to procure energy storage, ensuring that these resources are sufficiently planned in the context of the distribution grid [42].

Presently, most DER installations are "invisible" to T&D operators. The lack of coordination among DER owners, distribution operators, and transmission operators makes system operations more difficult, even as system operators remain responsible for the reliability and quality of electric service for all consumers. Likewise, utilities miss an opportunity to use DER, with the proper attributes, to support the grid. The expected services rendered from distributed storage in California are provided in Table 1. However, an integrated grid is required to enable many of these services, making integration beneficial to the entire system, not only to customers who own DER.

Category	Storage "End-Use"
Describes the point of use in the value chain	Describes the use or application of storage
Transmission/Distribution	Peak shaving
	Transmission peak capacity support (upgrade deferral)
	Transmission operation (short-duration performance, inertia, system reliability)
	Transmission congestion relief
	Distribution peak capacity support (upgrade deferral)
	Distribution operation (voltage/VAR support)
Customer	Outage mitigation: micro-grid
	Time-of-use (TOU) energy cost management
	Power quality
	Back-up power

Table 1: Expected T&D and Customer Services from Distributed Storage in California [43]

Realize the Benefits of Distributed Energy Resources

An Integrated Grid that enables a higher penetration of DER offers benefits to operators, customers and society. These examples illustrate the diverse nature of these benefits:

- **Provide distribution voltage support and ride-through** – DER can provide distribution grid voltage and system disturbance performance by riding through system voltage and frequency disturbances to ensure reliability of the overall system, provided there are effective interconnection rules, smart inverters, or smart interface systems.

- **Optimize distribution operations** – This can be achieved through the coordinated control of distributed resources, and using advanced inverters to enhance voltage control and to balance the ratio of real and reactive power needed to reduce losses and improve system stability.

- **Participate in demand response programs** – Combining communication and control expands customer opportunities to alter energy use based on prevailing system conditions and supply costs. Specifically with respect to ancillary services, connectivity and distribution management systems facilitate consumer participation in demand response programs such as dynamic pricing, interruptible tariffs, and direct load control.

- **Improve voltage quality and reduced system losses** – Included in this are improved voltage regulation overall and a flatter voltage profile, while reducing losses.

- **Reduce environmental impact** – Renewable distributed generation can reduce power system emissions, and an integrated approach can avoid additional emissions by reducing the need for emissions-producing backup generation. Also contributing will be the aggregation of low-emissions distributed resources such as energy storage, combined heat and power, and demand response.

- **Defer capacity upgrades** – With proper planning and targeted deployment, the installation of DER may defer the need for capacity upgrades for generation, transmission and/or distribution systems.

- **Improve power system resiliency** – Within an integrated grid, distributed generation can improve the power system's resiliency, supporting portions of the distribution system during outages or enabling consumers to sustain building services, at least in part. Key to doing this safely and effectively is the seamless integration of the existing grid and DER.

44

Figure 12 illustrates a concept of an integrated grid with DER in residences, campuses, and commercial buildings networked as a distributed energy network and described in a recent EPRI report [44].

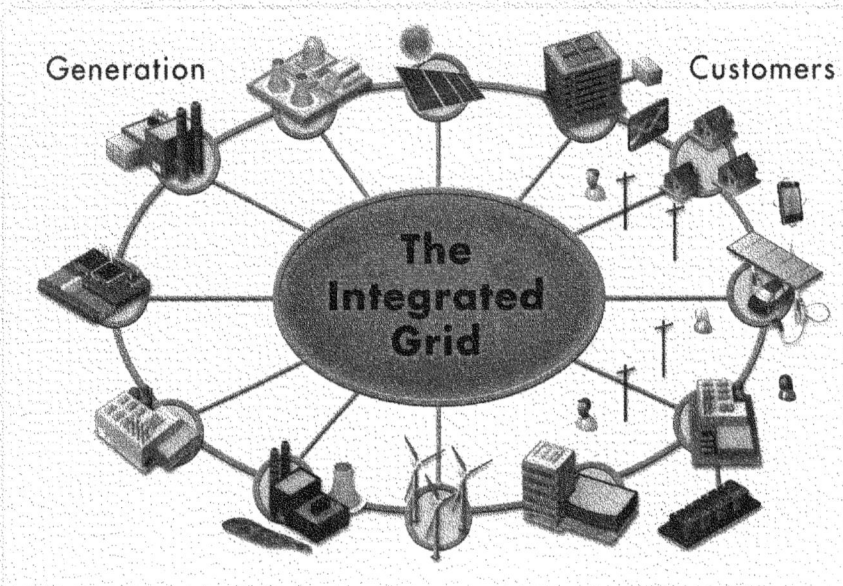

Figure 12: Creating an Architecture with Multi-Level Controller [44]

45

Grid Modernization: Imperative for the Integrated Grid

Grid modernization of the distribution system will include re-conductoring, and augmenting its infrastructure along with deploying smart technologies such as distribution management systems (DMS), communication, sensors, and energy storage is a key component of moving to the Integrated Grid. It is anticipated that this combination of infrastructure reinforcement and smart technology deployment can yield the lowest-cost solution for a given penetration level of DER in a feeder.

Table 2 shows a menu of technology options for the distribution system operator side, the consumer side and the integration of the two [45], [46] that will enable a distribution feeder to reliably integrate greater DER penetration. The solutions, which have been outlined and evaluated by others in the industry, are organized as follows:

* System Operator solutions are those actions that the DSO could take to bolster the performance and reliability of the system where DER deployment is growing.
* DER-Owner solutions are those that could be employed at the customer end of the system through installation of technology or operational response measures.
* Interactive solutions are those that require close coordination between the System Operator and DER-Owner and generally provide the operator the ability to interact with the DER-Owner's system to help maintain reliable system operation.

A comprehensive understanding of each approach is beyond the scope of this paper but is an important element of EPRI's proposed work. Assuming that any grid investment will be paid for by customers it is important to determine if, and under what situations, such investments may prove cost-effective and in the public interest.

The coordinated demonstration of each option outlined in Table 2 across different types of distribution system feeders can help provide a knowledge repository that stakeholders can use to determine the prudence of the various investments needed to achieve an integrated grid. Such demonstrations also can provide information essential for all stakeholders regarding rules of engagement among DER owners, DSOs, TSOs, and ISOs.

No one entity has the resources to conduct the demonstrations and the associated engineering analysis to document costs, benefits, and performance of all technology options across all types of distribution feeders. EPRI proposes using its collaborative approach globally to develop a comprehensive repository of data and information that can be used to move toward the Integrated Grid.

System Operator Solutions	Interactive Solutions	DG-Owner Solutions
Network Reinforcement	Price-Based Demand Response	Local Storage
Centralized Voltage Control	Direct Load Control	Self-Consumption
Static VAR Compensators	On-Demand Reactive Power	Power Factor Control
Central Storage	On-Demand Curtailment	Direct Voltage Control
Network Reconfiguration	Wide-Area Voltage Control	Frequency-based Curtailment

Table 2: Technology Options [45], [46]

Action Plan

The current and projected expansion of DER may significantly change the technical, operational, environmental, and financial characteristics of the electricity sector. An integrated grid that optimizes the power system while providing safe, reliable, affordable, and environmentally responsible electricity will require global collaboration in the following four key areas:

1. Interconnection Rules and Communications Technologies and Standards

  * *Interconnection rules* that preserve voltage support and grid management
  * *Situational awareness* in operations and long-term planning, including rules-of-the-road for installing and operating distributed generation and storage devices
  * Robust *information and communication technologies*, including high-speed data processing, to allow for seamless interconnection while assuring high levels of cyber security
  * A *standard language and a common information model* to enable interoperability among DER of different types, from different manufacturers, and with different energy management systems

2. Assessment and Deployment of Advanced Distribution and Reliability Technologies

  * *Smart inverters* that enable DER to provide voltage and frequency support and to communicate with energy management systems [1]
  * *Distribution management systems and ubiquitous sensors* through which operators can reliably integrate distributed generation, storage and end-use devices while also interconnecting those systems with transmission resources in real time [2]
  * *Distributed energy storage and demand response,* integrated with the energy management system [3]

3. Strategies for Integrating DER with Grid Planning and Operation

  * *Distribution planning and operational processes* that incorporate DER
  * *Frameworks for data exchange and coordination* among DER owners, DSOs, and organizations responsible for transmission planning and operations
  * Flexibility to *redefine roles and responsibilities* of distribution system operators and independent system operators

4. Enabling Policy and Regulation

  * *Capacity-related costs* must become a distinct element of the cost of grid-supplied electricity to ensure long-term system reliability
  * *Power market rules* that ensure long term adequacy of both energy and capacity
  * *Policy and regulatory framework* to ensure costs incurred to transform to an integrated grid are allocated and recovered responsibly, efficiently, and equitably
  * *New market frameworks* using economics and engineering to equip investors and other stakeholders in assessing potential contributions of distributed resources to system capacity and energy costs

Next Steps for EPRI

In order to provide the knowledge, information, tools that will inform key stakeholders as they take part in shaping the four key areas supporting transformation of the power system, EPRI has begun work on a three-phase initiative.

Phase I – Develop a Concept Paper

This concept paper was developed to align stakeholders on the main issues while outlining real examples to support open fact-based discussion. Input and review was provided by various stakeholders from the energy sector including utilities, regulatory agencies, equipment suppliers, non-governmental organizations (NGOs), and other interested parties. The publication of this paper will be followed by a series of public presentations and additional topical papers of a more technical nature that will more completely analyze various aspects of the Integrated Grid and lessons learned from regions where DER penetration has increased.

Phase II – Develop an Assessment Framework

In this six-month project, EPRI will develop a framework for assessing the costs and benefits of combinations of technology that lead to an Integrated Grid. Such a framework is required to ensure consistency in the comparison of options and to build a resource library that will inform the Phase III demonstration program.

In order to organize a comprehensive framework, EPRI will analyze System Operator, DER Owner, and Interactive Options listed in Table 2. Since each country, state, region, utility, and feeder may have differing characteristics that lead to different optimized solutions, efforts will be made to ensure that the framework is flexible enough to accommodate these differences.

Additionally, a testing protocol will be developed in support of the Phase III global demonstration program to ensure that a representative sample of systems and solutions will be tested.

Phase III – Conduct a Global Demonstration and Modeling Program

Phase III will focus on conducting global demonstrations and modeling using the analytics and procedures developed in Phase II to provide data and information that stakeholders will need for the system-wide implementation of integrated grid technologies in the most cost effective manner.

Using the Phase II framework and resource library, participants in Phase III can combine and integrate their various experiments and demonstrations under a consistent protocol. However, it is neither economic nor practical for an individual distribution system operator to apply all the technological approaches across different types of distribution circuits. Therefore, Phase III, planned as a two-year effort, will present the opportunity for utilities globally to collaborate to assess the cost, benefit, performance and operational requirements of different technological approaches to an integrated grid.

48

Demonstrations and modeling projects in areas where DER deployment is not expected near term will use the analytics and procedures developed in Phase II to ensure that results can provide data and information that utilities will need for planning investments in the system-wide implementation of integrated grid technologies.

With research organizations and technology providers working with distribution companies on individual demonstration projects, EPRI can work to ensure that findings and lessons learned are shared, and to consolidate the evaluations of the different approaches. The lessons learned from the real life demonstrations will be assembled in a technology evaluation guidebook and information resources and analysis tools.

New technologies for grid modernization will continue to evolve as the transformation to an integrated grid continues in this decade and beyond. The effort outlined in Phase II and Phase III will not be a one-time event but will set the stage for ongoing technology development and optimization of the integrated grid concept. As new technology evolves, a comprehensive framework for assessment of the technology as outlined in Phases II and III can support prudent investment for grid modernization using a solid scientific rigor of assessment before system-wide deployment.

An integrated grid that optimizes the power system while providing safe, reliable, affordable, and environmentally responsible electricity will require global collaboration.

Outputs from the Three-Phase EPRI Initiative

Taken together, Phases II and III will help identify the technology combinations that will lead cost-effective and prudent investment to modernize the grid while supporting the technical basis for DER interconnection requirements. Also to be developed are interface requirements that help define the technical basis for the relationship between DER owners, DSOs, and TSOs or ISOs. The information developed, aggregated, and analyzed in Phases II and III will help identify planning and operational requirements for DER in the power system and inform policymakers and regulators as they implement enabling policy and regulation. The development of a consistent framework backed up with data from a global technology demonstration and modeling program will support cost effective and prudent investments to modernize the grid in order to effectively integrate large amounts of DER into the existing power system.

A key deliverable from the Phase II and III efforts will be a comprehensive guidebook, analytical tools, and a resource library for evaluating combinations of technologies in distribution system circuits. In order to maximize the value of these deliverables, EPRI will seek to partner with organizations that are leading integrated grid-style analyses and demonstration projects to ensure that all have access to the full database of inputs and outputs from these important projects even if they were not directly involved in the technical work. Key components of the guidebook, analytical tools, and resource library will include:

- Comprehensive descriptions of technological approaches and how they can be applied in a distribution system
- Modeling tools and approaches required to assess the performance of the technical solutions
- Operational interface that will be required between DER owners and DSOs
- Analytics to assess the hosting capacity of distribution circuits
- Analytics to evaluate technology options and costs to support greater penetration of DER
- Analytics to characterize the value of integrated grid approaches beyond increasing feeder hosting capacity

A collaborative approach will be essential to develop the comprehensive knowledge repository of costs, benefits, performance and operational requirements of the multitude of technical approaches that can be implemented in a given distribution feeder for a specific level of DER integration. The guidebook, analytical tools and resource library will build on prior work of EPRI and other research organizations to develop a portfolio of solution options outlined in Table 2. They will also use the DOE/EPRI cost/benefit framework for evaluating smart grid investments as part smart grid demonstrations around the world [47].

The Integrated Grid

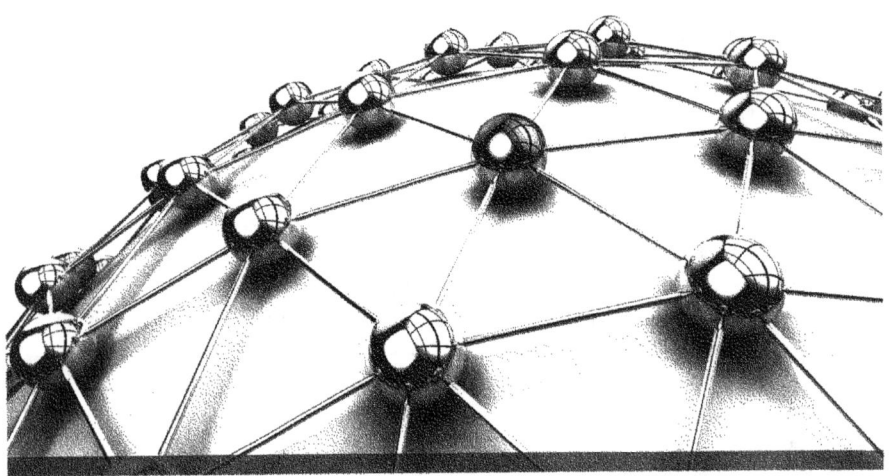

Conclusion

Changes to the electric power system with the rise of DER have had a substantial impact on the operation of the electric power grid in places such as Germany and Hawaii. As consumers continue to exercise their choice in technology and service, technologies improve in performance and cost, and federal and regional policy incentives are passed, DER could become even more pervasive.

DER deployment may provide several benefits, including reduced environmental impact, deferred capacity upgrades, optimized distribution operations, demand response capabilities, and improved power system resiliency. The successful integration of DER depends pivotally on the existing electric power grid, especially its distribution systems which were not designed to accommodate a high penetration of DER while sustaining high levels of electric quality and reliability. Certain types of DER operate with more variability and intermittency than the central power stations on which the existing power system is based. The grid provides support that balances out the variability and intermittency while also providing other services that may be difficult to replicate locally.

An integrated grid that optimizes the power system while providing safe, reliable, affordable, and environmentally responsible electricity will require global collaboration in the following four key areas:

1. Interconnection Rules and Communications Technologies and Standards

2. Deployment of Advanced Distribution and Reliability Technologies

3. Strategies for Integrating DER with Grid Planning and Operation

4. Enabling Policy and Regulation

In order to provide the knowledge, information, tools that will inform key stakeholders as they take part in shaping the four key areas supporting transformation of the power system, EPRI has begun work on a three-phase initiative:

- Phase I – Align stakeholders with a concept paper (this document)
- Phase II – Develop a framework for assessing the costs and benefits of combinations of technology that lead to an Integrated Grid
- Phase III – Initiate a worldwide demonstration program to provide data to those seeking to implement Integrated Grid solutions.

The initiative will help identify the technology combinations that will lead to cost effective and prudent investment to modernize the grid while supporting the technical basis for DER interconnection requirements. It will develop interface requirements to help define the technical basis for the relationship between DER owners, DSOs, and TSOs or ISOs. Finally, the information developed, aggregated, and analyzed in Phases II and III will help identify planning and operational requirements for DER in the power system while supporting the robust evaluation of the capacity and energy contribution from both central and distributed resources.

The development of a consistent framework supported by data from a global technology demonstration and modeling program will support cost effective, prudent investments to modernize the grid and the effective, large-scale integration of DER into the power system. The development of a large collaborative of stakeholders will help the industry move in a consistent direction to achieve an Integrated Grid.

52

References

[1] "Common Functions for Smart Inverters, Version 2," EPRI, Palo Alto, CA, 2012. 1026809.

[2] "Integrating Smart Distributed Energy Resources with Distribution Management Systems," EPRI, Palo Alto, CA, 2012. 1024360.

[3] "2012 Grid Strategy: Distribution Management System (DMS) Advanced Applications for Distributed Energy Resources," EPRI, Palo Alto, CA, October 2012. 1025572.

[4] "Key Facts About the Electric Power Industry," Edison Electric Institute, Washington, DC, 2013.

[5] "Electric Power Annual," Energy Information Administration (EIA), Washington, DC, 2012.

[6] "IEEE Distribution Reliability Benchmarking 2011 Results," IEEE Distribution Reliability Working Group, 2013.

[7] "Renewable Energy Medium-Term Market Report," International Energy Agency (IEA), Paris, 2013.

[8] "Recent Facts About Photovoltaics in Germany," Fraunhofer Institute, Freiburg, September 2013.

[9] "Annual Updates and Trends Report," Interstate Renewable Energy Council (IREC), Chicago, 2013.

[10] "Combined Heat and Power: A Clean Energy Solution," U.S. Department of Energy (DOE), Washington, D.C., August 2012.

[11] W. Yan et. al., "Operation Strategies in Distribution Systems with High Level PV Penetration," in ISES Solar World Congress, Kassel, 2011.

[12] B. I. Craciun et al., "Overview of Recent Grid Codes for PV Power Integration," in 13th International Conference on Optimization of Electrical and Electronic Equipment (OPTIM), 2012.

[13] K. Schmitz and C. Weber, "Does One Design Fit All? On the Transferability of the PJM Market Design to the German Electricity Market," Essen, 2013.

[14] P. Tielens and D. Van Hertem, "Grid Inertia and Frequency Control in Power Systems with High Penetration of Renewables," in Young Researchers Symposium in Electrical Power Engineering, Delft, 2012.

[15] "Grid Impacts of Distributed Generation with Advanced Inverter Functions: Hosting Capacity of Large-Scale PV Using Smart Inverters," EPRI, Palo Alto, CA, 2013. 3002001246.

[16] J. Smith, M. Rylander, and W. Sunderman, "The Use of Smart Inverter Controls for Accommodating High-Penetration Solar PV," in Distributech Conference and Exhibition, San Diego, CA, 2013.

[17] "Power Generation Systems Connected to the LV Distribution Network," VDE-AR-N 4105, Offenbach, August 2011.

[18] "Technical Guideline: Generating Plants Connected to the MV Network," BDEW, Berlin, 2013.

[19] "European Network Code Development: The importance of network codes in delivering a secure, competitive and low carbon European electricity market," ENTSO-E, March 2013.

[20] "Network Code for Requirements for Grid Connection Applicable to all Generators," ENTSO-E, March 2013.

[21] "Distribution Grid Study," Deutsche Energie-Agentur (DENA), Berlin, 2012.

[22] Jeffrey Sparshot. (2013, September) About 1 in 3 Households Has No Landline Phone. [Online]. http://blogs.wsj.com/economics/2013/09/05/about-1-in-3-households-has-no-landline-phone/

[23] "Benefits Provided to Distributed Generation by a Parallel Utility System Connection," EPRI, Palo Alto, CA, 2003. 1001668.

[24] "Operation and Application of Variable Frequency Drive (VFD) Technology," Carrier Corporation, Syracuse, NY, 2005.

[25] T. D. Kefalas and A. G. Kladas, "Grid Voltage Harmonics Effect on Distribution Transformer Operation," in *7th Mediterranean Conference and Exhibition on Power Generation, Transmission, Distribution and Energy Conversion (MedPower 2010)*, 2010.

[26] "Reciprocating Engines for Stationary Power Generation," EPRI, Palo Alto, CA, 1999. TR-113894.

[27] "Annual Energy Outlook 2012," Energy Information Administration (EIA), Washington, DC, April 2012. DOE/EIA-0383.

[28] "California Net Energy Metering (NEM) Draft Cost-Effectiveness Evaluation," Energy and Environment Economics (E3), San Francisco, CA, 2013.

[29] "2010 PJM Market Highlights: A Summary of Trends and Insights," PJM, Valley Forge, PA, 2011.

[30] J. C. Bonbright, A. L. Danielsen, and D. R. Kamerschen, *Principles of Public Utility Rates*. Public Utilities Report, Inc., 1988.

[31] A. E. Kahn, *The Economics of Regulation: Principles and Institutions*. MIT, 1988.

[32] A. Faruqui, R. Hledik, and B. Neenan, "Rethinking Rate Design: A Survey of Leading Issues Facing California's Utilities and Regulators," Demand Response Research Center, 2007.

[33] "DS3: System Services Review – TSO Recommendations," EirGrid and SONI, Dublin, June 2012.

[34] E. Ela, M. Milligan, and B. Kirby, "Operating Reserves and Variable Generation," NREL, Golden, CO, 2011. NREL/TP-5500-51978.

[35] M. Hummon et al., "Fundamental Drivers of the Cost and Price of Operating Reserves," NREL, Golden, CO, 2013. NREL/TP-6A20-58491.

[36] "Flexibility Requirements and Metrics for Variable Generation," North American Electric Reliability Council (NERC), Princeton, NJ, August 2010.

[37] "Fast Facts: What the Duck Curve Tells Us About Managing a Green Grid," California ISO, Folsom, CA, 2013.

[38] "Germany's New Energy Policy," Federal Ministry of Economics and Technology (BMWi), Berlin, 2012.

[39] E. Lannoye, "Quantifying and Delivering Flexibility in Power Systems," University College Dublin, 2013.

[40] A. Mills and R. Wiser, "An Evaluation of Solar Valuation Methods Used in Utility Planning and Procurement Processes," Lawrence Berkeley National Laboratory, December 2012.

[41] "2011 ERP All-Source Solicitation: Renewable Resource RFP," Public Service Company of Colorado, Denver, CO, 2011.

[42] "Order Instituting Rulemaking Pursuant to Assembly Bill 2514 to Consider the Adoption of Procurement Targets for Viable and Cost-Effective Energy Storage Systems," California Public Utilities Commission (CPUC), 2013.

[43] E. Malashenko et. al., "Energy Storage Framework Staff Proposal," California Public Utilities Commission (CPUC), April 2012.

[44] "ElectriNet," EPRI, Palo Alto, CA, 2009. 1018848.

[45] M. Vandenbergh et. al., "Technical Solutions Supporting the Large Scale Integration of Photovoltaic Systems in the Future Distribution Grids," in *22nd International Conference on Electricity Distribution (CIRED)*, 2013.

[46] "Prioritisation of Technical Solutions Available for the Integration of PV into the Distribution Grid," PV Grid, 2013.

[47] "Estimating the Costs and Benefits of the Smart Grid," EPRI, Palo Alto, CA, 2011. 1022519.

Appendix A – Cost Calculations

Generation, Transmission, Distribution vs. Cost of Energy and Capacity

Generation, Transmission, Distribution Breakdown are provided from EIA estimates in ($/kWh) assuming an average customer usage of 982 kWh/month.

Generation is broken down into two components (Energy and Capacity) based on PJM market estimates of the price breakdown: "2010 PJM Market Highlights: A Summary of Trends and Insights." 2011. http://www.pjm.com/~/media/documents/reports/20110513-2010-pjm-market-highlights.ashx

Of which, 80% was estimated as energy-related, while the other 20% was attributed to capacity.

Distribution and Transmission are estimated based on the following breakdown from SCE (E3 NEM Effectiveness Report): http://www.ethree.com/documents/CSI/CPUC_NEM_Draft_Report_9-26-13.pdf

Among the appendices, Southern California Edison's (SCE) implied transmission and distribution (T&D) costs were provided. When those costs were scaled back to national average values, the percentages are provided below:

SCE Implied Cost breakdowns (when scaled to $40/month)

Cost Breakdown	$/Month	Fixed %	Variable %	Fixed ($)	Variable ($)
Customer	$14.29	100%	0%	$14.29	$-
Distribution	$15.71	90%	10%	$14.14	$1.57
Sub-transmission	$4.29	60%	40%	$2.57	$1.71
Transmission	$5.71	100%	0%	$5.71	$-
TOTAL	$40.00			$36.71	$3.29

Thus the variable (energy-based) T&D costs were taken at $3/month.

Cost of Off-Grid Residential Solutions

Cost figures reflect the additional cost to take a residence that produces 100% of its energy locally (from PV) and turn it into a self-sufficient entity that can operate without a grid connection.

These costs include:

* Extra PV panels (beyond the annual kWh requirement)
* Battery storage
* Charge controller
* Backup generator

Which are then amortized across the lifetime of the project (20 years)

Software Package: HOMER Energy (Hourly energy profile simulator)
Locations: St. Louis, MO and San Francisco, CA
Analysis includes appropriate incentives Federal ITC and net-energy-metering

Location	St. Louis, MO	San Francisco, CA
Load Profile (OpenEI)	12MWh/yr.	7.67MWh/yr.
Real Interest Rate	3.5% (5.5% APR – 2% inflation)	
Project Lifetime	20 years (no salvage)	
PV System (Array + Inverter) Installed Cost	$3-$4/W installed (after incentive) [2013] $1.50-$2/W installed [2020]	
Battery Cost	$450-$550/installed kWh [2013] $200-$300/installed kWh [2020]	
Generator	$400/kW	
System Controller	$600/kW	
System O&M	$32/kW/yr PV system O&M + $0.50/hr generator O&M + $3/battery/yr	

© 2014 Electric Power Research Institute (EPRI), Inc.

56

57

The Electric Power Research Institute, Inc. (EPRI, www.epri.com) conducts research and development relating to the generation, delivery and use of electricity for the benefit of the public. An independent, nonprofit organization, EPRI brings together its scientists and engineers as well as experts from academia and industry to help address challenges in electricity, including reliability, efficiency, affordability, health, safety and the environment. EPRI also provides technology, policy and economic analyses to drive long-range research and development planning, and supports research in emerging technologies. EPRI's members represent approximately 90 percent of the electricity generated and delivered in the United States, and international participation extends to more than 30 countries. EPRI's principal offices and laboratories are located in Palo Alto, Calif.; Charlotte, N.C.; Knoxville, Tenn.; and Lenox, Mass.

Together... Shaping the Future of Electricity

3002002733   February 2014

Electric Power Research Institute
3420 Hillview Avenue, Palo Alto, California 94304-1338 • PO Box 10412, Palo Alto, California 94303-0813 USA
800.313.3774 • 650.855.2121 • askepri@epri.com • www.epri.com

The CHAIRMAN. Thank you, Dr. Howard. We will be sure to have some. Thank you for the model here of your integrated grid.

Let's go to Dr. Littlewood. Welcome.

STATEMENT OF DR. PETER LITTLEWOOD, DIRECTOR, ARGONNE NATIONAL LABORATORY

Dr. LITTLEWOOD. Good morning, Chairman Murkowski, Ranking Member Cantwell and members of the Committee. Thank you for the opportunity to appear before you.

A reliable, efficient, and secure electrical grid is, of course, essential to the United States' prosperity, competitiveness, and innovation. As we've just heard, our existing grid struggles to accommodate the new economy that is emerging as the nation and the world shift to clean energy and to digital technology.

Much of the grid was designed and built using technologies and organizational principles developed decades ago to serve vertically integrated markets with large scale generation sources located dozens of miles from consumers. They were designed to use centralized control schemes that deliver a one-way flow of power to customers. The emerging national economy needs a flexible modern grid that accommodates two-way flow of electricity and information, provides strong protection against physical and cyber risks and the impacts of natural disasters, and integrates widely varying, widely distributed energy sources such as solar and wind that produce electricity intermittently depending on the weather.

Argonne National Laboratory is a DOE Office of Science lab with a long and distinguished history in power R & D. We recently joined 13 other DOE laboratories in the Grid Modernization Laboratory Consortium. I'll refer to this as the GMLC, a new initiative started last November by Energy Secretary Ernest Moniz.

The GMLC is developing a vision and a plan for moving forward and has already identified three specific goals that can be achieved through a coordinated national effort to modernize the grid and to have in operation by 2025 which would yield a 10 percent reduction in the societal cost of power outages, a 33 percent reduction in the cost of utility's reserve margins, and a 50 percent reduction in the cost of integrating energy distribution within the grid.

The GMLC estimates conservatively that achieving these three goals would save the nation's economy $7 billion a year. In addition, a coordinated national grid modernization effort would help ensure that the future grid is a flexible platform for innovation by entrepreneurs and others who can develop tools and services that empower consumers and businesses helping them make informed energy decisions.

The GMLC has identified significant opportunities and needs in six broad technical areas that are critical to the establishment of the future grid. Sensing and measurement to report real time data from across the grid.

Devices and integrated systems that can talk to each other over the Internet allow customers to make decisions about their energy use and implement those decisions remotely. For example, Argonne is helping to make hybrid and electric vehicles future grid friendly by working with the automotive and electrical supply industries to develop and test systems and technologies that let electric vehicles communicate with grid operators and to enable ''smart'' charging and discharging, depending on the condition of the grid.

Advanced energy storage systems are critical to integrating intermediate, intermittent electricity sources within the grid. The Secretary of Energy recently stressed the need for us to consider grid and storage as an integrated activity. An excellent example of this integration is DOE's Joint Center for Energy Storage Research which leverages and expands current investments in energy storage research and is a vital component of the overarching grid strategy.

We need to work on system operations and power flow, because today's grid relies primarily on control rooms and centralized operations.

The future grid must also be secure and resilient in the face of disruptive threats that range from natural disasters to terrorist attacks.

Today's grid is largely reactive, responding to failures after they happen. Although advanced control systems have improved automatic outage responses, they can expose the grid to new cyber risks as digital components proliferate and create new points of entry.

As I mentioned earlier, advances in energy storage are critical to integrating clean, renewable energy sources such as wind and solar into our electricity supply and also to supporting fluctuating demand on microgrids. Breakthroughs in battery technology are needed to reduce our dependence on petroleum through a broader use of electric vehicles.

Grid modernization will provide the nation with a reliable and secure system that delivers increasingly clean energy to business and residences in ways that optimize customers' ability to control how, when, and where they consume energy. Grid modernization also assures a system that remains resilient to a range of threats and vulnerabilities. Success will not only save the nation billions of dollars annually, it would also create new suites of advanced technology.

Thank you for your time and attention to this critically important topic.

I'd be pleased to respond to any questions you may have.

[The prepared statement of Dr. Littlewood follows:]

Testimony of Dr. Peter Littlewood
Director, Argonne National Laboratory
before the
U.S. Senate Committee on Energy and Natural Resources
March 17, 2015

Chairman Murkowski, Ranking Member Cantwell, and members of the Committee, thank you for the opportunity to appear before you today to discuss the state of technological innovation related to the electric grid.

A reliable, efficient, and secure electrical grid is essential to the United States' prosperity, competitiveness, and innovation. It delivers the power that drives our economy, lights our homes and enables the near ubiquity of the Internet in our daily lives. It is essential for meeting the needs of our growing digital society and a clean energy economy. To achieve a reliable, resilient, and clean electric power sector, our nation must pursue the vision of a Future Grid in which all the parts communicate with one another and deliver reliable, affordable and clean electricity to consumers where they want it, when they want it, and how they want it.

While modern communication networks would have been unimaginable to Alexander Graham Bell, the current electrical grid would be much more familiar to Thomas Edison. That reflects the persistence of expensive physical investment and is a reminder that the grid we build today will be with us for many decades. As we exploit the convergence of information networks with physical transport networks for infrastructure – not only the electrical grid, but also gas, water, and transportation – we must build in the flexibility to support unheralded developments both in technology and in the needs of society. A future grid will likely need to support distributed generation from renewables, support an electrified transportation network, entwine with other physical distribution networks, and be embedded in a complex information-rich, sensorized environment.

This will not be an easy task, but it is a challenge we must meet head on. It will require investment, cooperation and technological innovation on a grand scale. It will require a concerted effort over decades, akin to the development of our interstate highway system – another driver of the United States' economic might. The competitive greatness of this country was built on the quality of its infrastructure, and we must once again rise to meet the challenge in order to ensure our future prosperity.

Our existing electrical grid struggles to accommodate the new economy that is emerging as the nation and the world shift to clean energy and digital technology. Much of the current electric grid was designed and built using technologies and organizational principles developed decades ago to serve vertically integrated markets with large-scale generation sources located dozens of miles from consumers; they were designed to use centralized control schemes that deliver a one-way flow of power to customers who have minimal opportunity to provide feedback. The emerging national economy needs a flexible modern grid that accommodates the two-way flow of electricity and information; provides strong protection against physical and cyber risks and the impacts of natural disasters; and integrates widely distributed, variable energy sources, such as solar and wind energy, that produce electricity intermittently, depending on the weather.

Our nation's electricity delivery systems face complex challenges as power companies, equipment manufacturers, federal and state regulators, electricity consumers, and other stakeholders evaluate alternative technologies, systems, and investment strategies for modernization. To succeed, grid modernization must incorporate intelligent technologies, next-generation components with "built-in" cybersecurity protections, advanced grid modeling and applications, and innovative control systems. The Electric Power Research Institute's 2011 report, *Estimating the Costs and Benefits of the Future Grid*, estimated that realizing these goals will require between 300 and 500 billion dollars of new investment over the next 20 years.

Today, we lack a good picture of how all the pieces of the grid fit and work together. The existing grid is not a single system, but a patchwork of thousands of independent generators, transmitters, and distributors that range from large regional utilities to rooftop solar collectors to backyard wind generators. The grand challenge is to understand the entirety of the grid and its interoperability, then to use game-changing technology to shape it effectively into that vision of a Future Grid that is both reliable and resilient.

Argonne National Laboratory is a DOE Office of Science laboratory with a long and distinguished history in power grid R&D. We recently joined 13 other DOE laboratories in the Grid Modernization Laboratory Consortium (GMLC), a new initiative started last November by Energy Secretary Ernest Moniz. The GMLC is developing a vision and a plan for moving forward and has already identified three specific goals that can be achieved through a coordinated national effort to modernize the grid:
- A 10 percent reduction in the societal costs of power outages.
- A 33 percent reduction in the cost of utilities' reserve margins, while maintaining reliability.
- A 50 percent reduction in the cost of integrating distributed energy sources with the grid.

The GMLC estimates that achieving these three goals would save the nation's economy an estimated $7 billion a year. In addition, a coordinated national grid modernization effort would help ensure the Future Grid is a flexible platform for innovation by entrepreneurs and others who can develop tools and services that empower consumers and businesses, helping them make informed energy decisions.

The GMLC has identified significant opportunities and needs in six broad technical areas that are critical to the establishment of the Future Grid:

- Sensing and measurement
- Devices and integrated systems
- System operations and power flow
- Design and planning tools
- Security and resilience
- Support for utilities and regulators

I agree with the GMLC list. I would also like to emphasize the need for advanced energy storage systems (included under integrated systems above) that are critical to integrating intermittent electricity sources with the grid and to building a national fleet of electric vehicles. The Secretary has recently stressed the need for us to consider grid and storage as an integrated activity. An excellent example of this integration is DOE's Joint Center for Energy Storage Research (JCESR), which leverages and expands current investments in energy storage research and is a vital component of an overarching grid strategy.

Sensing and measurement

Sensing and measurement are critical to enabling the millions of pieces of the Future Grid to communicate and interact effectively. A Future Grid needs miniature, low-cost sensing and measuring devices to gather and report real-time data from as many components as we can monitor. The ability to monitor and understand the state of the grid and all its parts in real time will provide enormous potential benefits. We could instantaneously detect and respond to changes in supply and demand by rerouting transmission or adjusting generation. We could not only detect system failures and outages instantly, but could develop modeling and simulation systems that predict future failures and allow us to stop them before they happen. By designing a system made of components that communicate with one another, we could create new self-healing systems that automatically adjust to minimize the scale and duration of outages and bring power back online quickly.

Many of these "distributed automation" systems have begun to appear in the field in recent years. In the past, when lightning tripped a switch at a transmission station, a worker had to visit the site to see if it was safe to reset the switch. Today, automated systems in the field communicate to reset switches automatically when it's safe.

Industry has been a leader in placing smart meters in homes and businesses, which presents new, as-yet-untapped opportunities to run the electrical grid more efficiently, more economically, and with less environmental impact. At the consumer level, we can let meters in homes and businesses save energy by communicating with utility systems to, for example, charge electric vehicles when electricity rates are low or to cut back on air conditioning and turn off unused equipment when rates are high.

Devices and integrated systems

Essential to this vision of the Future Grid is the need to develop and integrate devices that talk to one another over the Internet, allow customers to make decisions about their energy use, and implement those decisions remotely. This movement is already well under way in the form of products that use smartphone apps to control home thermostats and turn lights on and off from the other side of the world. Industry is also pursuing this vision by integrating systems and devices in our homes and automobiles to save time, money, energy, and lives. Examples include smart thermostats that sense when no one is home and turn down the temperature, car systems that reset home thermostats so the house is warm when the driver arrives, smoke alarms that flash bedroom lights when triggered in the middle of the night, and carbon monoxide detectors that communicate with smart light bulbs that flash red to alert home owners. The Future Grid promises to raise this vision to an entirely new level and unleash new realms of entrepreneurial opportunity.

Argonne is helping make hybrid and electric vehicles Future-Grid friendly by working with the automotive and electric supply industries to develop and test systems and technologies that let electric vehicles communicate with grid operators to enable 'smart' charging and discharging, depending on the condition of the grid. Smart charging will ensure that electric vehicles recharge their batteries when they can take advantage of excess power, low-cost electricity, or power from solar or wind. Smart discharging will return electricity from electric vehicle batteries to the grid when customers benefit, such as during a power outage. The long-term goal is to develop standard, industry-wide systems and interoperable connections that let anyone plug in and charge any car, anytime, anywhere in the world. Achieving this goal requires a coordinated world-wide effort and close collaboration with industry to develop, test, and validate devices and communications protocols that are secure, reliable, and robust.

System operations and power flow

The Future Grid will need advanced control technologies that enhance its reliability and optimize its transmission and distribution systems. While utilities install smart meters and similar devices around the nation, the research community lacks the analytic and diagnostic capabilities to take full advantage of data from these devices. Today's grid relies primarily on control rooms and centralized operations. The Future Grid calls for distributed controls that use computers to analyze data in real time and shift the flow of electrical power to meet changing conditions.

Distributed controls will give customers greater decision-making power over how and when they consume electricity.

We also need systems that are predictive and can alert human operators to developing concerns before they result in failures or outages. Today's grid is largely reactive, responding to failure and outages after they happen. Developing predictive systems and algorithms for the Future Grid will require collaboration in both fundamental and applied research by computer scientists, systems analysts, and engineers in a variety of specialized disciplines.

An underlying challenge is the grid's increasing use of microprocessors. While they provide many benefits, microprocessors make a system more complex and can slow it down. If a system grows too complex; it can behave in unexpected ways. We need fundamental research in the mathematical and computing sciences to understand this phenomenon and mitigate it.

Design and planning tools

We need a new generation of design and planning tools to accommodate the changing nature of the grid. The existing tools were created for a grid in which electricity was generated, transmitted, and delivered in one direction only–from the utility to the customer. This one-way model is breaking down with the rapid spread of distributed electricity generation based on renewable energy and the ability of customers to sell electricity back to utilities. Driven by rapid improvements in green energy generation, this practice will continue to grow, and we will need analytical methods to optimize its integration into the Future Grid.

Better design and planning tools are also needed to calculate accurate cost-benefit tradeoffs for the two-way flow of electricity between utilities and customers and to ensure reliable design and deployment of intermittent or variable electricity sources, such as wind and solar, which are not deployable on demand because their availability depends on the weather. To capture the complexity, uncertainty, and dynamics of growing renewable generation, we need new computational tools, methods, and software libraries that improve analysis and design capabilities a thousand-fold.

Security and resilience

The Future Grid must also be secure and resilient in the face of disruptive threats that range from natural disasters to terrorist attacks. Security and resilience need to be designed into all Future Grid systems and components from the start. Too much of the safety and resilience in today's grid was added as an afterthought—necessarily so, since much of today's grid was built decades ago. For tomorrow's grid we need to build security and resilience into both emerging and legacy grid technologies.

Resiliency research is a particular strength at Argonne National Laboratory, for the grid and for the rest of the nation's infrastructure. The grid is arguably our most important infrastructure and certainly one of the most complex.

Argonne is currently developing a concept for a national Resilient Design User Facility, which would give researchers from industry, universities, government, and national laboratories access to Argonne's physical science and computing facilities to test and simulate the resiliency of key infrastructure systems. The goal is to develop new knowledge, tools, and technologies to help communities mitigate the impacts of disasters and recover more quickly.

The starting point for infrastructure resiliency analysis is always the impact of a potential disaster on the local electrical grid. Those impacts are then analyzed to understand how they will cascade out to the region and nation, and what additional threats come into play as the impact spreads. For example, any prolonged power disruption is a danger to local economies, public health, and safety. Power disruptions can quickly cascade across interdependent infrastructures and adjacent regions, triggering disruptions in transportation services, wastewater processing, and other critical services. Although advanced control systems have improved automated outage responses, they can also expose the power grid to new cyber risks as digital components proliferate and create new points of entry.

Key threats to grid security include electromagnetic pulses (EMP) and geomagnetic disturbances (GMD) that could injure the grid so badly that outages last weeks or months and the damage takes years to fully repair. The Security and Resilience Committee of DOE's Grid Modernization Laboratory Consortium considers EMP and GMD threats to have low probabilities, but extremely high potential impacts. Historically, utilities have been most concerned with physical threats to infrastructure and more recently with cyber threats. The Security and Resilience Committee sees cybersecurity becoming an increasingly critical challenge as we move to the Future Grid with millions of components communicating in real time and more operating decisions made by computer.

We also need to develop methods for prioritizing which grid infrastructure assets need protection. Large transformers at substations are particularly vulnerable assets because they are custom-made and their replacement lead time is typically 12 to 18 months. The Department of Homeland Security has spearheaded work to develop modular transformers with standardized designs that can be quickly assembled and delivered. We might further minimize replacement time by developing standardized transformers in key high-voltage classes. A good deal of research has been done on solid-state transformer design that could aid this effort, but the focus has been on small-scale, low-voltage transformers. More research and development is needed to scale up to the larger transformers of interest to utilities.

The DOE labs, specifically Pacific Northwest National Laboratory, National Renewable Energy Laboratory, Los Alamos National Laboratory, Sandia National Laboratories, and Argonne National Laboratory, have developed the beginnings of a foundational key to the secure grid—a National Power Grid Simulator. Such a simulator would serve two functions. First, it would analyze the state of the grid in real time to provide broad situational awareness and to communicate important information to the grid's various operators. Faced with a threat to the grid, the real-time simulator would help operators understand the scope and progression of events, and would suggest control and mitigation strategies. Second, the simulator would include an offline component that would operate as a national user facility where researchers from industry, universities, government, and national laboratories would run virtual experiments to test potential vulnerabilities and mitigations, develop responses to various types of attack, and understand the impacts of new technologies—a capability of particular importance to the Future Grid and all the new technologies it will bring online.

Support for utilities and regulators

Utilities and regulators need better decision-making tools to provide the best information possible to guide and coordinate their planning and investment decisions. As the grid becomes smarter, better integrated, and more complex, the importance of good, rapid decision making will extend beyond the local level and have increasing regional and national impact.

The rapid growth of distributed energy sources, such as solar energy and wind energy, is changing the game in many ways. In the previous century, an electrical utility's biggest operating challenge might have been how to provide power for all the region's air conditioners on the hottest day of the year—a relatively easy challenge to anticipate. But how do you anticipate and quickly ramp up supply when a cloud bank suddenly blocks the sun and cuts off all the electricity the grid is receiving from solar collectors on neighborhood rooftops?

Better decision-making tools will help the states adjust their regulatory models to align utilities' interests with grid modernization and clean-energy goals. These tools will also provide methods for evaluating distributed energy technologies and services so all stakeholders understand them better and can make informed decisions on grid investments and operations.

Advanced battery research

Advances in energy storage are critical to modernizing the nation's aging electrical grid—and integrating clean, renewable energy sources such as wind and solar power into our electricity supply. Breakthroughs in battery technology are also needed to reduce our dependence on petroleum through broader use of electric vehicles.

Across our energy economy, effective storage of energy holds the key to the flexible energy sourcing and delivery required to diversify our energy portfolio, renovate our energy infrastructure, and alleviate the growing environmental costs and risks of continued reliance on fossil energy as our primary energy source. For the grid, advanced battery technologies are the key to storing energy from intermittent sources so we can release it later when we need it. The best solar tracking technologies have average capacity factors only just above 20 percent, and even in the windy plains, the median annual capacity factor of wind generators was only about 33 percent last year. (Plant capacity measures how much electricity a generator actually produces compared to the maximum it could produce at continuous full power operation during the same period.)

Fundamental research at our national laboratories and universities has yielded significant improvements in batteries and energy storage over the past 20 years. But we are still far short of comprehensive solutions for the grid and transportation.

Benefits of grid modernization

Grid modernization will provide the nation with a reliable and secure system that delivers increasingly clean electricity to businesses and residences in ways that optimize customers' ability to control how, when, and where they consume energy. Grid modernization also assures a system that remains resilient to a range of threats and vulnerabilities. Success will not only save the nation billions of dollars annually, it will also create new suites of advanced technology, such as:

- A new grid architecture that enables controllability across a system that consists of multiple microgrids, millions of distributed energy sources, and a wide variety of end-use devices, most of which are still emerging.
- Next-generation sensing and data management platforms that make the full system transparent and enable quick, adaptive responses over wide areas.
- New control theory and algorithms that speed system restoration by basing decisions on real-time analysis of real-time measurements.
- Contingency tools that predict outages in real-time in the face of threats and go beyond today's contingency tools, which consider only single-point failures, to analyze the impacts of multiple failures.
- New devices and methods to manage the flow of power transmission and distribution and reduce the need to expand transmission systems.
- Integration of high-performance computing to mitigate the uncertainty inherent in a system as complex as the Future Grid.
- Real-time control systems based on ultra-fast (less than one second) measurement and estimation of the state of the grid.

- Validation of algorithms and computer models to make the system work.
- Demonstrations of developing technology and systems, in cooperation with utilities, to verify developing concepts and strategies, such as reconfiguring grid connections in real time to minimize the size and duration of an outage.

A focused national effort to modernize the grid will fundamentally alter the way our electric grid and transportation systems function, enabling the creation of entirely new approaches to satisfy the increasing electricity needs of our country. It will inevitably create new technologies, new businesses and entrepreneurial opportunities, and new high-value jobs with the potential to power the American economy for decades to come.

Thank you for your time and attention to this critically important topic. I would be pleased to respond to any questions that you might have.

###

The CHAIRMAN. Thank you, Dr. Littlewood.
Next let's go to Dr. Taft, please. Good morning.

STATEMENT OF DR. JEFFREY TAFT, CHIEF ARCHITECT FOR ELECTRIC GRID TRANSFORMATION, PACIFIC NORTHWEST NATIONAL LABORATORY

Dr. TAFT. Thank you, Senator Murkowski and Ranking Member Cantwell and the Committee, for inviting me here today.

I'd like to offer three main points in my discussion. First, we're re-engineering the grid with advanced technologies, some of which you've heard about, to support many new capabilities. Second, these changes are actually challenging our existing grid structure and the methods that we use to operate it. Third, there are certain key technologies that can help resolve the widening gap in our ability to manage the grid reliably as we go forward.

In the U.S. our modernization efforts with advanced technology support many new goals and emerging trends that were not envisioned for the 20th Century grid. It was thought of as a one-way electricity delivery channel with large centralized generation to passive users who had little or no choice in their energy sources, and it had surprisingly little in the way of sensing and measurement to guide operation of that grid.

Our present modernization efforts are driving new technologies at an unprecedented pace, and some of these new goals that we have include expanding the diversity in consumer choice in electricity sources including distributed and clean generation sources.

The emergence of what we call prosumers. That's a combination of a customer who has both the ability to produce and consume energy and so thereby put energy back into the grid at times.

The ability of non-utility assets such as ordinary buildings to provide services to the grid and cooperate in managing grid operations. The convergence of fuel networks, transportation networks, and even social networks with the grid itself.

And then, of course, a desire for greatly improved reliability, resilience, and security for the grid.

These changes, some of which are actually occurring virally rather than being planned, are modifying the characteristics, the behavior, and in some cases the very structure of our grid and vastly increasing its complexity of a system that's already extremely complex.

The forgoing are causing this widening gap in the ability of us to manage this grid as it evolves. As that gap widens, the ability of utilities to manage the grid reliably is increasingly challenged. The uneven penetration of new technologies mixed with legacy systems only increases that challenge. We need new methods and new tools for operating the grid in addition to having these new technologies, and we need some new kinds of components to help with the grid as well.

Among all of the very valuable technologies that are being applied to the grid, a few stand out in particular as being key to resolving this widening gap between the grid as it is changing and our ability to operate it. These technologies are crucial to the future of the grid regardless of how modernization proceeds.

In particular they are sensing and data analysis—the electronic measurement and then the processing of the information that results from those measurements to extract useful information on very large scales requires new kinds of analysis tools for the utilities.

High voltage power electronics, adjustable electronics allow us to control grid power flows in ways that are more sophisticated than our present on/off, electric-mechanical switches.

Fast and flexible bulk electric storage which can act as the buffer that evens out variations in power fluctuations caused by various diverse energy sources.

And finally, advanced planning and control methods. We need new methods that will require the next generation of high performance computing coupled with new control mathematics.

The last three of those have so much potential for positive impact that we view the combination of storage, power electronics, and advanced control to be a new general purpose grid component as fundamental as a power transformer or a circuit breaker.

The Grid Modernization effort that you've already heard about, launched by Secretary Moniz, is an important effort to systematically address all of these emerging challenges. It's designed to leverage the broad assets of the national laboratory system and to deliver an integrated plan that connects all of the grid efforts in the Department of Energy. It also recognizes the importance of partnering with industry, the states and regional stakeholders in addressing these challenges going forward.

Thank you, and I'd be happy to answer any questions.

[The prepared statement of Dr. Taft follows:]

Dr. Jeffrey Taft, Chief Architect for Electric Grid Transformation, Pacific Northwest National Laboratory

Testimony of Dr. Jeffrey Taft
Chief Architect for Electric Grid Transformation
Pacific Northwest National Laboratory
before the
U.S. Senate Committee on Energy and Natural Resources
March 17, 2015

I would like to thank the Chair, Senator Murkowski, and the ranking member, Senator Cantwell, for inviting me here today.

My name is Jeffrey Taft, and I am the Chief Architect for Electric Grid Transformation at the Pacific Northwest National Laboratory. In my statement here today, I will offer three main points:

1. We are re-engineering the grid with advanced technology to support many new capabilities.

2. These changes are challenging our existing grid structure and grid management tools.

3. Certain key technologies could help resolve the widening gap in our ability to manage and operate the 21st Century grid reliably.

In the US, we are modernizing the grid with advanced technologies to support many new goals and emerging trends that were not envisioned in the original 20th Century grid.

The 20th Century model for the grid was a one-way electricity delivery channel from large centralized generation to passive users who have no choice of electric energy sources and with surprisingly little in the way of sensing and measurement to guide operation.

Present grid modernization efforts are driving new technologies into the grid at an unprecedented pace to serve a variety of new goals and emerging trends not contemplated for the 20th Century grid, including:

- expanded diversity and consumer choice in electricity sources, including distributed and/or clean generation such as distributed solar photovoltaics, wind, and energy storage,

- emergence of "prosumers" (customers who are both energy producers and consumers)

- ability of non-utility assets such as ordinary buildings to provide services to the grid and cooperate in managing grid operations

- convergence of fuel, transportation, and social networks with the grid

- desire for greatly improved reliability, resilience, and security for the grid

These changes, some of which are occurring virally rather than being planned, are actually modifying the characteristics, behavior, and even the very structure of the grid, and are vastly increasing the complexity of the already complex US power system.

The forgoing is causing a widening gap between the real grid and the one for which existing grid management methods and tools were designed.

New technologies and new goals are reducing the effectiveness of standard methods for operating and protecting the grid. As the gap widens between the emerging grid and traditional grid control tools, the ability of utilities to manage the grid reliability is increasingly challenged. Further, uneven penetration of new technologies, mixed with legacy systems increases the challenge. New methods and tools for grid control are needed, as well as new kinds of grid components and an architectural framework to coordinate overall integration and operation.

Among all the very valuable technologies being applied to the grid, a few stand out as key to resolving the widening gap between existing grid management tools/methods, and the needs of the 21st Century grid.

Certain technologies will be crucial to the future of the grid, regardless of how modernization proceeds. They are:

- Sensing and data analysis – electronic sensing and automated information extraction that will require new data collection and analysis tools

- High voltage power electronics – adjustable electronics for controlling grid power flows to replace today's on/off electromechanical switches

- Fast flexible bulk electric energy storage – can act as the buffer that evens out various power fluctuations and mismatches that can occur with diverse energy sources

- Advanced planning and control methods and tools – new approaches using advanced control methods suitable for the modern grid that will require next generation high performance computing coupled with new control mathematics

The last three have so much potential for positive impact that we view the combination of storage, power electronics, and advanced control to be a new grid component, as fundamental as a power transformer or circuit breaker.

The Grid Modernization cross-cut initiative launched last November by Secretary Moniz at DOE is an important effort to systematically address these emerging challenges. The initiative is designed to leverage the broad assets of the national laboratory system and deliver an integrated plan that connects all grid efforts at the Department of Energy. It also recognizes the importance of partnering with industry, the states and regional stakeholders in addressing these challenges going forward.

Thank you. I would be happy to address any questions that you may have.

Additional Written Testimony

Key Emerging Trends in the US Electric Utility Industry
A number of inter-related trends in the US utility industry are beginning to reach scales at which they may impact grid operations or interact with one another.

Increasing Data Volumes from the Grid
While much of the public discussion around increasing volumes of grid data has focused on meter data, the really large volumes are in fact coming --and will continue to grow--from newer instrumentation on both transmission and distribution grids. Already the more than 1,000 Phasor Measurement Units

(PMU's) on the US transmission grid produce vast volumes of data[1], and the number of PMU's is expected to grow significantly in coming years (PNNL expects that US transmission PMU data flow will eventually reach 50,000 PMU's and 1.5 Petabytes/year). Early adopters are the reliability coordinators and system operators (ISO/RTO), with transmission system operators following. Early work on applications of PMU's at the distribution level is being done, but no significant penetration exists to date.

PMU's are devices that measure voltage and current at different points across the grid as often as from 30 to 60 times per second. These measurements are also time-stamped, or synchronized, by GPS technology, which means that by comparing the measurements at any given moment operators can get an unprecedented picture of system conditions in near real-time. A handful of utilities and federal power marketing entities pioneered the early deployment of PMU networks, for purposes of wide-area situational awareness on the transmission system, particularly in response to cascading outage events in the West in 1996 and again in 1999. The Bonneville Power Administration became the first utility to deploy such a network in 2000[2]. Public/private partnerships were key to accelerating deployment of these networks after the Northeast/Midwest blackout of 2003—the largest in US history--and a number of investments pursuant to the American Recovery and Reinvestment Act (ARRA) of 2009 successfully leveraged additional industry investment[3].

The vast amounts of data now being generated from PMU's are due to that fact that these are streaming devices--much like video--in that, they produce streams of data that are used at multiple destinations. It's expected that similar technology is about to start penetrating distribution grids, which will have orders of magnitude more streaming sensors than the transmission system. As interest in asset monitoring continues to increase, vast new volumes of asset health and operational data will be generated, some to be used in real time, some to be stored and analyzed later. The newer protection and control systems needed for advanced grid functionality, such as integration of distributed generation and responsive loads at increasing scale will generate enormous volumes of sensor data that must be transported, processed, and consumed in real time, and potentially stored for offline analysis. All told, the utility industry will experience an expansion of data collection, transport, storage, and analysis needs of several orders of magnitude by 2030. Part of this growth is due to the next item in this list. New approaches for utility data acquisition, transport, storage, and analytics processing are needed, and new operational paradigms such as Cloud storage and computing will play roles as utility business models change.

Faster System Dynamics

The implementation of new grid capabilities has brought with it great increases in the speed with which grid events occur. This is especially true on distribution grids, although the trend exists for transmission as well. In the last century, aside from protection, distribution grid control processes operated on time scale stretching from about five minutes to much longer, and human-in-the-loop was (and still is) common. With the increasing presence of technologies such as solar PV and power electronics for inverters and flow controllers, active time scales are moving down to sub-seconds and even to

---

[1] North American Synchrophasor Initiative, "PMUs and Synchrophasor Data Flow in North America as of March 19, 2014;" available at:
https://www.smartgrid.gov/sites/default/files/doc/files/naspi_pmu_data_flows_map_20140325.pdf
[2] Virden et al, "Next Steps in Grid Modernization: Early Returns on U.S. Investment, and New Innovations in Electric Infrastructure Policy & Technology"; (April 2012).
[3] For example, see: Vanzandt and Nokes, "A Western Partnership Succeeds In Enhancing Grid Reliability"; (August 2014); available at: http://smartgrid.ieee.org/august-2014/1131-a-western-partnership-succeeds-in-enhancing-grid-reliability

milliseconds. The presence of significant levels or penetration of solar PV on a feeder (where prosumers may inject power into the feeder) has led to voltage stability problems, according to initial reports from Hawaiian Electric Company (HECO) and San Diego Gas & Electric (SDG&E). Conventional feeder control has been too slow to compensate, so each utility has applied fast power electronics in the form of DSTATCOM devices to stabilize voltages. As this fast automatic control has become necessary, the need to obtain data on the same times scales on which control must operate has arisen. Consequently, there is a sort of double impact: there are many more new devices to control on the system--and much faster dynamics for each device--leading to vast new data streams and increasing dependence on ICT for data acquisition and transport, analysis, and automated decision-making.

Hidden Feedbacks and Cross-coupling

As more advanced grid applications and systems are developed and deployed, more system interactions are emerging. These interactions are inevitable, although it seems that many applications have been developed to execute specific functions—without reference to broader system implications. These interactions occur and will continue to occur because the grid itself constitutes a hidden coupling layer for all grid systems. The coupling occurs due to the electrical physics of the grid, and in most cases this coupling propagates at nearly the speed of light. Such coupling can cause effects ranging from reduced effectiveness of smart grid functions, up to and including wide area blackout. Coupling exists because of electrical connectivity: for example, devices connected to the same feeder share voltages. In instances in which Demand Response (DR) is operated independently of voltage regulation, sudden changes in load change the conditions under which voltage regulation settings were created. This, in turn, leads to the settings becoming inappropriate on very short notice – too short for the relatively slow conventional voltage regulation methods to compensate effectively.

This applies to both commercial DR and aggregated residential DR, although most commercial DR has been operated so slowly that in the past this was not much of a problem, until more fully automated DR became available. The net impacts of this particular interaction can include[4]:

- Reduction of the effectiveness of DR by as much as 15 percent;

- Voltage violations on the affected feeder; and

- Feeder level circuit breaker trips.

This situation is further complicated by the fact DR applications may be developed outside the context of utilities' distribution management systems; and that, in some cases, third-parties are bypassing utility systems altogether, to aggregate DR through direct interaction with customers.

Other interactions have the potential to create wide area blackouts if they should occur during times of low stability margin operation. As smart grid functions become more complex, it is to be expected that more interactions will become manifest. Generally, effects of such interactions will not be important at the scale of pilot projects and demonstrations, but will become significant as penetrations pass tipping points that are becoming apparent from experience at several utilities.[5,6] In such cases, the correlation and

---

[4] Medina, et al, "Demand Response and Distribution Grid Operations: Opportunities and Challenges," IEEE Trans. On Smart Grid, September, 2010, pp 193-198

[5] Thomas Bialek, "Renewable Impact on Electric Planning," available online: http://www1.eere.energy.gov/solar/pdfs/highpenforum2013_1_3_sdge.pdf

concentration of assets involved in these new grid applications—within and potentially outside utilities' control—will determine the operational consequences of such interactions.

RPS and VER Penetration

The trend of converting from traditional thermal generation to renewables such as solar and wind (known as Variable Energy Resources [VER]) has been driven by a combination of factors over the past decade, including Federal tax policy and state Renewable Portfolio Standards (RPS). Since VER is not dispatchable in the same way as traditional generation, operational challenges arise for a system originally designed around the concepts of power balance and load-following generation control. Solutions to these problems involve new types of grid components such as energy storage, but also greatly expanded measurement, data transport, analytics, and control.

Bifurcation of the Generation Model

Similarly, the VER/RPS trend is shifting the model of central station generation connected to transmission, to a mix of that and distributed generation connected to distribution grids. This split or bifurcation changes grid operations drastically, introducing multi-way real power flows and other effects not included in original grid design assumptions. In addition, distributed generation may be able to offer services *back* to the grid operator, such as reactive power regulation—a shift in paradigm that can provide operational benefits if the appropriate incentives are put in place.

Responsive Loads

Demand response has been used by the utilities for decades, mostly in conjunction with commercial and industrial customers, and mostly in a non-automated fashion. More recently, efforts have been made to create automatically responsive loads at the commercial building level, at the residential level, and even at the individual appliance level. With the rise of advanced commercial building controls, behind-the-meter storage, wide area communications, bulk power markets, and evolving approaches to "transactive" load coordination and control, the concept of building-to-grid integration is moving to a bidirectional multi-services model. This suggests it is possible that a grid/buildings convergence is forming[7]. Ultimately, this convergence would result in an extension of the grid--involving a class of assets *not* owned by the utilities. In this scenario, the observability and controllability issues resident in the operation of today's distribution grids will also extend to include grid-connected, responsive loads.

Changing Fuel Mix

The change from traditional thermal generation to renewable sources is one shift that's been underway for some time; but it is also the case that retirements of coal and potentially nuclear plants will manifest in new grid operating characteristics. In addition, this trend is surfacing in its effect on utility planning – for example, in some regions, gas pipeline planning and build-out has to a significant degree displaced transmission line planning and build-out. At present, the industry lacks the tools required for *joint* electric and gas system planning. Moreover, Information and Communication Technology platforms do not yet commonly exist within utilities to support interactions with both electric and natural gas markets.

---

[6] Martin Lamonica, "Why is Hawaii Scaling Back on Solar?" Greenbiz.com, Jan 28, 2014, available online: http://www.greenbiz.com/blog/2014/01/28/solar-hawaii-utilities-scaling-back

[7] See Hagerman et al, "Buildings to Grid Technical Opportunities"; available at: http://energy.gov/sites/prod/files/2014/03/f14/B2G_Tech_Opps--Intro_and_Vision.pdf

Evolving Industry/Business Models and Structure

A number of key stakeholders believe that the penetration of new functions at the distribution level, along with responsive loads and distributed generation, is causing the original mode of distribution operations to become inadequate. Proceedings in Hawaii, New York and California are all aimed at reconsidering the roles and responsibilities of distribution grid operators as is much thought leadership in the industry at large[8,9,10].

Evolving Control System Needs

Utility controls systems have traditionally been centralized, with hub and spoke communication to remote subsystems and equipment, as needed. As the various trends cited above have emerged, the need for changes in control system structure has become apparent. Specifically, control systems must change from being centralized, to a hybrid of central and distributed control. Distributed control is distinguished from *decentralized* control in the following important way: decentralized control involves moving some control to remote locations; but the remote elements perform controls tasks in isolation, with perhaps some coordination from a centralized supervisory element. Distributed control includes those aspects, but also is distinguished by the following: the decentralized elements cooperate on solving a common problem. This aspect creates new requirements for communication, methods for coordinating the elements and converging on a common solution.

Interdependence on Information and Communication Technology

Interdependence of electric and ICT infrastructure has increased in recent decades, and recent trends in the utility industry suggest an even tighter coupling of these networks in coming years. While cyber vulnerability must remain a focus of federal research, development and information-sharing efforts with industry, the convergence of these networks also holds substantial promise as a platform for energy innovation, leading to potential new value streams and enhanced system resilience. The pace at which this convergence occurs and new services and operational methods emerge will turn on a number of factors, including regulatory structures that set the framework within which utilities and grid operators prioritize infrastructure investment decisions.

In assessing the challenges and opportunities presented by the enhanced interdependence of grid and ICT infrastructure, it is key to understand the ways in which utilities might use different classes of data, the characteristics (such as latency) that determine the operational and business value of that data, the implications for communications network investments, and required evolution in analytics, visualization and software tools that will help unlock new services and bolster desired system attributes such as resilience. Grounded in this understanding, a handful of key priorities emerge as potentially appropriate initiatives designed to convene relevant stakeholders, provide tools and methods that help inform industry investment strategies, and accelerate the pace at which innovations are brought to market. In particular:

- Leadership in convening industry stakeholders for purposes of developing a *reference architecture for control systems*-- extensible across electric and ICT networks--is a key first step in enabling the

---

[8] Lorenzo Kristov and Paul De Martini, "21st Century Electric Distribution System Operations," available online: http://smart.caltech.edu/papers/21stCElectricSystemOperations050714.pdf

[9] Hawaii PUC, "Exhibit A: Commission's Inclinations on the Future of Hawaii's Electric Utilities," available online: http://puc.hawaii.gov/wp-content/uploads/2014/04/Commissions-Inclinations.pdf

[10] NYS Department of Public Service staff, "Reforming the Energy Vision," available online: http://www3.dps.ny.gov/W/PSCWeb.nsf/96f0fec0b45a3c6485257688006a701a/26be8a93967e604785257cc40066b91a/$FILE/ATTK0J3L.pdf/Reforming%20The%20Energy%20Vision%20(REV)%20REPORT%204.25.%2014.pdf

kinds of innovations that will enhance grid observability and controllability in coming decades. A reference architecture is a technology neutral framework applicable to complex systems such as the grid, which takes a disciplined approach to characterizing system components, structures and attributes. Such an architecture helps identify potential gaps in technology and operations, assists in defining key system and component interfaces and provides context for interoperability and standards-setting activities.

- Exploration of mechanisms and tools relevant to *ensuring ICT network investments are sufficient* to enable enhanced grid management functions at the distribution level. For example, while wireless mesh networks built to support AMI deployments are more affordable than optical fiber or other advanced wireless technologies, certain characteristics may render them insufficient for system restoration and resilience functions in an outage or emergency scenario. In addition, early indications suggest meter communication networks have often been designed only to support consumers' usage reporting and thus lack the bandwidth and latency capabilities needed to support operation as a grid sensor network. Networks are increasingly integral to modern power grid operations, yet most power grid simulation and design tools today lack means to include communications-related elements. Measures to accelerate integration efforts and move them into use by utility planners and design engineers may help better inform grid operators' investment decisions. Moreover, certain regulatory reforms and/or tax incentives to encourage appropriately scaled investments may warrant consideration.

- And finally, *acceleration of ongoing research and development efforts to develop new grid management tools*, linking capabilities in high-performance computing and advanced power systems engineering with software developers and utility/grid operators. While it's reasonable to expect the commercial marketplace to *eventually* solve issues associated with emerging software needs of the utility industry, it is unclear this will take place in time to keep pace with the changing operational landscape of the grid, particularly at the distribution level. That's because software developers face a classic "chicken and egg" scenario. The market for these solutions is relatively thin (confined to the number of utilities in North America), which leads to conservatism in investing in new products for control systems that, in essence, might also replace existing product lines. Utilities, in turn, may agree with an assessment of their changing needs—but don't commit to buying new solutions until they have been well tested and demonstrated. Within this context, DOE can play a key, ongoing role in bringing together the ecosystem of stakeholders required to accelerate the path for new products from the laboratory to control rooms, in a way that unlocks new value streams and bolsters system attributes such as enhanced reliability and resilience.

Architectural View of the Grid

The grid may be viewed as a complex network of structures that has evolved over the past century, driven by a patchwork of regional economic prerogatives, diverse business models and variable regulatory structures. A number of current trends including the convergence of electricity and natural gas infrastructures, and the bifurcation of generation—the emerging split of generation between bulk transmission-connected generation and smaller distribution-connected resources—are adding additional complexity, as well as providing potential opportunity to create new value streams and enhance system resilience.

Below, selected views of selected present grid structures provide a number of key insights relevant to emerging trends, specifically with respect to industry structure, business/value streams, electric/power system structure, and control/coordination frameworks.

Industry Structure

Geographic-based structures have shaped the evolution of the electric power industry over the past century. However, the deployment of more non-utility assets interacting with the grid and emergence of merchant and prosumer-controlled distributed energy resources operating as a set or group despite wide geographic dispersal can erode the concept of a geographically bounded customer.

A review of industry structures shows that distribution operations are disconnected from the rest of the system in a control and coordination sense. In certain contexts, however, system operators at the wholesale level had already begun bypassing distribution utilities to directly engage distributed energy resources in the last few years. Recent court rulings and industry deliberations on the future of distribution have already opened up reconsideration of roles and responsibilities for ensuring system reliability, especially at the distribution level and have implications for grid control and coordination structure. However, many state and local laws and regulation would have to be changed.

Business/Value Stream Structure

Modeling the accrual of value streams within industry structures helps illustrate the kinds of business ecosystem partnerships required to realize such value. Regulatory variables figure prominently in determining which entities can realize such value, and what forms these values (products or services) may take.

Low-growth value streams are those most directly connected to provision of electricity as a regulated commodity; whereas potential high-growth streams are tethered to customer/prosumer products, devices, and services. Once again, what value streams are regulated, by whom, and under what terms, will bear on the distribution of these opportunities in what is essentially a zero-sum situation, and what entities are positioned to capture shares of the sum.

Electric System Structure

The structure of the grid determines important system properties and basic limits. For example, in major cities, the structure of dense underground urban mesh underlying the distribution system limits any services that buildings might supply to grids to the local feeder secondary, except for those that reduce net load. In these contexts, distributed generation and storage cannot push power back into the distribution feeders, and thus cannot push power to the grid. Furthermore, tripping of multiple network protectors will cause a portion of the secondary mesh to island. Since the network protectors are not coordinated, the extent of the island is unpredictable. Where fuses are used in the secondary, some of these may blow, requiring truck rolls to replace before normal operation can be restored.

The enablement of two-way flows within distribution systems in the face of such structural limitations can have costs that go beyond those related to new premises equipment and software. Some amount of change at the utility level may be needed just to unblock the potential for certain building-to-grid energy/power services.

While basic coupling occurs electrically at multiple levels in the grid, coupling can and does occur in other ways, some of which can be quite subtle. Coupling can occur through controls, markets, communications networks, fuel systems, loads, and social interactions of customers/prosumers. Unsuspected coupling is a hazard of increasing grid complexity.

The list of interactions between system elements is growing as the penetration of new devices and functionality increases. Responsibilities for reliability management have historically been established hierarchically, starting with wholesale generation/transmission treated in a semi-integrated fashion, but then separately at a lower level within distribution—where reliability requirements have historically been assigned to single regulated entities. Two-way power flows within distribution systems will require greater focus on making more explicit shared responsibilities for reliability management (and supporting investments) between distribution system operators and loads/producers within that distribution system.

Another structural consideration relates to system inertia and coupling of generators with droop control through the transmission system, which is crucial to proper grid operation. The implications for system inertia associated with replacing traditional forms of central station generation with DG and variable resources are not thoroughly understood. This is particularly the case in the loosely coupled Western Interconnection. Exploring methods for measuring—and potentially predicting—system inertia associated with existing operations as well as in context of a changing generation mix may provide key insights for policymakers and regulators concerned with system reliability. At present, this may require additional R&D efforts. In addition, such methods would be useful in the development of joint planning tools, which likewise do not yet exist for purposes of enhancing industry and policymakers' understanding of emerging infrastructure interdependencies (such as electricity and natural gas). Meanwhile, efforts underway in ERCOT to consider inertia-related grid services merit careful attention. Novel configurations of assets at the distribution level (including storage) may ultimately be leveraged to help provide such services—but once again, regulatory friction associated with determining which entities are eligible to provide such services, and allocation of costs and benefits, may once arise under current law.

Control/Coordination Framework

The inclusion of markets inside closed loop grid controls means that markets could contribute to control instability. The problem will worsen with additional entities in the loop and the presence of faster dynamics and diverse sources of net load volatility.

Consider the isolation of distribution control and coordination from the rest of the grid in the light of regulatory structure, namely the Federal regulation of the bulk power system, versus State and local regulation of distribution grids. Note that regulatory structure, industry structure and control/coordination structure are currently aligned—but this alignment is with a control structure that is increasingly problematic as the grid changes due to emerging trends.

In particular, the changing nature of system dynamics, implications of DER deployment at increasing scale, new technologies and models of consumer engagement are putting pressure on regulatory boundaries drawn over the past century. Current academic and industry literature suggests a consideration of a new, Distribution System Operator (DSO) model, though this thinking is very new and includes a highly varied set of topics.

Grid Architecture 2030

Looking forward using grid architecture principles and methods, it is possible to derive a number of preliminary insights:

Buildings: Buildings are significant users of electricity. Today, they exist primarily as passive loads, but hold promise for potentially providing services back to the grid in a transactive mode. The key grid-side factors limiting the expansion of building-to-grid services are not interoperability or interface standards (important though these are) or quantification of value streams. Instead, they are structural limitations to the distribution grid (such as those previously discussed in context of dense urban mesh), and current lack of a coordination mechanism on the grid side that extends across the grid/building boundary.

Storage: Storage is unique in that it can be capable of taking energy or power from the grid, adding energy or power to the grid, and supplying a wide range of grid services on short (sub-second) and long (hours) time scales. It can supply a variety of services simultaneously. There is an emerging sense that the combination of fast bilateral storage, flexible grid interface mechanisms, and advanced optimizing control is _a general purpose grid element as fundamental as power transformers and circuit breakers._ One of the most significant impacts of storage will be the ability to _decouple generation and load volatilities._ Since it is known that the impact of storage can be location-dependent, there is a need for new planning tools and procedures to make use of storage as a standard grid component, and to optimize storage location and size.

Whole Grid Coordination (Laminar Coordination Framework): Coordination is the means by which distributed elements are made to cooperate to solve a common problem—in this case, grid control. It is clear that existing grid coordination has gaps and lacks a rigorous basis—and that the gap is widening, with respect to grid behavior and desired capabilities. Where the grid is concerned, a structure that accommodates multiple simultaneous approaches to control is likely required. **_Local optimization inside global coordination_** is a principle for a mix of centralized and distributed control that provides properties such as boundary deference, control federation, disaggregation and scalability.

ISO's/RTO's and DER Dispatch: In certain (but not all) markets today, DER is being dispatched by Independent System Operators or Regional Transmission Operators, which retain system balancing and reliability responsibilities at the transmission level, and also operate wholesale markets. The ISO/RTO approach has led to several problems that have caused the industry to seek alternative arrangements. For example, letting an ISO/RTO handle DER causes a bypassing of distribution operators, which introduces ambiguity in the responsibility for distribution reliability, compromising the ability of the distribution operator to manage its assets and operations. A recent 3rd Circuit Court of Appeals ruling, a position statement by PJM, and proceedings in California and New York are addressing these issues. Second, as the number of devices that can participate in the markets and grid operations grows, a scaling problem arises in terms of communications, as well as in the complexity and computational requirements associated with control mechanisms (and associated latency requirements).

DSO Structure: While motivated by the need to clarify and simplify responsibility for distributed reliability, the emerging thinking around a Distribution System Operator (DSO) model appears entirely consonant with a laminar coordination structure. Since the laminar structure was motivated by the need for whole grid coordination with a rigorous basis for predicting properties such as scalability, it is reasonable to expect that the DSO model can share those properties that derive from such structure. If the DSO were to be implemented as an independent DSO (IDSO), then the IDSO may have issues of economy of scale sufficient to be viable and related cost problems.

Power Electronics/AC Power Flow: There are several means to adjust power flows in AC power systems, including phase shift transformers, variable frequency transformers, and various forms of power electronics. Power electronics get attention as edge connection tools, in the form of inverters for solar PV and storage, but can be used internally in the grid for power flow control.

Flexible Electric Circuit Operation: Adjustable flow control can be used to provide flexibility in electric circuit operation. It can also be used to cut or limit the effect of some kinds of constraints that exist in present circuits, such as unwanted cross feeder flows or unscheduled flows to the transmission system. Meshing provides more paths for power flow (with flow controllers directing the "traffic"), such that it becomes possible to make more effective use of storage and distribution level DG. That means the cost effectiveness of such assets is enhanced two ways: better sharing of the assets, and enablement of new value streams and innovations. At present, distribution grids suffer from poor observability given their lack of sensing capability. Additional efforts to develop observability strategies and tools for design of distribution sensor networks would further enhance flexible circuit operations.

The CHAIRMAN. Thank you, Dr. Taft.

Ms. Barton, welcome.

STATEMENT OF LISA BARTON, EXECUTIVE VICE PRESIDENT, TRANSMISSION, AMERICAN ELECTRIC POWER

Ms. BARTON. Good morning, Chairman Murkowski, Ranking Senator Cantwell, members of the Committee and fellow panelists. I appreciate the opportunity to speak with you here today.

My name is Lisa Barton. I'm the Executive Vice President of Transmission for American Electric Power. I also serve on the Board of Directors of Reliability First Corporation.

There are three key points that I'd like to leave you with today for your consideration.

The first is that the grid is a natural enabler of new technologies. When I think about the grid and what it does for new technologies, I make the analogy between its and our robust data network. Today we have the ability with smart devices at any time, nearly anywhere, to pull out that device and be able to do tasks that ten years ago we never would have dreamed possible. The high voltage electric grid is similar and plays a similar role in society. It provides the backbone that supports diverse generation and distributed energy technologies.

The AEP grid is designed to accommodate two way power flows, so it is different from the distribution system, and it's a point worth noting.

It is also designed to withstand component failures, designed to withstand the loss of a large generator and the loss of one or more large transmission lines, so it is an extraordinarily robust system. The grid ensures that electricity is delivered in a cost effective, efficient, and reliable manner whether it's produced at a large power plant or on the roof of a house.

The second point that I'd like to convey is the importance of ensuring a reliable and resilient grid for our economy and for our national security.

The grid has evolved over time from a fragmented group of small, inter-dispersed systems to one that is a networked, a very strong networked system. It has been touted as the largest synchronist machine on the Earth. We own the system that many nations aspire to own. Therefore it is vital that we do not casually discount the power of the grid, the existing fleet of generation, and the tens of thousands of utility workers, particularly line workers, who stand ready to address system needs during times of system emergencies.

The grid serves as the foundation of our economy, prosperity, and national security. As policy makers examine and evaluate the purpose of the grid and the potential for new technology such as wind, solar, distributed generation, battery technologies, and microgrids it's important to recognize that the benefits to consumers will be maximized by combining the strengths of these resources. It's imperative that we view these resources as complementing each other as opposed to competing with each other.

Finally, it's imperative that we support diversified technological solutions. To maximize the benefits to the grid, we really need to avoid picking winners and losers and instead allow the market to identify the best solution for the particular circumstance.

With respect to AEP's territory, one of the unique things that we have is a unique perspective given the vastness of the territory. We have 11 states where we have transmission facilities, and we're also constructing new projects in two additional states. Our service territory touches Mexico and goes as far north as Michigan.

As noted in my testimony there are different solutions that are in place to fix the problem at hand. We do not have the same system in West Virginia as we do in Texas and therefore the technologies that we employ in the system itself are very different in each of those areas. There is no one size fits all solution with respect to the grid.

Imagine a world where say, three years ago, we chose one application or one technology over another. We would have missed out on a much more advanced and better solutions, so it's important that we don't make those same mistakes with respect to the grid. By supporting a diverse set of generation resources continuing to invest in the backbone of the grid, we will ensure that energy is available, affordable, and reliable while enabling customers to optimize cutting edge technologies for their benefits.

I thank you for the opportunity to speak with you here today, and I welcome any questions.

[The prepared statement of Ms. Barton follows:]

Testimony of
Lisa M. Barton
Executive Vice President, Transmission
American Electric Power

on

The State of Technological Innovation Related to the Electric Grid

Before the
Committee on Energy and Natural Resources
United States Senate

March 17, 2015

Good morning Chair Murkowski and Ranking Member Cantwell, members of the Committee and fellow panelists. My name is Lisa M. Barton, and I am Executive Vice President of Transmission for American Electric Power (AEP). I also serve on the board of directors of Reliability First Corporation, the regional reliability organization for a 13-state region stretching from Wisconsin to Virginia. Thank you for inviting me to testify today.

AEP is one of the largest electric utilities in the United States, delivering electricity to more than 5.3 million customers in 11 states. AEP owns nearly 38,000 megawatts (MW) of generating capacity in the U.S., and owns and operates the nation's largest electricity transmission system, with 40,000 miles of transmission lines. AEP's network includes more 765-kilovolt (kV) extra-high voltage transmission lines than all other U.S. transmission systems combined. AEP's transmission system directly or indirectly serves about 10 percent of the electricity demand in the Eastern Interconnection, and approximately 11 percent of the electricity demand in ERCOT, the transmission system that covers much of Texas. AEP's headquarters are in Columbus, Ohio.

Today's hearing seeks to evaluate the state of technological innovation related to the electric grid. In my testimony, I will describe AEP's experience with a number of advanced grid technologies. I hope to leave you with three key messages today:

- **A robust grid is a critical enabler of generation diversity, new storage and demand-side technologies**. Just as the nation's robust data network serves as a foundation to modern communications and provides an enabling function for various technologies across the communication sector, the nation's high-voltage electric grid serves a similar role with respect to enabling diversity in generation and distributed energy technologies. The electric grid aggregates generation and demand-side technologies and ensures that resources, from whatever source derived, are delivered to customers in a cost-effective, efficient and reliable manner.

- **Maintaining a reliable and resilient grid is critical to economic and national security**. The reliability provided by the integrated electric grid serves as the foundation of our economy; it provides stable electricity service necessary for our economic well-being. As policy makers examine and evaluate the potential for new electricity technologies such as solar, wind, distributed generation, grid-scale battery technologies and micro grids, it is critical to appreciate how the system works to maintain the high level of reliability and affordability that we currently enjoy. Integrated appropriately, advancements in generation, grid and end-use technology will over time serve to strengthen the robustness of the network by providing greater diversity in resources and better responsiveness in the grid itself, supporting reliable delivery of power to all consumers.

- **To maximize the beneficial impact of new technologies, policymakers should avoid picking winners and losers and allow the market to identify the best solutions for a particular circumstance**. Today, we are seeing new technology being applied and implemented all along the value chain from generation, through transmission and distribution, to homes and businesses. Technologies that are targeted to address deficiencies in reliability, improve system efficiencies, reduce the cost to consumers or diversify generation resources should be applied as the specific system needs and circumstances dictate. The choice among competing technologies should be driven by relative performance and cost.

I.   **The Grid of the Future Will Be More Flexible and Adaptable for the Benefit of Consumers and the U.S. Economy**

A key element of any "utility of the future" model will be a modern, efficient grid that not only handles new generation and end-use technologies but also enhances the efficiency of the existing grid. To succeed in the future, our industry not only must continue doing what it does today in terms of enhancing and improving reliability and connectivity, it also must enable the integration of new technologies. As our customers are able to more fully utilize the electric grid as a technology integration network, they will realize its full value.

The electrification of the U.S. started with smaller generators and isolated utilities; over time, these resources were networked together through the increasingly integrated grid that we know today. The integrated grid that has developed over the last century served to increase reliability, resiliency and efficiency. No longer did a neighborhood need to rely on one local source of power - it could rely on larger, more efficient units that were more cost-effective and diversified when integrated together. Diversification in the source, location and type of generation today continues with the integration of wind and solar generation and micro grids.

The transmission system evolved over time to be the aggregator of generation and an efficient distributor of energy to load centers. While originally designed for the movement of conventional power resources to distribution systems, the grid has evolved to integrate variable energy resources such as wind and solar. For example, the AEP 765-kV and 345-kV extra-high voltage transmission lines initially were planned to integrate AEP-owned generation with AEP load. Today, those lines provide 9,300 MW of transfer capacity from western PJM and MISO to eastern PJM, supporting all types of resources, including wind. In the future, the grid will need to evolve further and be more intelligent and responsive, better able to manage two-way flows of power and information. This will support further integration of distributed energy resources – such as rooftop solar and micro grids – as well as larger-scale intermittent generation resources such as wind and solar in a cost-effective and reliable manner.

A smarter grid will be especially important to accommodate the coming changes in how and when customers use electricity, providing consumers with more information and choices with respect to energy consumption. The demands on today's grid have changed from a few decades ago, and the demands in the future will continue to be shaped by consumer consumption patterns, which are in turn shaped by new technologies such as smart appliances, plug-in electric vehicles, and customers managing their electric use with mobile devices. The continued evolution of the grid to incorporate new technologies is essential and will provide for a more flexible, resilient and interactive grid to advance evolving societal needs.

II. Reliable, Affordable Electric Service Depends on a Robust and Resilient Backbone Transmission Grid

New electric technologies such as micro grids, distributed generation, demand response and other localized solutions will complement rather than replace conventional generation and the transmission network that brings that power to homes and businesses every day. Today's system provides a tremendous amount of resiliency whereby customers are no longer exposed to outages associated with the loss of a single transmission line or a generating station; rather the grid ensures reliable power even during times when there are maintenance or storm-related outages of major system components.

AEP strongly supports implementation of technologies that, working together, improve the resiliency, functionality, reliability and operability of the grid. When all is said and done, maintaining an adequate level of generation resources, in combination with a robust grid, is necessary to ensure reliability is maintained.

The importance of maintaining the reliability of service provided by the system as policies change was recently demonstrated by Germany's efforts to promote renewable generation, where implementation had the unintended consequence of forcing conventional generation out of the mix to the detriment of reliability. As a consequence, Germany's Federal Ministry of Economics & Technology issued a New Energy Policy in 2012 recognizing that "Conventional power stations will remain indispensable to our electricity supply in the years ahead. This is because they can do what most renewable energy sources cannot: provide a reliable supply of power precisely when it is needed."[1]

---

[1] Federal Ministry of Economics and Technology, Germany's New Energy Policy – Claims and Challenges (July 2012), *available at* http://www.diha.al/fileadmin/ahk_albanien/Dokumente/Germanys_New_Energy_Policy.pdf.

III.   New Environmental Requirements Will Require Additional Transmission Investments

One of the largest drivers of transmission investments is large scale changes in the location and type of generation resources on the system. As units retire and new resources are added to the system, investments in transmission are needed to ensure that the grid continues to function as the reliable aggregator of generation.   It is clear that the implementation of the Environmental Protection Agency's Clean Power Plan will require significant transmission construction to interconnect the new natural gas generation that will be required, to interconnect location-constrained resources such as wind and centralized solar power, and to preserve grid stability given the retirement of coal-fueled units. To maintain the level of reliability and system resiliency we enjoy today, the retired generation cannot simply be replaced with distributed generation alone; transmission investments will be needed.

While constructing this new transmission infrastructure provides an opportunity to expand and update the current grid, this work, like all electric infrastructure development, takes time. It will require significant lead times to obtain approvals, permits, and rights-of-way, and to complete construction. Let me bring to your attention an excellent law review article written by two former Federal Energy Regulatory Commission regulators on the difficulties surrounding transmission development, *Regulatory Federalism and Development of Electric Transmission: A Brewing Storm.*[2] The article proposes a balanced approach to state-federal and state-to-state jurisdictional disagreements that often impede transmission development.

IV.   AEP's Experience with Key Grid Technology Innovations

Like any design challenge, aligning the solution with the problem encountered is essential. AEP's service territory covers 11 states, and we own transmission assets in 13 states. The diversity of our service territory has been the mother of invention and many innovative technologies have been utilized by AEP to tailor the solution to the need. Innovations that we have advanced include:

A. BOLD™ - Moving More Electric Power over Greater Distances with Fewer Losses

Breakthrough Overhead Line Design™ (BOLD™) is a new type of transmission line developed by AEP that utilizes a more compact and efficient configuration, featuring a single arch shaped tubular cross arm that supports the circuits. As a result, BOLD™ towers are 33% shorter than the standard transmission pole and utilize less right-of-way than higher voltage facilities that would transport the same amount of power. It offers a high capacity, reliable solution to many of the nation's transmission challenges. BOLD™ can effectively deliver large blocks of power over long distances, connecting

---

[2] James J. Hoecker & Douglas W. Smith, Regulatory Federalism and Development of Electric Transmission: A Brewing Storm?, 35 Energy L.J. 71 (2014), *available at* http://www.felj.org/sites/default/files/docs/elj351/16-71-Hoecker-Smith_Final%205.13.14.pdf.

remote renewable generation projects to the power grid and load centers while boosting the load capacity of extra-high-voltage lines by 50 percent or more and requiring no more right-of-way than traditional 345-kV lines. This technology can also be used to increase capacity in the same right-of-way when replacing existing older lines with BOLD™ technology.

The BOLD™ design can mitigate the need for additional transmission lines. Furthermore, BOLD™ offers lower magnetic field strengths, reduced energy losses and greater aesthetic appeal. This technology is best applied in areas where there are right-of-way constraints, the need to increase the amount of power being moved or where there is a desire not to introduce higher voltages. AEP is piloting this technology in Indiana with a project that will be in service in 2016.

B. Variable Frequency Transformer

The Variable Frequency Transformer (VFT) is a controllable bi-directional power flow control device used to transmit electricity between two systems similar to a back-to-back high-voltage direct current (HVDC) converter. The VFT is based on the combination of hydro generator and transformer technologies. AEP deployed the first application of the VFT in the United States to address reliability problems in a load pocket near Laredo, Texas. The VFT was utilized by AEP to tie the Mexico and Texas grids together asynchronously in support of the Laredo load pocket. The transmission into the Laredo area, at the time, was no longer capable of supporting the load and the condition was made worse with the shutdown of the generation within the load pocket. The long-term solution is ultimately the completion of a new 345-kV transmission line into the area, but a short-term solution was also required. The VFT technology provides a controlled transmission path between the U.S. and Mexican electrical grids, improving reliability and permitting power exchanges, which was not possible with conventional technologies.

C. Phase Shifting Transformers

The Phase Shifting Transformers (PST) are a special type of transformer that is connected in series with the transmission line to control the flow of power between the sending end and receiving end of a line. Today, AEP employs PSTs to balance the power flows on the transmission system to avoid thermal overloads on transmission lines and more efficiently and effectively utilize the capacity of the grid. This not only improves the reliability of the system, but it also provides the system operator with additional flexibility to manage the maintenance of the system. Use of this technology is applicable in areas where the transmission system is weak and there is a need to change power flows to better balance flows across large geographic areas. Using this technology allows for larger transmission investments to be deferred into the future, giving AEP the flexibility to utilize capital more efficiently in other areas of the transmission system.

D. Drop-in Control Modules

AEP has designed and expanded the use of prefabricated drop-in control modules (DICM) to meet rising customer demand, especially from the oil and gas industry, and to provide grid reliability as coal units retire. A DICM is a factory-built module made to meet AEP specifications that houses all the communication and control technology for the station. It can be placed into service in eight to ten weeks, which is half the time needed to build a traditional control room. Although prefabricated control rooms are not new, AEP designed its units to be more flexible and expandable. Our extensive use of DICMs began in 2011, and we are currently seeking a patent for a DICM expansion concept. They are reliable, save time and have proven to be a cost-efficient solution. Since 2011, AEP Transmission has installed more than 200 DICMs throughout our service territory. Our goal is to install 2,000 over the next 20 years. Use of this technology addresses aging infrastructure, future reliability compliance, physical security and situational awareness.

E. Energy Storage-Battery Technology

In 2006, AEP was the first utility in North America to deploy a megawatt-scale sodium sulfur (NaS) battery at its Chemical Station in Charleston, WV. In 2009, AEP's Electric Transmission Texas installed two 2.4-megawatt NaS batteries in Presidio, Texas, to provide transmission backup in the event of a transmission line outage. Presidio is a small, remote community bordering Mexico along the Rio Grande River – the only load at the end of a single radial transmission line. Previously, when Presidio's line encountered an outage, the town had an immediate blackout and its only alternative electricity could come from Mexico.

Energy storage via battery technology has long been viewed as a game changer for the electricity industry, if it could be implemented cost-effectively. Currently, the electricity infrastructure is built to address peak loads, recognizing significant fluctuations in energy consumption throughout the day and throughout the year. If cost-effective energy storage devices such as batteries became commercially available, they could effectively change the existing planning parameters and applications of our assets in a profound manner. For example, if storage technology can be cost-
effectively combined with intermittent renewable generation such as wind and solar, it can mitigate the problem of output variability when renewable generation portfolios are significant. In most circumstances, the cost of energy storage currently exceeds what the market will support. Storage costs need to come down, and that can only happen with increased research and development, and greater market penetration. Deployment of energy storage can also face regulatory barriers, because it offers multiple types of services. For instance, a battery project may be a low-cost alternative to a transmission/distribution upgrade, may have value in energy markets, and may provide ancillary services, but capturing this range of values may not be practical given regulatory treatments.

F. Volt Var Optimization – an Intelligent Energy Efficiency Opportunity

Volt-Var Optimization (VVO) is technology that manages voltage as power moves from a substation to household appliances. VVO reduces demand and energy by 2-3 percent and sometimes more on a reliable basis at the customer's meter, which lowers customer bills. It does not require any changes to a consumer's equipment or behavior. Verifiable benefits are realized immediately upon deployment. This cost-effective technology provides for flexible implementation, because it can be used on a stand-alone basis or as part of a larger smart grid deployment. The technology is not proprietary; several suppliers manufacture the necessary equipment and software.

VVO increases the efficiency of the distribution system, reduces the need for distribution capital investments, and reduces air emissions associated with avoided energy production. It also provides a platform for future grid modernization that will deliver greater visibility and control of distribution system operations and improved reliability with relatively small incremental investment. In order to be successfully deployed, investment in this technology needs to be supported though the state regulatory processes.

G. Solar Pilot by AEP's Indiana Michigan Power Company

On February 4, 2015, the Indiana Utility Regulatory Commission approved the application of AEP's Indiana Michigan Power Company (I&M) to build, own and operate a Clean Energy Solar Pilot Project. The pilot project will consist of four to five separate solar facilities totaling nearly 16 MW, most of which will be on or near existing and future substation properties. Locating them in this way helps to minimize the cost of delivering the energy to the transmission grid. The addition of zero-carbon solar also meets the increasing interest of customers who want to use more renewable energy to meet their needs.

H. AEP Ohio gridSMART® Demonstration Project

This demonstration project tested a number of energy-saving programs. The heart of the gridSMART® Demonstration Project is the smart meter, a digital electric meter equipped with two-way communications technology. With the installation of the smart meter, AEP Ohio was able to develop and offer many innovative customer services and programs. Because smart meters are able to communicate in real-time, the company is better equipped to detect power outage locations, improve reliability, and provide faster response to certain customer service requests such as meter reading and service connections.

One project goal was to develop programs that would help customers manage their electricity use and save money. Some of the programs AEP Ohio tested include SMART Shift, a time-of-day rate plan that helped customers save money by moving electricity use to off-peak times, and SMART Cooling^{SM}, an air-conditioning conservation program that helped reduce peak demand during the summer months. In addition to smart meters, the company was able to test other smart grid technologies, such as:

- Distribution Automation Circuit Reconfiguration (DACR) – A system that allows the automatic re-routing of electricity during service interruption, limiting the scale of outages.

- Volt Var Optimization – Technology that manages voltage as power moves from substations to household appliances.

- Smart Appliances – Clothes washers, dryers, refrigerators and other appliances that can work with smart meters to respond to high energy demand and operate all or parts of the appliance when costs are lower.

V. Conclusion

Whatever the technology future holds for the electricity sector, a robust integrated electric grid will be essential to providing reliable and affordable delivery of electricity to the nation's consumers. Ongoing investments will be needed to reshape the grid in light of changes to the generation mix driven by regulatory requirements and market forces. These enhancements will further the robustness of the grid and enable greater implementation of advanced grid technologies, which should be incorporated wherever they can cost-effectively enhance the flexibility and reliability of the grid.

Thank you for the opportunity to address you on these important issues. I would be happy to respond to any questions.

The CHAIRMAN. Thank you, Ms. Barton. And finally, let's go to Ms. Edgar. Welcome.

STATEMENT OF HON. LISA EDGAR, PRESIDENT, NATIONAL ASSOCIATION OF REGULATORY UTILITY COMMISSIONERS AND COMMISSIONER, FLORIDA PUBLIC SERVICE COMMISSION

Ms. EDGAR. Thank you. Good morning, Chairman Murkowski, Ranking Member Cantwell, Committee Members. My name is Lisa Edgar, and I have the honor of serving as President of the National Association of Regulatory Utility Commissioners, also known as NARUC. I'm also a member of the Florida Public Service Commission, and my comments today reflect both responsibilities.

Thank you for the opportunity to testify on issues regarding technological innovation and the electric grid. I applaud the Committee for holding this hearing and for recognizing the role and advances the states have made to improve electric service. For state utility regulators ensuring the safe, reliable, and affordable delivery of essential utility service is our most pressing duty.

Today's hearing is very timely. Coast to coast change is happening all around the electricity industry from smart grid deployments to energy efficiency and distributed generation projects, state public utility commissions are on the front lines and pursuing new and innovative programs across the country. As you've heard today DG can offer economic, reliability, and environmental benefits to consumers who are able to access and use them. When combined with smart meters and other resources DG can significantly change how some consumers use and consume electricity.

These resources may also transform our current utility construct in ways we haven't even imagined. Yet it is important to remember that while consumers come in all shapes and sizes from residential to large industrial, the expectation and the need for affordable, reliable service is the same no matter who is producing or delivering their electricity.

To be sure, states are leading the way in implementing DG programs. At last count over 43 states and the District of Columbia had adopted net metering policies. In addition numerous states have deployed smart and distributed resources.

For example, in Florida, my state, one of our utilities recently installed a smart grid system called Energy Smart Florida. This program installed over four and a half million smart meters. Now, of course, this did not come for free or without controversy, but it is a concrete step to keep Florida on the path for a smarter, nimbler and more reliable grid.

But while DG can have multiple benefits it also brings idiosyncrasies and challenges that cannot be ignored. Solar and wind resources are not dispatchable. So if they are needed at a time that the sun or the wind isn't producing those contributions are limited. Solar and wind also need support to operate under many different scenarios and configurations.

Likewise small, backup fossil units can have worse air emission profiles than utility scale units. Grid operators generally don't control these resources, and it is hard to predict when they will come into the system, where and for what time period. These advantages and tradeoffs must be better understood and balanced while making policy and cost allocation decisions.

In our view experience has demonstrated that states pursuing these initiatives at their own pace is good policy. We understand the value these technologies bring, but we also recognize the challenges associated with integrating them into the grid. Utilities are required to provide electricity 24 hours a day, seven days a week. We need to ensure that the core responsibilities of reliability and affordability are maintained while taking into account our local and regional differences.

At NARUC our members have passed resolutions recognizing the many collaborative efforts between regulators, consumer advocates, utilities and other key stakeholders to address the potential for DG and other technologies. These efforts continue to be important as we work to better evaluate the benefits and the costs as these technologies grow throughout the country.

In addition hearings like this help ensure that necessary consumer protections are maintained.

It remains our responsibility to facilitate the continued provision of safe, reliable, resilient, secure, cost effective, and environmentally sound, energy services at fair and affordable rates. As regulators part of our job is to bring certainty into this fast changing and uncertain dynamic to ensure safety, reliability, customer affordability, customer satisfaction, environmental sustainability, and financial viability. Our unique reality is that we must regulate in the public interest for consumers and communities while these systems are in transformation.

Given our statutory responsibility many of these decisions are best addressed at the state level.

Thank you for the opportunity to continue a dialogue and so that we all better understand the opportunities and the challenges. Thank you.

[The prepared statement of Ms. Edgar follows:]

BEFORE THE
UNITED STATES SENATE

COMMITTEE ON ENERGY AND NATURAL RESOURCES

TESTIMONY OF THE HONORABLE LISA EDGAR
PRESIDENT, NATIONAL ASSOCIATION OF REGULATORY UTILITY
COMMISSIONERS
COMMISSIONER, FLORIDA PUBLIC SERVICE COMMISSION

ON BEHALF OF THE
NATIONAL ASSOCIATION OF REGULATORY UTILITY COMMISSIONERS

ON

"The State of Technological Innovation Related to the Electric Grid"

March 17, 2015

National Association of
Regulatory Utility Commissioners
1101 Vermont Ave, N.W., Suite 200
Washington, D.C. 20005
Telephone (202) 898-2200, Facsimile (202) 898-2213
Internet Home Page http://www.naruc.org

THE HONORABLE LISA EDGAR
PRESIDENT, NATIONAL ASSOCIATION OF REGULATORY UTILITY
COMMISSIONERS
COMMISSIONER, FLORIDA PUBLIC SERVICE COMMISSION

Good morning Chairman Murkowski, Ranking Member Cantwell, and Members of the Senate Energy and Natural Resources Committee. My name is Lisa Edgar and I have the honor of serving as President of the National Association of Regulatory Utility Commissioners. Thank you for your interest in hearing from the States on these issues.

NARUC is a non-profit organization founded in 1889. Our members are the public utility commissions in all 50 States and the U. S. territories. NARUC's mission is to serve the public interest by improving the quality and effectiveness of public utility regulation. Our members regulate the retail rates and services of electric, gas, water, and telephone utilities. We are obligated under the laws of our respective States to assure the establishment and maintenance of essential utility services as required by public convenience and necessity and to assure that these services are provided under rates, terms and conditions of service that are just, reasonable, and non-discriminatory.

I am also a member of the Florida Public Service Commission and have served in that capacity since 2005. My comments are reflective of both responsibilities and will focus on issues within the purview of State utility regulators.

Thank you for the opportunity to testify today on issues regarding technological innovation and the electric grid. I applaud the Committee for holding today's hearing and for recognizing the advances States have made and will continue making to improve electric utility service to customers. For economic utility regulators in each State across the country, ensuring the safe, reliable and affordable delivery of essential utility service is our most pressing duty. At NARUC, this has been our cause for the last 126 years, and it will only grow in importance in the future.

Today's hearing is timely. Coast to coast, change is happening all around the electric utility industry. From smart-grid deployments to energy efficiency and distributed generation projects, State public utility commissions are on the front lines in pursuing new and innovative changes across the country.

Distributed generation technologies can be defined as non-centralized sources of electricity generation interconnected to the distribution system and located at or near customers' homes and businesses. These kinds of resources include solar devices, energy storage, fuel cells, micro turbines, small wind turbines, backup generation, and much more. DG can offer economic, reliability, and environmental benefits to consumers who are able to access and use them.

When combined with smart meters and other advanced resources, distributed generation can revolutionize how some consumers use and consume electricity. These resources will also transform our current utility construct in ways we probably haven't quite imagined. Yet, it is important to also remember that consumers come in all shapes and sizes, from residential to large industrial, but their expectation of affordable, reliable utility service is the same, no matter who is producing or delivering their electricity.

States are leading the way in implementing DG programs to provide options, while still assuring reliable service at fair rates. By the end of 2013, 43 States and the District of Columbia had adopted net metering policies which credit consumers for excess electricity generated and exported to the electric grid. Indeed, several States—such as New York, Minnesota, New Jersey, Georgia, California, Texas, and elsewhere—have all deployed varying degrees of smart and distributed resources. As an example, in my State of Florida, one of our utilities recently installed a smart-grid system called Energy Smart Florida. This program installed around 4.5 million smart meters. This did not come for free or without controversy. However, it is a concrete step to keep Florida on the path for a smarter, more nimble, reliable, and efficient grid.

As I described earlier, distributed generation can have multiple benefits for consumers and the grid. More importantly, however, DG including solar, has idiosyncrasies and challenges that should not be ignored. For example, solar and wind resources are not dispatchable, so if they are needed at a time the sun or wind isn't producing, those contributions to power supply are significantly limited. Solar and wind also need grid support to operate under many different scenarios and configurations. Likewise, small backup fossil units can have worse air emissions profiles than their utility-scale brethren.

Grid operators generally don't control these resources, and it is hard to predict when they will come in, where and for what time period. While these effects can be manageable to a degree, they should not be dismissed, as they come with economic and operational costs. The advantages and trade-offs of distributed resources must be better understood and balanced while making public policy and cost allocation decisions.

As States have encountered and embraced these challenges, experience has demonstrated that individual States pursuing these initiatives at their own pace works. More understanding and buy-in, consumer protections, and public education occur. These policy discussions do at times become heated, even at the State level, with media and advocates on every side sometimes utilizing sound bites rather than informative discourse.

Let me be clear, State regulators understand the value innovative technologies can bring, but we also understand there are challenges associated with integrating these technologies. Utilities are required to provide electricity at all times, 24 hours a day, seven days a week. State commissions and legislatures need to be able to determine the best way to proceed to ensure that the core responsibilities of reliability and affordability are maintained while taking into account state and regional differences.

Change is here in the electric utility industry. With this comes innovation and enthusiasm, but also challenges at both the state and federal levels. States are working to determine which technologies work best for their ratepayers and the specifics of their respective systems. At NARUC, our members have passed resolutions recognizing numerous collaborative efforts between regulators, consumer advocates, utilities and other key stakeholders to address the potential for DG and other new technologies and the impact on the electricity grid. These dialogues are so important as they allow all of us to better evaluate system-wide benefits and costs as use of these technologies continues to grow throughout the country.

In addition, hearings such as this and ongoing discussions will help us all ensure that necessary consumer protections are maintained and that consumers of all types have the information needed as they consider whether to invest in DG and other technologies and services. It remains our responsibility to facilitate the continued provision of safe, reliable, resilient, secure, cost-effective, and environmentally sound energy services at fair and affordable rates as new and innovative technologies are added to the energy mix, and to engage fully and effectively at both the State and federal levels on technology policy considerations.

As State utility regulators, part of our job, more so than in the past, is to help bring some certainty into this fast changing and uncertain dynamic, to ensure safety, reliability, customer affordability, environmental sustainability and financial viability. Our unique reality is that we have to regulate, in the public interest, for consumers, short term and long term, while our systems are in transformation.

Given our statutory responsibilities over the various components of the electric system, many of these decisions may best be made at the State level. NARUC and State Commissioners look forward to an open dialogue with Congress in an effort to better understand both the opportunities and the challenges presented by existing and future innovative grid technologies.

Thank you again for your attention to these issues.

The CHAIRMAN. Thank you, Ms. Edgar.

Thank you to all of the panelists this morning for your comments and for beginning what, I think, is going to be a very, very interesting discussion.

Ms. Edgar, you summed it up there when you were talking about your responsibility as a state regulator to ensure the safety, reliability, customer affordability, environmental sustainability, and financial viability and do this all at the same time where every day you have to do the basic function of making sure that it happens. You have to be able to perform while at the same time you're pushing out and you're really leading.

Ms. Barton, you mentioned that we are the envy of the world with our grid and what we have, yet we are certainly not sitting still here every day.

We are seeing advances that lead us to be able to do more, quicker, faster, and smarter. But this, I think, is our real challenge here, and I would like to propose to anybody that wants to jump in here.

You have got, almost, a do no harm type of philosophy that you have to get up every morning with, while you're pushing out at Argonne with new technologies. Those that are implementing and integrating need to make sure that you're able to do so without being so out of the way that things don't function anymore.

The question I have is whether or not we can keep going at the rate that we are with the innovation and still keep the commitment to reliability, affordability, the financial security, and the environmental security. Is this sustainable? And I hope you say, yes. [Laughter]. Let's start with you, Dr. Howard.

Dr. HOWARD. Okay. The answer is absolutely yes. The hallmark of this industry has been innovation and implement that while at the same time you absolutely maintain affordability and reliability, and that is the keystone of this industry.

The CHAIRMAN. What do you think is the biggest impediment to doing all of this? What is your anchor here?

Dr. HOWARD. Well——

The CHAIRMAN. Your rock?

Dr. HOWARD. We're advancing technologies and that's coming along. We need to continue to make investments in research so that we can continue to advance these technologies and many of which we talked about here today, power electronics, energy storage and reconductoring with advanced cables and so on. So there's a variety of different technologies and we need to make sure we continue to lead in innovation, that we can implement these innovations at the regional level because what happens in one part of the country is going to be different in another part of the country. So there's a lot of regionality issues. And we need to make sure that we continue to focus on not just an interconnected system which the system is now but an integrated system from end to end.

These are very tough challenges, but I think it comes down to research, development and innovation that we're so good at in this industry.

The CHAIRMAN. So just in terms of keeping up as these new technologies are being developed, being incorporated, you have your integrated grid here. It's being used in ways for which it was not initially designed. We have now recognized that you've got this flow both ways.

Ms. Edgar, this is probably best directed to you so how do pricing mechanisms then need to be adjusted so the value that's generated from both the grid and the distributed generation is captured? Because if you've got a system that was in place and this is your base and you've got all these changes, is that keeping up with the technology?

Ms. EDGAR. Thank you, Madam Chair, for the question. As you're well aware, customers all across the country have invested a great deal of money, money from their utility bills, money from their monthly family bills, into the system that we have. A concern that I believe needs to be part of the discussion is that those customers continue to get value from those investments as we are integrating new technologies and new systems into the larger system.

The CHAIRMAN. Are we getting that yet?

Ms. EDGAR. In some areas, yes and perhaps in some areas, less so. One of the discussions here today has been about how to maximize efficiencies by using new technologies, and that's an important balance in all of these discussions. As is how do we use traditional rate making, innovative rate making, how do we attract investment while giving good value to customers and continuing to have that reliable, resilient system?

The CHAIRMAN. Senator Cantwell.

Senator CANTWELL. Thank you, Madam Chair.

I think I am going to try to get three things in, since I don't know what is going to happen with votes.

Dr. Taft, could you tell us what you think the three federal priorities should be to accelerate the transition that's already happening?

Dr. Littlewood, how far do you think we are from game changing storage?

Ms. Edgar, isn't the grid investment really part of our security? Ms. Barton mentioned it as well.

When you talk about grid security or any security, it is always about hardening the targets and redundancy. And basically continuing the development of smart grid gives us an incredible amount of redundancy, juxtaposed to some of the other ideas that are being talked about in other committees. Investing in smart grid will help us with our security.

Dr. TAFT. I'll start. We need to be able to take advantage of the start we already have in adding instrumentation and measurement capability to the grid. A lot of work has already been done at the transmission level with the deployment of phasor measurement units. More work like that is being done at the distribution level. It produces enormous amounts of data.

Going forward one of the things we need to do is be able to process that data to extract the useful information and connect that to decision and control.

So it's really a three step process: measure, analyze and then control. We need more advanced capabilities for that control including distributed control as well as centralized control, and for certain key technologies like storage we need to continue driving the cost down so those become very practical.

Lastly, I'll say because of the complexity and the growth in complexity we need better ways to look at and manage and understand that.

We need architectural tools that help people. All kinds of stakeholders appreciate that you have this large tapestry and if you pull on the thread on one end something happens on the far edge. We see that in the New York REV process, for example, where they're working with and struggling with all the complexities. So we need tools to help them visualize and understand the consequences of changes too.

Senator CANTWELL. Dr. Littlewood, how far away are we from game changing storage?

Dr. LITTLEWOOD. Because game changing storage is really a research activity, it's rather difficult to predict that. Let me first say that conventional storage technologies are improving slowly, a few percent a year, and have been doing so for a long time. Of course, if you project forward the time scale where they will have impact in this business it's probably another 20 years at that rate. So it isn't good enough. So that's why, of course, there are some major research efforts sponsored by Department of Energy and other agencies of the government to try and break out of that.

What I will say is that the space, intellectually, for being able to do that is very large. And I am confident that on some time scale, which unfortunately I can't give you, we'll make the discoveries needed to get forward and do that.

I would also say that there's a point where you're already beginning to see incorporation of storage within microgrids. I think the experiments that we'll be doing in the United States where microgrids are gradually being introduced in various areas and we learn how to integrate storage, wind, solar, small scale energy generation in rural areas will propel this much faster than it has been at the moment.

Senator CANTWELL. Thank you.

Dr. LITTLEWOOD. I'm actually very bullish on this. I think there are great opportunities that will make a big difference.

Senator CANTWELL. Great. Thank you. Ms. Edgar, under 30.

Ms. EDGAR. Thank you, Senator Cantwell. Coming from Florida, clearly you can recognize that resiliency and redundancy are issues that are very important to us. When I first came to the Public Service Commission the issues that we were dealing with on a day to day basis were because over the course of two years Florida had eight hurricanes and two named storms hit our coastline and even into the central areas. We had areas of the state that had large outages, three, four, five times over the course of two years.

That resiliency and redundancy is near and dear to the issues that we deal with. Changes with technology and incorporating new technology certainly will help us withstand cyber security risks, physical security risks, weather risks, but it also, as we make these changes, brings in new technology issues that need to be addressed. And one of the things, as a state regulator, that we want to do is make sure that no customer group is left behind and that there is transparency as we have the discussions about investment and cost allocation.

Senator CANTWELL. Thank you.

Ms. EDGAR. Thank you.

The CHAIRMAN. Senator Barrasso.

Senator BARRASSO. Thank you very much, Madam Chairman.

Dr. Littlewood, in your testimony you state that our electric grid must be secure and resilient in the face of disruptive threats that range from natural disasters to terrorists attacks. You explain resiliency research is a particular strength at the Argonne National Laboratory, and go on to say that Argonne and other labs have developed the beginnings of a foundational key to secure the grid, the national power grid, simulator. The simulator would run virtual experiments to test potential vulnerabilities and to develop responses to various types of attacks.

In January the Department of Energy's Inspector General released a report examining how FERC, the Federal Energy Regulatory Commission, conducted a similar process. In 2013 former FERC Chairman, Jon Wellinghoff, directed FERC staff to identify critical electric substations by location. Chairman Wellinghoff also directed FERC staff to create failure scenarios to simulate the impact of the loss of these substations. The Chairman then decided to share this information with individuals and entities outside of the federal government. The Inspector General actually found that Chairman Wellinghoff made the decision to share the information without determining whether the information was classified and did so even though FERC staff expressed concern that the information, ''could provide terrorists and other adversaries with data they might use to disable portions of the grid.''

Now personally I find these actions deeply troubling, but I'm even more troubled that emails between the Chairman and FERC staff on this issue were missing from the Chairman's email account.

I understand that Argonne National Laboratory handles sensitive information on a regular basis. What steps do you, as the Director of Argonne, take to safeguard this information?

Dr. LITTLEWOOD. Sir, Argonne has a prominent role, actually, in cyber security in a large number of areas, so we actually work very hard to make sure that all transactions like this are appropriately vetted and are not shared inappropriately.

I share your concern, Senator, about the risks of potentially advertising the vulnerabilities of our network.

I also think that we need to work, in fact, to develop cyber security protocols for how the future grid will operate. And this is a very important part, a kind of hidden part, of making sure that any developments that we move forward with on the future grid are, in fact, robust against terrorist acts and terrorist intrusions and other such things.

Senator BARRASSO. One other thing. I noticed in your testimony you compare the electric grid with the interstate highway system.

Dr. LITTLEWOOD. Right.

Senator BARRASSO. The comparison might suggest that Washington should assume virtually all of the responsibility for funding the electric grid, but today the vast majority of investment for the electric grid actually comes from entities other than the federal government. Are you suggesting that the federal government be the principle source of funding the electric grid or how do you view that whole thing?

Dr. LITTLEWOOD. I certainly didn't intend to suggest that, and it's none of my business, actually, to decide what the federal government should be doing in that area. I was simply pointing out the physical scale of the infrastructure.

As I think I said earlier with regard to microgrids, one of the advantages the United States has over much more centrally planned economies is the ability to do experiments and actually to develop infrastructure in a way that's responsive to the market and enables us to take great advantage of the intellectual and scientific and engineering strengths that we have.

So one of the challenges, I think, for the industry and we discuss this a great deal, is developing business models which are appropriate to be able to deal with that. Notwithstanding as we've already heard from EPRI something like $300 to $500 billion of investment are expected to be required over the next 20 years.

Senator BARRASSO. Thank you. Thank you, Madam Chairman.

The CHAIRMAN. Senator Warren.

Senator WARREN. Thank you, Madam Chair.

Last month Energy Secretary Moniz came before this Committee to discuss the Department's budget request, and during that hearing we had a chance to talk about the power grid and specifically how to make sure that the grid can reliably distribute electricity throughout the country.

Well, our power grid has been reliable for so long that, for the most part, no one even thinks about it. The grid is aging. Much of the technology was developed by Thomas Edison and much of the structure was built shortly after World War II, and now this aging grid faces new challenges, particularly with the rise of extreme weather events that threaten to shut down parts of the grid during weather emergencies.

During our earlier hearing, Secretary Moniz said that, ''Distributed generation and microgrids are themselves a resiliency tool,'' meaning that in addition to the environmental benefits a power grid with more wind and more solar spread across multiple locations can actually help protect against extreme weather.

Dr. Littlewood, your testimony discusses how resiliency research is a specialty at Argonne National Laboratory which you run. Can you explain to us in more detail how it is that distributed generation itself can produce system wide resiliency?

Dr. LITTLEWOOD. Thank you very much, Senator. That's a very prescient and precise question.

The grid is a very complicated object, so even now when it's being driven in a one-way mode it interacts in very complicated and very subtle ways. And actually we do not understand it.

If you look at it as a basic principle of resilience, you look for individual units that can continue to fight the war after the command or control has somehow been taken out. So that's one very basic principle about how you build a resilient network. It cannot be a single connected object. It has to be a web in the same way that the Internet is a web. So by having microgrids that can operate independently but still communicate, you have the opportunity in the case of a natural disaster to be able to reroute energy around affected areas to be able to make sure that the other infrastructure, which may be affected in the area of a natural disaster, is also not impacted.

One of the dangers in major disasters, of course, is that you lose first the grid, then the Internet, then water, then the ability to distribute many, many different kinds of things. And the very basic principles of resilience tell you that one monolithic big thing is a dangerous thing to be working with.

So the other side of that, the need that actually my colleagues have commented on for the ability to model, understand and work the operation of the grid in real time. And so that's the high level infrastructure which is necessary to monitor it, the sensors, the data, the integration, the high performance computing.

Senator WARREN. Thank you very much.

The Energy Secretary and the technical experts at the Department are not the only folks who made this connection. A recent analysis from the World Bank showed that when there's more diversity of energy sources, including more renewable energy that's connected to the grid, the grid becomes more resilient. Similarly a Massachusetts Climate Change Adaptation report from 2011 also recommended diversifying energy supplies as a strategy to make our system less prone to failures.

Extreme weather events are on the rise. Every year these events stress the capacity of our power grid. If our technical experts are telling us that plugging more renewable energy into the system can help protect the grid in the face of these extreme weather threats, then we should make it a priority. Whatever you think about climate change, we all have an interest in keeping the lights on. Thank you.

The CHAIRMAN. Senator Risch.

Senator RISCH. Thank you, Madam Chair.

Ms. Barton and Ms. Edgar, this question is for you. On the Intelligence Committee, we spend hours and hours and hours listening to and wringing our hands about the cyber threats. There's a lot of us who are convinced that that's going to be the next big one that hits the country.

Can each of you please explain how this plays out in the real world? What challenges do you face in the real world dealing with this? Ms. Barton, why don't you start with you?

Ms. BARTON. Certainly. The cyber threats are something that the utility industry takes extremely seriously. We have a number of people who basically sit behind a desk, 24/7, and monitor all of the threats that are going on, the attacks that are happening, whether it be through fishing attacks and so forth or direct attacks with respect to the system itself.

The ability of this industry to work with the federal government and other service providers has been outstanding in terms of being able to mitigate any threats. We have a number of people on our staff who have a secret clearance so they can get access to all kinds of different information.

We have also paid a lot of attention with respect to physical security and our control houses and making sure that they are much more resilient than they have been in the past. In the end system redundancy is diversity and redundancy are really at the heart of reliability. So from a physical standpoint one of the things that we are doing is working with the RTOs who are responsible now for planning and determining what new transmission needs to be built, we are really encouraging them to take an accelerated view with respect to transmission investments so that the system itself can be more reliable by diversifying.

Senator RISCH. Thank you. I appreciate that, and it is interesting to hear you talk about your relationship with the federal government. As you know, we're struggling with that when you have the issues of privacy versus the issues of what's got to be done in order to stop this, so I appreciate your input on that.

Ms. Edgar.

Ms. EDGAR. Thank you, Senator, and I'll approach the question from perhaps a slightly different angle.

First off, I will say that, again, the issues of cyber security are truly something that, as regulators, do keep us up at night. None of us wants to get that call from the governor that says, the system is down. Somebody has hacked it. What are you doing to fix it?

However, I can tell you I recently attended a meeting that's an ongoing collaborative effort between the Department of Energy, the Department of Homeland Security and EEI. I found it to be a very progressive and very forward thinking collaborative effort between the private industry, state regulators and the federal agencies, so I can tell you that a lot is going on that should give us all reassurance.

However, as a state regulator one of the things that we struggle with, again, is the cost. Because much of this information is and necessarily so, confidential, it is difficult when utilities come to us, as a state regulator and economic regulator, and we are asked to approve these investments and put them into the rate base, into the monthly bills of consumers. Yet, because of the security issues, we cannot really closely evaluate where that money is going and what for. That is something, at the state regulatory level, that is still a puzzle that we are needing to work through.

Senator RISCH. That's interesting. I think most consumers really aren't fully aware of the challenges the utilities face. We all take for granted when we flip the switch, the light goes on. And when that doesn't happen, the reasons for it are complicated and could be a real danger to the security of the United States. So, thanks, thanks for all of that.

My time is almost up.

Dr. Taft, I'm aware of the Pacific Northwest smart grid demonstration project that you're working on, and I have a couple of questions regarding that. Since my time is up and with respect to the other people, would it be alright if I submit those in writing and you could respond?

Dr. TAFT. Absolutely. We will be very happy to respond.

Senator RISCH. Thank you very much, Dr. Taft. I appreciate it.

Senator BARRASSO [presiding]. Thank you, Senator Risch. To members of the panel, we're in the middle of roll call voting on the Floor of the Senate. You'll see some of the Committee members leave and then come back.

And with that, Senator Franken, you're next.

Senator FRANKEN. Thank you. I'm sorry. I'm going back and forth to the HELP Committee, health hearings.

Dr. Howard, I like distributed energy, and I like distributed generation. It makes our grid more resilient, allows critical infrastructure, like hospitals and military bases and others, to stay online during an outage which is very important. That's why I've always supported the increased deployment of things like combined heat and power and district energy, which we have in St. Paul, burns clippings and tree limbs and stuff like that then provides electricity for downtown St. Paul and heats and cools the buildings. So I like distributed energy.

So I want to ask the panelists about another aspect of distributed energy and that's grid scale storage. I apologize if this has been discussed already, but I think this is a real game changer because it will allow us to deploy renewables, more renewables, wind and solar, when needed. It will help stabilize the grid, and it will improve our resiliency by making sure that we have electricity available on demand and in case of an outage.

Can the panelists in the national labs please describe some of the recent advances in grid scale storage and what your visions are for deploying this technology in the future?

Dr. TAFT. So let me start, Senator Franken. One of the things that we have been doing is looking at storage both in terms of the core technology, the fundamentals of how the storage devices work, but also how they can be used as system components. And you mentioned grid scale storage in particular. The ability to use storage in multiple modes is one of the things that makes it most interesting. There are a variety of functions that it can supply that are useful to the grid, and unlike most components, storage can be used in multiple ways simultaneously.

So one of the things that we have been working on is how you actually use that in a multi mode fashion because that helps spread the cost over more capabilities and make it more cost effective.

The reason that we think of storage, power electronics and advance control as a new grid component is because of those new capabilities and so many of them as opposed to other components that have one function. That's why we think that will become a general purpose tool for buildings and grids of the future, and so we focus a lot on how to apply it as a system component. I'll stop there and let my colleague take over.

Dr. LITTLEWOOD. Thank you. I agree with those comments. Let me focus a little bit on the technology side of that.

Since the need for grid scale storage has only recently emerged, it turns out that there's a substantial vacuum in technology and science which is needed to support that. Unlike mobile storage which has been driven by the consumer electronics industry for the last few years, this is a recently emerging driver. And that's one of the reasons that the DOE has been heavily investing in fundamental research on the basic science to be able to do this with the main goal of taking prototypes that we already know and understand and have opportunities to work with and drive the cost down so that they are appropriate for the grid.

I actually think that on grid scale storage, at least electrical storage, we will make some very substantial advances within the next few years of driving costs down because this is an area which has been underexploited for quite some time.

Senator FRANKEN. I agree, and I'm glad you're doing this.

Let me ask about like electric vehicle storage. I mean, there you have like at night, wind blows, wind. Electric vehicles can store at night. Drive, less gas. Then during the day the sun is shining so we get solar, and what are the benefits of this approach in terms of reducing our carbon emissions?

Dr. LITTLEWOOD. I think they're actually very substantial over the long term.

In the short term, it actually involves getting, you know, the electric vehicles to a stage where they are widely adopted. And that's largely, again, a matter of cost in it, cost to do this.

Senator FRANKEN. Right.

Dr. LITTLEWOOD. It also involves, I think, doing experiments with microgrids because this is the first opportunity to do so.

I think there are big opportunities in rural areas and there are big opportunities in certain subdivisions to design microgrids that are able to take solar, wind and electrical storage and use them effectively. I think a number of utilities are actually exploring models by which they can introduce those and that involves developing new business models which go along with the technologies as they're being invented.

Senator FRANKEN. I know I'm out of time, but this is exploding. I mean, this is happening faster than anyone could have conceived. If you look at the explosion in cell phones and think about that technology and think about what we're going to be doing with all of this stuff. This is a revolution, right? And it's good, right? Except for Ms. Barton, right? It's good for you too?

Ms. BARTON. Diversity in the generation portfolio is always a good thing.

Senator FRANKEN. Okay, good. I'm sorry I went over my time.

Senator BARRASSO. Thank you, Senator Franken. Senator Daines.

Senator DAINES. Thank you, and thanks for holding this hearing today.

I represent the State of Montana. Certainly innovation and reliability in the electric grid is an important topic for our consumers back home. And thank you for what the witnesses have relayed today regarding the challenges that the modern electric grid faces.

But too often the change in generation sources is brought by misguided federal policies. And somebody from a state like Montana we have truly an all of the above energy portfolio. The sun shines a lot. The wind blows a lot. We have an abundance of coal, oil, natural gas and hydro, but Montana consumers rely on over half of their electrical supply from coal. Montanans are severely concerned about the impacts of the EPA's regulation on their energy prices, transmission capacity, grid reliability and the very jobs that depend on coal production in Montana.

The Crow Tribe, for example, their unemployment rate today is 50 percent. If we lost the coal mining jobs on the Crow Reservation, their unemployment rate goes to somewhere north of 80 percent.

The tax revenue of coal production in Montana is $118 million a year. That is how we fund our infrastructure, our teachers and our schools. These proposed EPA clean power plant rules could threaten operations at Colstrip.

In fact it threatens coal-fired plants in other states like Michigan. In fact it's coal for Montana that powers those plants in Michigan, that supplies electricity for the auto manufacturing industry there in Detroit.

My questions surround the impact of clean power plan on electricity reliability.

Ms. Barton, Nick Akins, the CEO of American Electric Power, has been outspoken on grid reliability. About a year ago he stated in testimony before this very Committee that the country, ''dodged a cannonball during the polar vortex.'' He also reiterated a point he made a month earlier that AEP used 89 percent of generation that will retire in 2015 to meet electricity demand during the polar vortex.

My question is this. Will the EPA 111D rule on existing power plants impact conventional sources of electricity like coal? And if so, how will that impact grid reliability?

Ms. BARTON. Thank you. Will the EPA 111D impact conventional generating units? Yes, it will.

One of the things that we have been doing is we have been studying the grid, in particular, studying the PJM system, to really understand and appreciate the impact of what some of these retirements are. I think it's important to note that when you talk about the grid, the grid is the combination of the generation, transmission and distribution system that is there in a given area. The grid in New York City is very different from the grid in Montana, the grid in Texas and so forth. And so it's very important that we appreciate the inner workings of the grid in those regional areas.

PJM, for example, has actually recently done a study and it has gotten some news attention. And that is an economic study. An economic study really is just looking to determine is there a sufficient amount of generation on the system. Their next stage is to look at a load flow analysis. The load flow analysis really looks at can you get power from that generation to your load centers. In that recent analysis there are basically five load pockets that did not have enough import capability in a post 111D world, to be able to function reliably.

So when you look at the impacts of really any government initiative, whether it be 111D or what have you, it's important to take a step back and really take a look at the reliability of the grid. We have incredible modeling technology. We have folks who basically are station operators who are looking at the system and appreciating what needs to happen to keep the system reliable at any given moment.

We then have a number of regional transmission planners who are looking at that system ten years into the future, five years into the future. What you end up having in some of these situations is a chicken and egg situation. You need to first identify well, what generation is retiring? Where is it retiring? What then changes are necessary with respect to the transmission grid to be able to maintain the reliable and stable system we have today?

Senator DAINES. Thanks for your insights into the supply chain. I appreciate that.

Commissioner Edgar, similarly, do the federal regulations like 111D help your Commission create certainty or do these regulations create uncertainty for consumers?

Ms. EDGAR. Certainly. Thank you, Senator.

Well at this point there is great uncertainty. As you're well aware, we are in the stage of, still in the stage of rulemaking or EPA is still in the rulemaking stage. Many, many discussions have occurred about what the proposed rule really means, what it requires, what costs will come from the changes that it may or may not require? But of course, it's still a proposed rule at this point in time.

And consumers, regulators, states and industries are continuing to make investment decisions in order to meet the needs of customers, short term and long term, not knowing what the final rule is going to be or even once it comes out what the implementation processes will be. So yes, there is certainly uncertainty now as that process moves forward.

Senator DAINES. Thank you.

Ms. EDGAR. Thank you.

The CHAIRMAN [presiding]. Thank you, Senator Daines.

You will see that members have fled for the votes, but they are on their way back. So as members come in I will turn to them, but let me ask another series of questions here.

I mentioned in my opening comments my interest in the potential that we have for microgrids and recognizing, as you pointed out, Ms. Barton, that there are great differences region to region and that's why there can't be a one size fits all application when it comes to our energy sources and how our grid functions here.

Dr. Littlewood, what kind of coordination goes on with our national labs and what the states are doing or even more locally? For instance, in a very small village in the Southwestern part of Alaska, not connected by road, there's a village by the name of Kongiganak, with their own little microgrid between three wind turbines, a battery storage unit and small heating units within homes which is perfect for them. They've really pioneered much of what they have provided for their community of several hundred.

How much coordination goes on between the labs and then what we're seeing within the states, whether through our universities or just from those that, by necessity, are pulling together these very small microgrids?

Dr. LITTLEWOOD. So one of the things that the labs are doing is trying to develop test beds that can be used for communities to do virtual experiments within the lab system to be able to design systems that could be exported to be used in small communities.

So I'll mention a program at the NREL, National Renewable Energy Lab, which is ESIF which actually provides a kind of basic test bed. There's similar work been going on at PNNL and at Argonne and at many of the labs.

So, of course what we can't do always is to connect with individual communities one by one, but we do work quite closely with a view, to get a view, both from regulators or from within individual states with those states running those kind of experiments.

I think the labs would say that they would, in fact, benefit from the attention coming from states who have particular problems and particular issues that they want to solve. For example, I think my colleagues would be interested to come and study while it's actually going on on the ground. And we do, in many places, like can't vouch for the fact that we've been to a particular village in Alaska, although I'm sure that there are some pretty creative ideas which are on the ground there already.

The CHAIRMAN. I think it would be helpful for all of us to have a better understanding in terms of what is happening in some pretty unique situations, because of necessity in so many of these areas, and then integrating that with our brilliant minds that we have in our laboratories.

Dr. Taft, you had mentioned four technologies that are key to modernizing the grid. And this is a sensing and data analysis, high voltage power electronics, fast and flexible bulk energy storage and then the advanced planning and control methods and tools. Which of these is most readily deployable? And as we are moving toward this modernization, of these four, where's our biggest challenge? So what can we do to deploy quickly? And again, what's our drag here?

Dr. TAFT. The utilities have had an opportunity recently to expand their capability to make measurement, especially at the transmission level, some at the distribution level. More is coming.

So we're at the point of needing to be able to manage and analyze that data. And just to give you a sense of how much more data we're talking about, an ordinary utility in the past might collect up enough data to fill up a book like a Tale of Two Cities, once every second. And what we're looking at now in the future is being able to fill up a book like War and Peace, 846 times a second.

So the need there is to be able to sift through all of that data and get the useful information and then act upon it. That's an immediate need because the sensor capability is coming online. The tools to do that in other industries exist but the tools to do it in the utility industry are somewhat lagging. So we need that. We need it immediately. We also need advanced controls.

And you talked about microgrids. You know, when we did the grid architecture work for DOE last year we looked at microgrids as a specific example, looked at NRCA work in that area because that more distributed architecture has lessons for us to learn. And so we started to gather that kind of information and make use of it and think about that in terms of how we would go forward.

The control systems for all of that are an area where we need considerable work. And there's a lag there because the industry doesn't invest a lot in research. They can't do a lot in research. So the vendors see a thin market and there's a nice role for the federal government to do the key development and demonstration of that capability so that it moves forward.

The storage technologies are moving forward commercially, but there's a great deal to be done to drive the costs down to make them more effective and to make them flexible.

So all four of those technologies are, kind of, in the same place in a sense of there's not one that's way ahead of the other as we, kind of, need all four. But that's why I talked about them as a group. That we need to advance all of those together and when you put together the power of electronic storage and control, in particular, you get this remarkably flexible tool that can be applied at the microgrid level, at the bulk system level, even at the individual building level, for example, if you think of a building as acting like a microgrid on its own.

It's imperative that we move those forward because they have so much flexibility to help us deal with all of these variations, fluctuations and complexity that we're faced with going forward.

The CHAIRMAN. Thank you.

Dr. Howard, you have said that a consumer with distributed energy resources is not necessarily a self-sufficient energy consumer. Talk a little bit about that because I think you have folks that think, well, wait a minute, I've got solar. I've got panels on my home, and I'm seeing those benefits. I am a self-sufficient energy consumer.

You say that they are not necessarily so. Speak to that and what would it take for an energy consumer to really become completely self-sufficient?

Dr. HOWARD. Okay. The answer to that question really gets to the issue of capacity and energy. A consumer that has, for example, solar PV, may at certain times have excess energy that is being produced. But when they don't have that excess energy they need to tap into the grid. And at that point they hope that the capacity is available to get access to that energy that they need.

And so it's really a two way street. And that's why it's important going forward that we really understand that a lot of these different technologies may be rich in energy, but you need to also tap into the power system to get access to that capacity that's needed so that when the sun doesn't shine, they have electricity now.

The CHAIRMAN. But hope doesn't necessarily——

Dr. HOWARD. That's right.

The CHAIRMAN [continuing]. Get you to the reliability——

Dr. HOWARD. That's right.

The CHAIRMAN [continuing]. Piece of it, so——

Dr. HOWARD. That's right. So you need devices that we referred to here like smart inverters or power electronics. You need sensors. Also energy storage can play a very important role in helping to certainly end those times when the sun isn't shining then you can either use energy storage or the biggest energy storage device is the grid itself and tap back into the grid to get the electricity that you need.

So it all has to work together. But as we look forward I think there's a whole new approach that we need to think about, not just energy, but also making sure that we have the capacity available at every single minute to provide the energy that's needed.

The CHAIRMAN. Well, let's talk about the capacity here. And you used the term capacity related costs because in my opening statement I, kind of, teased and said I wanted to know what's our time line for development here and what are those costs.

I'm pretty sure that Senator Cantwell used the same number that I did that we're looking at a cumulative investment of between $300 and $500 billion over the next 20 years to really modernize our grid here.

When we're talking about this capacity related cost is this the number that we're talking about here?

Dr. HOWARD. Yeah, yeah, that's right. We, EPRI, did a report in 2011, and it's called ''Estimating the Costs and Benefits of the Future Grid.'' I will make sure that that report is available for this Committee.

The CHAIRMAN. Okay.

[The information referred to follows:]

EPRI | ELECTRIC POWER
RESEARCH INSTITUTE

Estimating the Costs and Benefits of the Smart Grid

A Preliminary Estimate of the Investment Requirements and the Resultant Benefits of a Fully Functioning Smart Grid

2011 TECHNICAL REPORT

Estimating the Costs and Benefits of the Smart Grid

A Preliminary Estimate of the Investment Requirements and the Resultant Benefits of a Fully Functioning Smart Grid

EPRI Project Manager
C. Gellings

EPRI | ELECTRIC POWER RESEARCH INSTITUTE

3420 Hillview Avenue
Palo Alto, CA 94304-1338
USA

PO Box 10412
Palo Alto, CA 94303-0813
USA

800.313.3774
650.855.2121

askepri@epri.com

www.epri.com

1022519
Final Report, March 2011

116

DISCLAIMER OF WARRANTIES AND LIMITATION OF LIABILITIES

Electric Power Research Institute (EPRI)

NOTE

117

Acknowledgments

The following organization prepared this report:

Electric Power Research Institute (EPRI)
3420 Hillview Ave
Palo Alto, CA 94304

Principal Investigator
C. Gellings

This report describes research sponsored by EPRI. The project manager would like to acknowledge the contributions of the following EPRI staff members:

Gale Horst – Project Manager, for contributions regarding Customer Costs

Mark McGranaghan – Vice President – Power Delivery and Utilization, for contributions regarding Distribution

Paul Myrda – Technical Executive, for contributions regarding Transmission and Substations Costs

Brian Seal – Senior Project Manager, for contributions regarding Home Area Network (HAN) & Automated Metering Infrastructure (AMI)

Omar Siddiqui – Director, for contributions regarding Customer Costs

This publication is a corporate document that should be cited in the literature in the following manner:

Estimating the Costs and Benefits of the Smart Grid: A Preliminary Estimate of the Investment Requirements and the Resultant Benefits of a Fully Functioning Smart Grid. EPRI, Palo Alto, CA: 2011. 1022519.

118

Donald Von Dollen – Senior Program Manager, for contributions regarding Overall Communication & IT Integration Costs

Mark Duvall – Director, for contributions regarding Overall Treatment Costs of Electric Vehicle & Plug-In Hybrid Electric Vehicle Costs

Andrew Phillips – Director, for contributions regarding Overall Transmission & Substation Costs

Daniel Rastler – Senior Project Manager, for contributions regarding Storage Costs

Bernard Neenan – Technical Executive, for contributions related to the overall cost framework

The project manager would also like to acknowledge the thoughtful contributions of the following reviewers:

Stuart Brindley and Mark Lauby – North American Electric Reliability Corporation

Tommy Childress and Ruben Salazar – Landis+Gyr

Lynn Coles, Henry Kenchington, and Eric Lightner – U.S. Department of Energy

David Curtis – Hydro One, Inc.

Paul DiMartini – Cisco Systems

Frank Doherty – Consolidated Edison Company of New York

James Alan Ellis and Ramesh Shankar – Tennessee Valley Authority

119

John Estey - S&C Electric

Ahmad Faruqui and Doug Mitarotonda – Brattle Group, Inc.

Joshua Hambrick – National Renewable Energy Laboratory

Tim Heidel – Massachusetts Institute of Technology

Louis Jahn and Eric Ackerman– Edison Electric Institute

Richard Joyce – GE Energy Services

William Mintz – Alabama Power Company

Mark Nealon – Ameren

Ernst Scholtz – ABB

Leslie Sibert – Georgia Power Company

Thomas Wick – We Energies

Ralph Zucker – BC Hydro

Abstract

The present electric power delivery infrastructure was not designed to meet the increased demands of a restructured electricity marketplace, the energy needs of a digital society, or the increased use and variability of renewable power production. As a result, there is a national imperative to upgrade the current power delivery system to the higher performance levels required to support continued economic growth and to improve productivity to compete internationally. To these ends, the Smart Grid integrates and enhances other necessary elements including traditional upgrades and new grid technologies with renewable generation, storage, increased consumer participation, sensors, communications and computational ability. According to the Energy Independence and Security Act of 2007, the Smart Grid will be designed to ensure high levels of security, quality, reliability, and availability of electric power; improve economic productivity and quality of life; and minimize environmental impact while maximizing safety. Characterized by a two-way flow of electricity and information between utilities and consumers, the Smart Grid will deliver real-time information and enable the near-instantaneous balance of supply (capacity) and demand at the device level.

The primary goal of this report, which is a partial update to an earlier report (EPRI 1011001), is to initiate a stakeholder discussion regarding the investment needed to create a viable Smart Grid. To meet this goal, the report documents the methodology, key assumptions, and results of a preliminary quantitative estimate of the required investment. At first glance, it may appear the most obvious change from the 2004 report is the significant increase in projected costs associated with building the smart grid. In actuality, the increased costs are a reflection of a newer, more advanced vision for the smart grid. The concept of the base requirements for the smart grid is significantly more expansive today than it was seven years ago, and those changes are reflected in this report.

Keywords
Smart grid
Smart grid costs and benefits
Functionality
Power delivery system
Transmission
Distribution

Table of Contents

122

List of Figures

125

126

List of Tables

127

128

Section 1: Executive Summary

This report documents the methodology, key assumptions, and results of a quantitative evaluation of the investment needed (costs) for an envisioned Smart Grid, and it represents a partial update to an earlier EPRI report (EPRI TR-1011001). It also offers a preliminary estimate of benefits of implementing a Smart Grid. This report is a framework for discussing possible levels of investment to achieve a fully functioning Smart Grid. It is not a definitive analysis of all attributes or costs of enhancing the power delivery system.

What is the Smart Grid?

The Smart Grid as defined here is based upon the descriptions found in the Energy Independence and Security Act of 2007. The term "Smart Grid" refers to a modernization of the electricity delivery system so that it monitors, protects, and automatically optimizes the operation of its interconnected elements – from the central and distributed generator through the high-voltage transmission network and the distribution system, to industrial users and building automation systems, to energy storage installations, and to end-use consumers and their thermostats, electric vehicles, appliances, and other household devices.

Background

The present electric power delivery infrastructure was not designed to meet the needs of a restructured electricity marketplace, the increasing demands of a digital society, or the increased use of renewable power production. In addition, investments in expansion and maintenance are constantly being challenged, and the existing infrastructure has become vulnerable to various security threats.

Figure 1-1 illustrates today's power system. As shown, it is primarily comprised of large central-station generation connected by a high voltage network or Grid to local electric distribution systems which, in turn, serve homes, business and industry. In today's power system, electricity flows predominantly in one direction using mechanical controls.

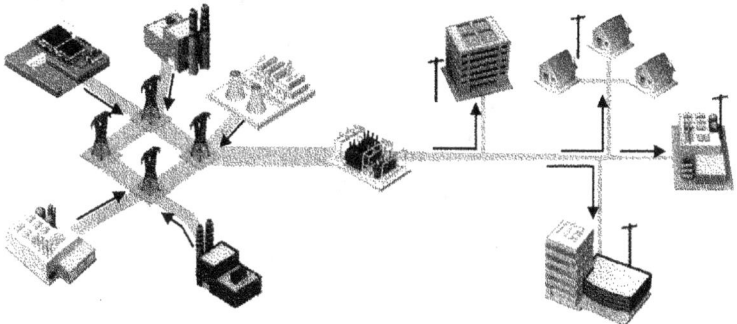

Figure 1-1
Today's Power System

Figure 1-2 illustrates the elements which will be part of a fully functional Smart Grid. The Smart Grid still depends on the support of large central-station generation, but it includes a substantial number of installations of electric energy storage and of renewable energy generation facilities, both at the bulk power system level and distributed throughout. In addition, the Smart Grid has greatly enhanced sensory and control capability configured to accommodate these distributed resources as well as electric vehicles, direct consumer participation in energy management and efficient communicating appliances. This Smart Grid is hardened against cyber security while assuring long-term operations of an extremely complex system of millions of nodes.

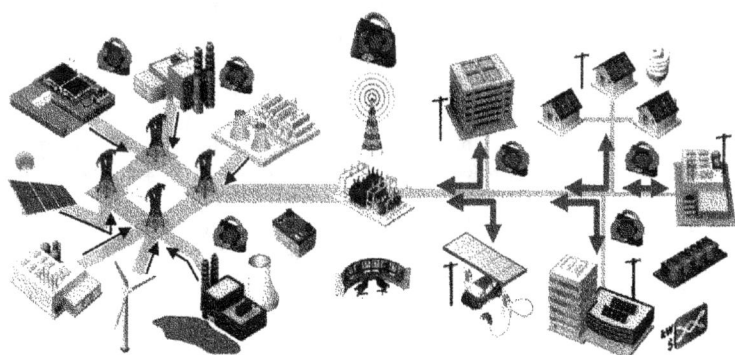

Figure 1-2
Tomorrow's Power System: A Smart Grid

As a result, there is a national imperative to modernize and enhance the power delivery system. The Smart Grid is envisioned to provide the enhancements to ensure high levels of security, quality, reliability, and availability (SQRA) of electric power; to improve economic productivity and quality of life; and to minimize environmental impact while maximizing safety and sustainability. The Smart Grid will be characterized by pervasively collaborative distributed intelligence, including flexible wide band gap communication, dynamic sharing of all intelligent electronic devices and distributed command and control. Achieving this vision will require careful policy formulation, accelerated infrastructure investment, and greater commitment to public/private research, development, and demonstration (RD&D) to overcome barriers and vulnerabilities.

Previous Studies

Previous EPRI studies have estimated both the costs and benefits of a Smart Grid. According to an earlier study, "The Power Delivery System of the Future," the Smart Grid would require $165 billion in net investment (over and above investment for load growth and that needed to maintain reliability), and lead to a benefit-to-cost ratio of 4:1 (EPRI 1011001).

The Smart Grid, combined with a portfolio of generation and end-use options, could reduce 2030 overall CO_2 emissions from the electric sector by 58% relative to 2005 emissions (EPRI 1020389). A Smart Grid would be capable of providing a significant contribution to the national goals of energy and carbon savings. One EPRI report (EPRI 1016905) estimated the emissions reduction impact of a Smart Grid at 60 to 211 million metric tons of CO_2 per year in 2030.

Other EPRI studies have estimated the cost of power disturbances across all business sectors in the U.S. at between $104 billion and $164 billion a year as a result of outages and another $15 billion to $24 billion due to power quality (PQ) phenomena (EPRI 1006274). The cost of a massive blackout is estimated to be about $10 billion per event as described in EPRI's "Final Report on the August 14, 2003 Blackout in the United States and Canada."

Purpose and Scope

The purpose of this study is to inform the public debate on the investment needed to create a fully functioning Smart Grid. For each key portion of the overall task, the project team selected methods based on the availability of credible information and the need to conduct a cost-effective and time-efficient study. The resulting estimates of costs remain highly uncertain and open to debate. This report is viewed as a starting point for discussion of possible levels of investment to bring the current power delivery system to the higher performance levels required for a Smart Grid.

In addition to welcoming and encouraging comments on this report, EPRI invites the participation of energy companies, universities, government and regulatory agencies, technology companies, associations, public advocacy organizations, and other interested parties throughout the world in refining the vision for the Smart Grid. Only through collaboration can the resources and commitment be marshaled to achieve the vision.

Summary of Results

Over and above the investment to meet electric load growth, Table 1-1 shows that the estimated *net* investment needed to realize the envisioned power delivery system (PDS) of the future is between $338 and $476 billion. The total value estimate range of between $1,294 and $2,028 billion; and when compared to the Future PDS cost estimate results in a benefit-to-cost ratio range of 2.8 to 6.0. Thus, based on the underlying assumptions, this comparison shows that the benefits of the envisioned Future PDS significantly outweigh the costs. At first glance, it may appear the most obvious change from the 2004 report (EPRI 1011001) is the significant increase in projected costs associated with building the smart grid. In actuality, the increased costs are a reflection of a newer, more advanced vision for the smart grid. The concept of the base requirements for the smart grid is significantly more expansive today than it was seven years ago, and those changes are reflected in this report. The project team has made every effort to capture a reasonable send-state of the Smart Grid in this report, rather than creating a snap shot that will change in another six or seven years.

Table 1-1
Summary of Estimated Cost and Benefits of the Smart Grid

	20-Year Total ($billion)
Net Investment Required	338 – 476
Net Benefit	1,294 – 2,028
Benefit-to-Cost Ratio	2.8 – 6.0

This indicates an investment level of between $17 and $24 billion per year will be required over the next 20 years. The costs cover a wide variety of enhancements to bring the power delivery system to the performance levels required for a Smart Grid. The costs include the infrastructure to integrate distributed energy resources (DER) and to achieve full customer connectivity, but exclude the cost of generation, the cost of transmission expansion to add renewables and to meet load growth, and a category of customer costs for smart-grid ready appliances and devices. Table 1-2 lists major components of the total cost. As highlighted in the body of the report, the wide range in these estimates reflects the uncertainty the industry currently faces in estimating these costs and the possible reductions which may or may not occur over time.

Smart Grid Costs

Included in the estimates of the investment needed to realize the Smart Grid, there are estimated expenditures needed to meet load growth and to enable large-scale renewable power production. As part of these expenditures, the components of the expanded power system will need to be compatible with the Smart Grid.

134

Table 1-2
Total Smart Grid Costs

Costs to Enable a Fully Functioning Smart Grid ($M)		
	Low	High
Transmission and substations	82,046	90,413
Distribution	231,960	339,409
Consumer	23,672	46,368
Total	**337,678**	**476,190**

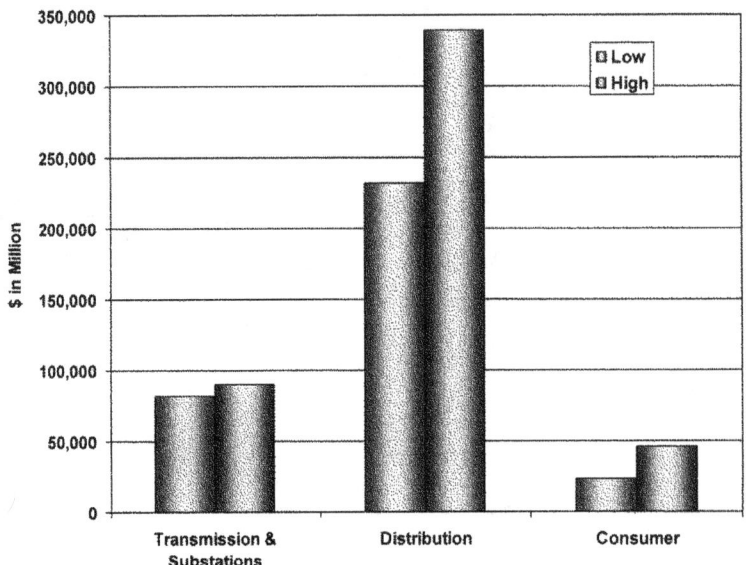

Figure 1-3
Total Smart Grid Costs

Table 1-3 summarizes one attempt to portray the consumer implications of the
EPRI estimate of Smart Grid costs.

In this table, the Smart Grid costs are allocated to classes by energy (which are
often what a regulatory body would mandate in the case of a regulated utility)
and then calculated in several ways: (1) total Smart Grid cost divided by the
number of customers for each class (a one-time payment proxy); (2) total cost per
customer per year by class for 10 years for a 10-year amortization of the Smart
Grid cost (in nominal, not present value, terms); and (3) the monthly equivalent

of the annual amortized cost. Finally, for the last value, the EPRI team calculated the corresponding percentage increase in the average customer monthly bill.

In practice, more complex cost allocation methods might well be applied that would shift cost among the class. This calculation assumes that the Smart Grid costs are equalized over customers across the country. However, the Smart Grid cost per costumer is likely to vary considerably, and therefore, the total estimated Smart Grid cost may be more concentrated in some areas, which would raise their cost per customer in those areas and reduce it elsewhere. These costs are modest when compared to the benefits the Smart Grid will yield. However, the challenge for all of those in the electricity sector will be communicating that the Smart Grid is indeed a good investment.

Table 1-3
Possible Consumer Implications of the EPRI Estimate of Smart Grid Costs

	Smart Grid Cost to Consumers – Allocated by Annual kWh (a)							
	$/Customer Total Cost (b)		$/Customer-Year, 10-Yr Amortization (c)		$/Customer-Month, 10-Yr Amortization (d)		% Increase in Monthly Bill, 10-Yr Amort (e)	
Class	Low	High	Low	High	Low	High	Low	High
	$/Customer	$/Customer	$/Cust/yr	$/Cust/yr	$/Cust/Month	$/Cust/Month		
Residential	$1,033	$1,455	$103	$145	$9	$12	8.4%	11.8%
Commercial	$7,146	$10,064	$715	$1,006	$60	$84	9.1%	12.8%
Industrial	$107,845	$151,877	$10,785	$15,188	$899	$1,266	0.01%	1.6%

(a) LOW refers to EPRI low estimate of $ total SG costs; HIGH is the other SG cost. Customer numbers by class (residential, commercial industrial) are for 2009 from EIA. SG costs are allocated to customer classes based on 2009 kWh sales (38 %residential; 37% Commercial; 25% industrial).

(b) Total SG cost divided by customers for each segment (residential +commercial+ industrial).

(c) Annual cost per customer per year for total SG cost spread out (amortized) equally over 10 years (nominal values).

(d) Annual cost per customer per month for total SG cost spread out (amortized) equally over 10 years (nominal values).

(e) Annual increase in monthly bill for based on (d).

◄ 1-7 ►

Smart Grid Benefits

The benefits of the Smart Grid are numerous and stem from a variety of functional elements which include cost reduction, enhanced reliability, improved power quality, increased national productivity and enhanced electricity service, among others. Table 1-4 and Figure 1-4 summarize these benefits. In general terms, the Smart Grid will assure that consumers are provided with reliable, high quality digital-grade power, increased electricity-related services and an improved environment. The Smart Grid will allow the benefits resulting from the rapid growth of renewable power generation and storage as well as the increased use of electric vehicles to become available to consumers. Without the development of the Smart Grid, the full value of a lot of individual technologies like Electric Vehicles, Electric Energy Storage, Demand Response, Distributed Resources, and large central station Renewables such as wind and solar will not be fully realized.

As detailed in Chapter 2, the benefits of the Smart Grid include:

- **Allows Direct Participation by Consumers.** The smart grid consumer is informed, modifying the way they use and purchase electricity. They have choices, incentives, and disincentives.

- **Accommodates all Generation and Storage Options.** The Smart Grid accommodates all generation and storage options.

- **Enables New Products, Services, and Markets.** The Smart Grid enables a market system that provides cost-benefit tradeoffs to consumers by creating opportunities to bid for competing services.

- **Provides Power Quality for the Digital Economy.** The Smart Grid provides reliable power that is relatively interruption-free.

- **Optimizes Asset Utilization and Operational Efficiently.** The Smart Grid optimizes assets and operates efficiently.

- **Anticipates and Responds to System Disturbances (Self-heal).** The Smart Grid independently identifies and reacts to system disturbances and performs mitigation efforts to correct them.

- **Operates Resiliently against Attack and Natural Disaster.** The Smart Grid resists attacks on both the physical infrastructure (substations, poles, transformers, etc.) and the cyber-structure (markets, systems, software, communications).

Table 1-4
Estimated Benefits of the Smart Grid

Attribute	Net Present Worth (2010) $B	
	Low	High
Productivity	1	1
Safety	13	13
Environment	102	390
Capacity	299	393
Cost	330	475
Quality	42	86
Quality of Life	74	74
Security	152	152
Reliability	281	444
Total	**1294**	**2028**

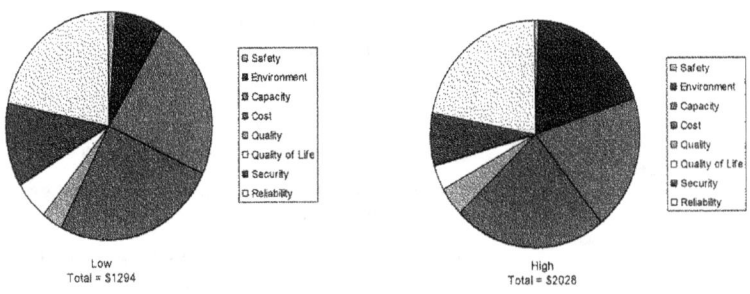

Figure 1-4
Estimated Benefits of the Smart Grid ($ in billions)

Cyber Security

Concern has arisen recently regarding the security of an information technology regime which would be integral with tomorrow's grid. Electric utilities have been incorporating cyber security features into their operations since the early 2000s. In recent years as the Smart Grid became increasingly popular, cyber security concerns have increased significantly. While there have to date been few reliable reports of cyber attacks on power systems, there is a great deal of concern that as the grid becomes smarter and more interactive, disruption of the reliability of U.S. electricity supply will become easier.

Cyber security is an essential element of the Smart Grid. It involves the protection needed to ensure the confidentiality and integrity of the digital overlay which is part of the Smart Grid.

The project team estimates for proper cyber security protection are included in the preceding estimates. An investment of approximately $3,729 million will be needed for the Smart Grid in addition to a related investment in information technology of approximately $32,258 million.

Section 2: Introduction

Smart Grid Vision

This section contains a definition of Smart Grid, an outline of benefits, Smart Grid characteristics, and challenges as contained in EPRI's report to the National Institute of Science and Technology (NIST). The vision is presented as it appeared in EPRI's report to NIST (EPRI, 2009).

What is the Smart Grid?

The Smart Grid definition is based upon the description found in the Energy Independence and Security Act of 2007. The term "Smart Grid" refers to a modernization of the electricity delivery system so it monitors, protects and automatically optimizes the operation of its interconnected elements – from the central and distributed generator through the high-voltage network and distribution system, to industrial users and building automation systems, to energy storage installations and to end-use consumers including their thermostats, electric vehicles, appliances and other household devices.

The Smart Grid will be characterized by a two-way flow of electricity and information to create an automated, widely distributed energy delivery network. It incorporates into the grid the benefits of distributed computing and communications to deliver real-time information and to enable the near-instantaneous balance of supply and demand at the device level.

Smart Grid Characteristics: Drivers and Opportunities

The definition of the Smart Grid builds on the work done in EPRI's IntelliGrid Program (intelligrid.epri.com), in the Modern Grid Initiative (MGI) (NETL, 2007), and in the GridWise Architectural Council (GWAC) (gridwise.org). These considerable efforts have developed and articulated the vision statements, architectural principles, barriers, benefits, technologies and applications, policies, and the frameworks that help define the Smart Grid.

Smart Grid Benefits

Smart Grid benefits can be categorized into 5 types:

- **Power reliability and power quality.** The Smart Grid provides a reliable power supply with fewer and briefer outages, "cleaner" power, and self-healing power systems, through the use of digital information, automated control, and autonomous systems.

- **Safety and cyber security benefits.** The Smart Grid continuously monitors itself to detect unsafe or insecure situations that could detract from its high reliability and safe operation. Higher cyber security is built in to all systems and operations including physical plant monitoring, cyber security, and privacy protection of all users and customers.

- **Energy efficiency benefits.** The Smart Grid is more efficient, providing reduced total energy use, reduced peak demand, reduced energy losses, and the ability to induce end-users to reduce electricity use instead of relying upon new generation.

- **Environmental and conservation benefits.** The Smart Grid facilitates an improved environment. It helps reduce greenhouse gases (GHG) and other pollutants by reducing generation from inefficient energy sources, supports renewable energy sources, and enables the replacement of gasoline-powered vehicles with plug-in electric vehicles.

- **Direct financial benefits.** The Smart Grid offers direct economic benefits. Operations costs are reduced or avoided. Customers have pricing choices and access to energy information. Entrepreneurs accelerate technology introduction into the generation, distribution, storage, and coordination of energy.

Stakeholder Benefits

The benefits from the Smart Grid can be categorized by the three primary stakeholder groups:

- **Consumers.** Consumers can balance their energy consumption with the real-time supply of energy. Variable pricing will provide consumer incentives to install their own infrastructure that supports the Smart Grid. Smart grid information infrastructure will support additional services not available today.

- **Utilities.** Utilities can provide more reliable energy, particularly during challenging emergency conditions, while managing their costs more effectively through efficiency and information.

- **Society.** Society benefits from more reliable power for governmental services, businesses, and consumers sensitive to power outage. Renewable energy, increased efficiencies, and Plug-In Electric Vehicle (PEV) support will reduce environmental costs, including carbon footprint.

A benefit to any one of these stakeholders can in turn benefit the others. Those benefits that reduce costs for utilities lower prices, or prevent price increases, to customers. Lower costs and decreased infrastructure requirements enhance the value of electricity to consumers. Reduced costs increase economic activity which benefits society. Societal benefits of the Smart Grid can be indirect and hard to quantify, but cannot be overlooked.

142

Other stakeholders also benefit from the Smart Grid. Regulators can benefit from the transparency and audit-ability of Smart Grid information. Vendors and integrators benefit from business and product opportunities around Smart Grid components and systems.

Modern Grid Initiative Smart Grid Characteristics

For the context of this section, characteristics are defined as prominent attributes, behaviors, or features that help distinguish the grid as "smart". The MGI developed a list of seven behaviors that define the Smart Grid. Those working in each area of the Smart Grid can evaluate their work by reference to these behaviors. These behaviors match those defined by similar initiatives and workgroups.

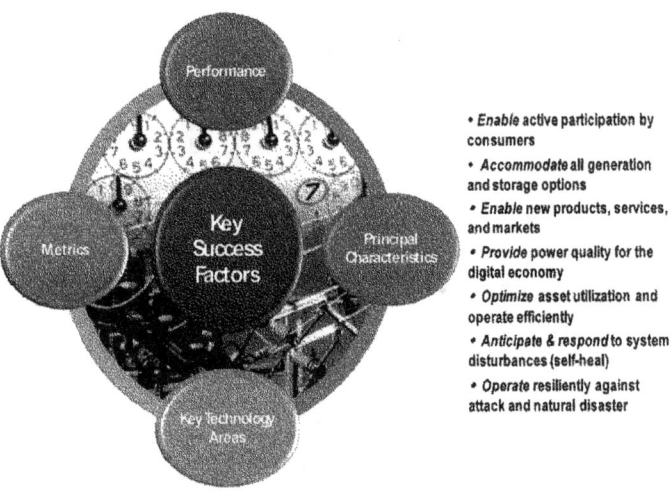

Figure 2-1
MGI's Principle Characteristics are Part of Their Smart Grid System Vision for Measuring Success (Source: EPRI Report to NIST, 2009)

The behaviors of the Smart Grid as defined by MGI are:

- **Enable Active Participation by Consumers.** The Smart Grid motivates and includes customers, who are an integral part of the electric power system. The smart grid consumer is informed, modifying the way they use and purchase electricity. They have choices, incentives, and disincentives to modify their purchasing patterns and behavior. These choices help drive new technologies and markets.

- **Accommodate All Generation and Storage Options.** The Smart Grid accommodates all generation and storage options. It supports large, centralized power plants as well as Distributed Energy Resources (DER). DER may include system aggregators with an array of generation systems or a farmer with a windmill and some solar panels. The same is true of storage, and as storage technologies mature, they will be an integral part of the overall Smart Grid solution set.

- **Enable New Products, Services, and Markets.** The Smart Grid enables a market system that provides cost-benefit tradeoffs to consumers by creating opportunities to bid for competing services. As much as possible, regulators, aggregators and operators, and consumers can modify the rules of business to create opportunity against market conditions. A flexible, rugged market infrastructure exists to ensure continuous electric service and reliability, while also providing revenue or cost reduction opportunities for market participants. Innovative products and services provide 3rd party vendors opportunities to create market penetration opportunities and consumers with choices and clever tools for managing their electricity costs and usage.

- **Provide Power Quality for the Digital Economy.** The Smart Grid provides reliable power that is relatively interruption-free. The power is "clean" and disturbances are minimal. Our global competitiveness demands relatively fault-free operation of the digital devices that power the productivity of our 21st century economy.

- **Optimize Asset Utilization and Operate Efficiently.** The Smart Grid optimizes assets and operates efficiently. It applies current technologies to ensure the best use of assets. Assets operate and integrate well with other assets to maximize operational efficiency and reduce costs. Routine maintenance and self-health regulating abilities allow assets to operate longer with less human interaction.

- **Anticipate and Respond to System Disturbances [Autonomously] (Self-heal).** The Smart Grid independently identifies and reacts to system disturbances and performs mitigation efforts to correct them. It incorporates an engineering design that enables problems to be isolated, analyzed, and restored with little or no human interaction. It performs continuous predictive analysis to detect existing and future problems and initiate corrective actions. It will react quickly to electricity losses and optimize restoration exercises.

- **Operate Resiliently against Attack and Natural Disaster.** The Smart Grid resists attacks on both the physical infrastructure (substations, poles, transformers, etc.) and the cyber-structure (markets, systems, software, communications). Sensors, cameras, automated switches, and intelligence are built into the infrastructure to observe, react, and alert when threats are recognized within the system. The system is resilient and incorporates self-healing technologies to resist and react to natural disasters. Constant monitoring and self-testing are conducted against the system to mitigate malware and hackers.

Smart Grid Challenges

The Smart Grid poses many procedural and technical challenges as we migrate from the current grid with its one-way power flows from central generation to dispersed loads, toward a new grid with two-way power flows, two-way and peer-

144

to-peer customer interactions, distributed generation, distributed intelligence, command and control. These challenges cannot be taken lightly; the Smart Grid will entail a fundamentally different paradigm for energy generation, delivery, and use.

Procedural Challenges

It the short term it will be useful to prioritize the challenges that the Smart Grid needs to overcome first as a foundation for what is to come. The industry should collaborate to segregate the challenges into buckets to test a hypothesis under which to move forward or so that addressing these challenges becomes more manageable.   To address this problem, EPRI is working with several members to develop roadmaps for achieving the promise of the Smart Grid including the necessary decision trees, off ramps and schedules. These are expected to become available during 2011. The procedural challenges to the migration to a smart grid are enormous, and all need to be met as the Smart Grid evolves:

- **Broad Set of Stakeholders.** The Smart Grid will affect every person and every business in the United States. Although not every person will participate directly in the development of the Smart Grid, the need to understand and address the requirements of all these stakeholders will require significant efforts by utilities, system operators, third party electricity service providers and consumers themselves.

- **Complexity of the Smart Grid.** The Smart Grid is a vastly complex machine, with some parts racing at the speed of light. Some aspects of the Smart Grid will be sensitive to human response and interaction, while others need instantaneous, intelligent and automated responses. The smart grid will be driven by forces ranging from financial pressures to environmental requirements.

- **Transition to Smart Grid.** The transition to the Smart Grid will be lengthy. It is impossible (and unwise) to advocate that all the existing equipment and systems to be ripped out and replaced at once. The smart grid supports gradual transition and long coexistence of diverse technologies, not only as we transition from the legacy systems and equipment of today, but as we move to those of tomorrow. We must design to avoid unnecessary expenses and unwarranted decreases in reliability, safety, or cyber security.

- **Ensuring Cyber Security of Systems.** Every aspect of the Smart Grid must be secure. Cyber security technologies and compliance with standards alone are not enough to achieve secure operations without policies, on-going risk assessment, and training. The development of these human-focused procedures takes time—and needs to take time—to ensure that they are done correctly.

- **Consensus on Standards.** Standards are built on the consensus of many stakeholders over time; mandating technologies can appear to be an adequate short cut. Consensus-based standards deliver better results over.

- **Development and Support of Standards.** The open process of developing a standard benefits from the expertise and insights of a broad constituency.

The work is challenging and time consuming but yields results more reflective of a broad group of stakeholders, rather than the narrow interests of a particular stakeholder group. Ongoing engagement by user groups and other organizations enables standards to meet broader evolving needs beyond those of industry stakeholders. Both activities are essential to the development of strong standards.

- **Research and Development.** The smart grid is an evolving goal; we cannot know all that the Smart Grid is or can do. The smart grid will demand continuing R&D to assess the evolving benefits and costs, and to anticipate the evolving requirements.

- **Having a Critical Mass.** It is unclear to the EPRI project team if the Smart Grid implementation is subject to considerations like those of critical mass needed, tipping points and penetration of implementation. There is some concern that early efforts must yield benefits in order to gather support for the development. That support may not accrue until a critical number of consumers are on board with the concepts. If everything the industry does in building the Smart Grid is amenable to a slow diffusion model for evolution as opposed to undertaking some elements in a concentrated way, those benefits may not be revealed quickly enough.

Technical Challenges to Achieving the Smart Grid

Technical challenges include the following:

- **Smart Equipment.** Smart equipment refers to all field equipment which is computer-based or microprocessor-based, including controllers, remote terminal units (RTUs), and intelligent electronic devices (IEDs). It includes the actual power equipment, such as switches, capacitor banks, or breakers. It also refers to the equipment inside homes, buildings and industrial facilities. This embedded computing equipment must be robust to handle future applications for many years without being replaced.

- **Communication Systems.** Communication systems refer to the media and to the developing communication protocols. These technologies are in various stages of maturity. The smart grid must be robust enough to accommodate new media as they emerge from the communications industries, while preserving interoperable, secured systems.

- **Data Management.** Data management refers to all aspects of collecting, analyzing, storing, and providing data to users and applications, including the issues of data identification, validation, accuracy, updating, time-tagging, consistency across databases, etc. Data management methods which work well for small amounts of data often fail or become too burdensome for large amounts of data—and distribution automation and customer information generate lots of data. Data management is among the most time-consuming and difficult task in many of the functions and must be addressed in a way that can scale to immense size.

- **Cyber Security.** Cyber security addresses the prevention of damage to, unauthorized use of, exploitation of, and, if needed, the restoration of

146

electronic information and communications systems and services (and the information contained therein) to ensure confidentiality, integrity, and availability.

- **Information/Data Privacy.** The protection and stewardship of privacy is a significant concern in a widely interconnected system of systems that is represented by the Smart Grid. Additionally, care must be taken to ensure that access to information is not an all or nothing at all choice since various stakeholders will have differing rights to information from the Smart Grid.

- **Software Applications.** Software applications refer to programs, algorithms, calculations, and data analysis. Applications range from low level control algorithms to massive transaction processing. Application requirements are becoming more sophisticated to solve increasingly complex problems, are demanding ever more accurate and timely data, and must deliver results more quickly and accurately. Software engineering at this scale and rigor is still emerging as a discipline. Software applications are at the core of every function and node of the Smart Grid.

Smart Grid Networking

The Smart Grid is a network of networks, including power, communications and intelligence. That is, many networks with various traditional ownership and management boundaries are interconnected to provide end-to-end services between stakeholders and in and among intelligent electronic devices (IEDs).

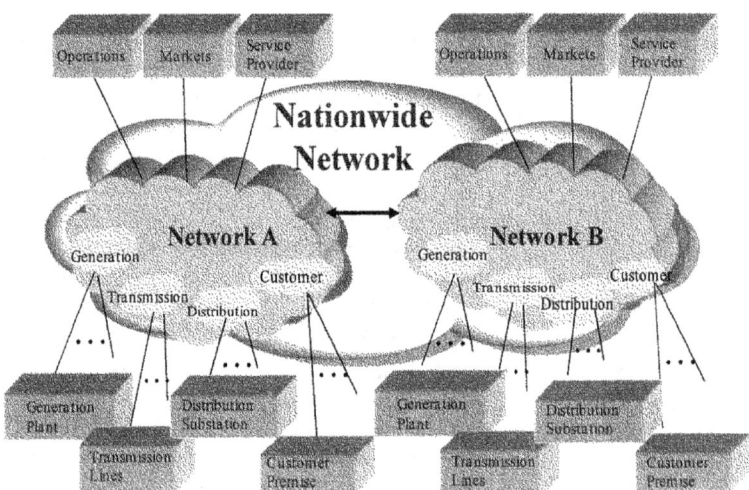

Figure 2-2
Smart Grid Networks for Information Exchange (Source: EPRI Report to NIST, 2009)

Figure 2-2 is a high level view of the information network for the Smart Grid. It handles the two-way communication between the network end points residing in their respective domains. By domain, we mean the unique distributed computing environments in which communicating end points can be found (see next section). Thus, any domain application could communicate with any other domain application via the information network, subject to the necessary network access restrictions and quality of service requirements.

The applications in each domain are the end points of the network as shown on the top and bottom of Figure 2-2. For example, an application in the Customer domain could be a smart meter at the customer premise; an application in the Transmission domain could be a phasor measurement unit (PMU) unit on a transmission line or in a Distribution domain at a substation; an application in the Operation domain could be a computer or display system at the operation center. Each of these applications has a physical communication link with the network. The smaller clouds within the network represent sub-networks that may be implementing unique functionality. The networking function in the Operations, Market, Service Provider domains may not be easily differentiated from normal information processing networks; therefore no unique clouds are illustrated.

This information network may consist of multiple interconnected networks as shown in Figure 2-2, where two backbone networks, A and B are illustrated. The physical links within these two networks and between the network and the network end points could utilize any appropriate communication technology currently available or yet to be developed.

Additional requirements for the information network are as follows:

- Management functionality for network status monitoring, fault detection, isolation, and recovery,
- Secure protocols to protect Smart Grid information in transit and authenticate infrastructure components,
- Cyber security countermeasures,
- Addressing capability to entities in the network and devices attached to it,
- Routing capability to all network end points,
- Quality of service support for a wide range of applications with different latency and loss requirements.

The Smart Grid Conceptual Model

The Smart Grid Conceptual Model is a diagram and description that are the basis for discussing the characteristics, uses, behavior, interfaces, requirements and standards of the Smart Grid. This does not represent the final *architecture* of the Smart Grid; rather it is a tool for describing, discussing, and developing that architecture. The conceptual model provides a context for analysis of interoperation and standards, both for the rest of this document, and for the development of the architectures of the Smart Grid. The top level of the conceptual model is shown in Figure 2-3.

Conceptual Model

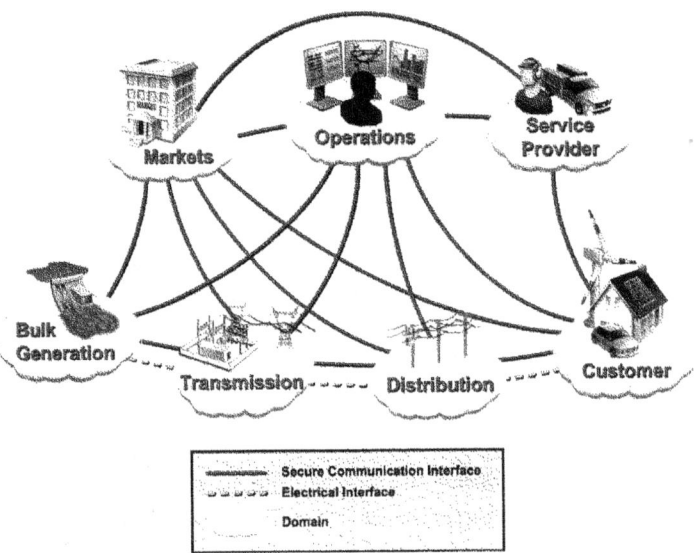

Figure 2-3
Smart Grid Conceptual Model – Top Level (Source: EPRI Report to NIST, 2009)

The conceptual model consists of several *domains*, each of which contains many *applications* and *actors* that are connected by *associations*, which have *interfaces* at each end:

- **Actors** may be devices, computer systems or software programs and/or the organizations that own them. Actors have the capability to make decisions and exchange information with other actors through interfaces.

- **Applications** are the tasks performed by the actors within the domains. Some applications are performed by a single actor, others by several actors working together.

- **Domains** group actors to discover the commonalities that will define the interfaces. In general, actors in the same domain have similar objectives. Communications within the same domain may have similar characteristics and requirements. Domains may contain other domains.

- **Associations** are logical connections between actors that establish bilateral relationships. At each end of an association is an *interface* to an *actor*.

- **Interfaces** show either electrical connections or communications connections. In the diagram, electrical interfaces are shown as yellow lines and the communications interfaces are shown in blue. Each of these interfaces may be bi-directional. Communications interfaces represent an information exchange between two domains and the actors within; they do not represent physical connections. They represent logical connections in the smart grid information network interconnecting various domains (as shown in Figure 2-2).

The domains of the Smart Grid are listed briefly in Table 2-1 and discussed in more detail in the sections that follow. In Figure 2-3, domains are shown as clouds.

Table 2-1
Domains in the Smart Grid Conceptual Model

Domain	Actors in the Domain
Customers	The end users of electricity. May also generate, store, and manage the use of energy. Traditionally, three customer types are discussed, each with its own domain: home, commercial/building, and industrial.
Markets	The operators and participants in electricity markets
Service Providers	The organizations providing services to electrical customers and utilities
Operations	The managers of the movement of electricity
Bulk Generation	The generators of electricity in bulk quantities. May also store energy for later distribution.
Transmission	The carriers of bulk electricity over long distances. May also store and generate electricity.
Distribution	The distributors of electricity to and from customers. May also store and generate electricity.

It is important to note that domains are NOT organizations. For instance, an ISO or RTO may have actors in both the Markets and Operations domains. Similarly, a distribution utility is not entirely contained within the Distribution domain – it is likely to also contain actors in the Operations domain, such as a Distribution Management System, and in the Customer domain, such as meters

Additional Challenges

Work Force Issues

The utility work force is undergoing a significant challenge. One-half of the 500,000 to 600,000 utility workers will be eligible to retire in the next five years. They need to be replaced with a trained and motivated work force. Introducing Smart Grid technologies requires employees with different skills to support the implementation, maintenance, and operation of the systems with digital components. Accomplishing this when it is already difficult to get highly skilled employees with technical experience will be challenging. In addition, these staffing requirements must be met at the same time as utilities reduce their work forces in order to hold the line on costs (Lave, 2007.) This report includes the labor costs associated with installation of Smart Grid devices and software as well

150

as the differential maintenance. However, it does not include costs for a workforce which generally has different skills involving more expertise in digital devices and communications.

Outage management systems can be used to save costs associated with incorrect outage reports by verifying power outages at customer facilities. (PECO estimates that it avoided 7,500 crew dispatches in 2005 because it was able to see that the customer-reported outage was inaccurate (PECO, 2006).

Regulatory Challenges to Achieving the Smart Grid

Smart Grid technologies offer unprecedented challenges to regulators in encouraging and adjudicating decisions regarding Smart Grid investments. Power systems have largely operated without "smart" technology for decades. In fact, many power systems operate at 99.999% reliability at the bulk transmission level. As long as reliability levels have been maintained (the lights were still on) and costs were low (rates have been essentially flat for decades), it is conceptually difficult to understand how a basket of digital-based technologies can improve the current value of today's power system.

New regulatory and business models are being considered which would offer a greater incentive for utilities to engage in energy efficiency. In some cases, they may be faced with reducing rates as part of seeking approval for Smart Grid investments and, subsequently, losing revenue. Regulated utilities are increasingly embracing energy-efficiency measures on the distribution system or those involving end-use customers.

An additional regulatory challenge is to understand the incremental value of Smart Grid investments. Often, for logical reasons, Smart Grid technologies are implemented in stages, with each stage requiring a business plan for regulators to approve. However, the benefits of many of the Smart Grid efforts come from the synergistic applications of a portfolio of Smart Grid technologies.

This issue was summarized nicely by the Illinois Commerce Commission in a recent report (ISSGC, 2010):

> "The issue of smart grid cost recovery has been a matter of controversy and litigation for several years. Disagreements exist about whether recovery of a utility's smart grid costs should be restricted to the 'traditional' rate-base method, or whether a 'non-traditional' method (e.g., 'rider' recovery) should be used.
>
> Some stakeholders are concerned that utility proposals for cost recovery of smart grid investments would lead to significantly higher monthly bills and a shift in the risk of investment from utilities to ratepayers. Others believe that non-traditional cost recovery would be essential to accelerate deployment of smart grid technologies."

Finally, regulators and utilities are accustomed to utility asset life in the range of 30 to 50 years and business plans and rate cases for regulated utilities are typically based on that supposition. When digital equipment is involved – particularly at the early stages of evolution – the depreciation rates for a significant amount of capital investment may be 5 to 15 years. Appropriate depreciation rates must be allowed in order to pay for asset renewal without increasing costs more than necessary.

In short, a forward view of technology must be embraced by utilities and regulators in order for the Smart Grid to be successful.

Drivers of Smart Grid Investment

A wide variety of policies and economic trends have begun to stimulate and drive U.S. investment in Smart Grid technology, including:

- **The Energy Independence and Security Act (EISA) of 2007** established national policy for grid modernization, created new federal committees, defined their roles and responsibilities, and provided incentives for investment.

- **The American Recovery and Reinvestment Act of 2009** provided more than $3.4 billion in stimulus funding for Smart Grid technology development and demonstration, plus $615 for Smart Grid storage. In October, 2009, 100 Smart Grid Investment Grants were awarded. These were 50/50 matching grants over a three-year period leading to an infusion of $7 to $8 billion in Smart Grid investments that are specifically targeted at projects that can be emulated by others.

- **Renewable portfolio standards** have been established in 30 states plus the District of Columbia stimulating rapid expansion of renewable technology and accelerating the need for Smart Grid technology for grid integration. EPRI's Prism analysis anticipates 135 GW of renewables by 2030 (EPRI 1020389). A number of states have also enacted policies to address specific environmental concerns.

- **Smart Grid interoperability standards,** called for by the EISA, moved forward with NIST's release in September, 2009, of a roadmap for interoperability standards. NIST's efforts were aided by EPRI's draft interim roadmap report released in August, 2009. While not a driver of Smart Grid investment of itself, these recommended standards facilitate Smart Grid deployment.

- **Critical Infrastructure Protection (CIP) Cyber Security Standards,** maintained by the North American Electric Reliability Corporation (NERC) and approved by the Federal Energy Regulatory Commission (FERC) in 2006, are intended to ensure the protection of the critical cyber assets that control or effect the reliability of North America's bulk electric systems. The CIP Cyber Security Standards are mandatory and enforceable across all users, owners, and operators of the bulk power system (LogRhythm, 2009).

- **Demand response programs** have accelerated. The Federal Energy Regulatory Commission's (FERC's) 2008 Demand Response (DR) and Automated Metering Infrastructure (AMI) Survey indicated that advanced metering penetration reached 4.7% of total meters, up from less than 1% in 2006, and that 8% of U.S. customers were currently involved in DR programs. DR continues to gain ground through state legislative initiatives and utility regulation. FERC's "A National Assessment of Demand Response Potential – Staff Report" in June 2009 and FERC's staff report "National Action Plan on Demand Response" in June 2010 highlight the potential (FERC, 2009 and 2010).

- **Market demand** for Smart Grid technologies has drawn the interest of many major information technology companies, from Cisco and Intel, to Google, IBM, and Microsoft which want to participate in one of the most attractive business opportunities of the future. The Smart Grid is viewed as the market equivalent of the Internet in terms of its trillion-dollar potential worldwide. It is the enabling infrastructure for the accelerated deployment of electric vehicles and plug-in hybrid electric vehicles. In addition, consumers will a driving force in shaping the nature of tomorrow's Smart Grid technologies as they respond to evolving offerings by equipment providers.

- **Venture capital** is also entering the Smart Grid domain in a large way, promising to bring faster and more concentrated technical innovation in the areas of Automated Metering Infrastructure (AMI), communications and network technologies. More than $1 billion in VC funding has been extended to key startups, typified by GridPoint and Silver Spring Networks (Green tech media, 2010).

- **Smart Grid roadmaps** are being developed by many electric utilities to optimize their investment strategy going forward. Objectives and starting points vary from company to company, and the optimal pathway difficult to assess. EPRI has been working with SCE, FirstEnergy, and SRP, among others, to create roadmaps for Smart Grid investment. What is critical today is reliable data on benefits from Smart Grid demonstrations. Every effort must be made to measure the actual benefits realized through Smart Grid demonstrations as projects are deployed.

- **National transmission corridors** have been identified. The Energy Policy Act of 2005 authorized the DOE to conduct national electric transmission congestion studies and to designate National Corridors if appropriate. In 2007, DOE designated a Mid-Atlantic National Corridor and a Southwest Area National Corridor (DOE, 2007).

- **Outage prevention** becomes increasingly important in an information-service-based economy. There have been five major blackouts in the last 40 years, three of which occurred in the last decade. The Northeast blackout of 2003 resulted in an estimated $7 to $10 billion in losses to the region. Less disruptive but more pervasive power quality problems are estimated to now cost the U.S. $119 to $188 billion per year (EPRI 1006274).

Previous Studies by EPRI

EPRI has been involved in collaborative, electricity-based innovation in what is now called the Smart Grid since the mid-1980s. EPRI has acted as the catalyst for a process of engagement and consensus building among diverse parties within and outside the electricity enterprise. This effort continues.

Previous EPRI studies have estimated both the costs and benefits of a Smart Grid.

- According to a 2004 study, "The Power Delivery System of the Future" will require $165 billion in net investment (over and above investment for load growth and correcting deficiencies), and lead to a benefit-to-cost ratio of 4:1. Benefits accrue from:

  - Reduced energy losses and more efficient electrical generation.
  - Reduced transmission congestion.
  - Improved power quality.
  - Reduced environmental impact.
  - Improved U.S. competitiveness, resulting in lower prices for all U.S. products and greater U.S. job creation.
  - Fuller utilization of grid assets.
  - More targeted and efficient grid maintenance programs.
  - Fewer equipment failures.
  - Increased security through deterrence of organized attacks on the grid.
  - Improved tolerance to natural disasters.
  - Improved public and worker safety.

- EPRI studies show the annual cost of power disturbances to the U.S. economy ranges between $119 and $188 billion per year (EPRI 1006274). The societal cost of a massive blackout is estimated to be in the order of $10 billion per event as described in a report published by the North American Electric Reliability Corporation titled "Final Report on the August 14, 2003 Blackout in the United States and Canada" (NERC, 2004)

- The Smart Grid is capable of providing a significant contribution to the national goals of energy and carbon savings, as documented in two recent reports.

  - One report by EPRI states that the emissions reduction impact of a Smart Grid is estimated at 60 to 211 million metric tons of CO_2 per year in 2030.

  - Another report by Pacific Northwest National Laboratory (PNNL) states that full implementation of Smart Grid technologies is expected to achieve a 12% reduction in electricity consumption and CO_2 emissions in 2030.

- And in another report, EPRI estimated that the Smart Grid, combined with a portfolio of generation and end-use options, could reduce 2030 annual CO_2 emissions from the electric sector by 58% relative to 2005 emissions (EPRI 100389).

EPRI Demonstrations

EPRI's Smart Grid Demonstration initiative involves a number of ongoing projects to demonstrate the potential for integrating distributed power generation, storage, and demand response technology into "virtual power plants." Demonstrations include both utility side and customer side technologies, and are intended to address the challenges of integrating distributed Energy resources (DER) in grid and market operations, as well as in system planning. The program addresses key industry challenges, such as:

- Demonstrating effective ways of integrating different forms of distributed resources.

- Demonstrating multiple levels of integration and interoperability among various components.

- Exploring existing and emerging information and communication technologies.

The demonstrations are taking place at a number of U.S. locations and will include a variety of feeder constructions, climate zones, and technologies. Individual demonstrations are focused on the integration of specific feeder types used in residential neighborhoods, in a mixture of residential and commercial customers, and in areas with mostly commercial customers.

Purpose of this Report

The primary purpose of this report is to initiate a discussion and debate of the investment needed to create a viable Smart Grid. To meet this objective, this report documents the methodology, key assumptions, and results of a preliminary quantitative estimate of the needed investment (cost). The report is a starting point intended to encourage further stakeholder discussion of this topic.

The complexity of the power delivery system and the wide range of potential technology applications and configurations to enhance its performance complicate the process of quantitatively estimating the needed investment. In addition, due to the various types of information available, complexity of subparts of the analysis, and uncertainties associated with estimating techniques, no single approach can be applied to all portions of the evaluation. Nevertheless, the debate over the appropriate level of power delivery system investment cannot be advanced without some preliminary estimate of costs. Hence, for each key portion of the overall task, the project team selected methods based on the availability of credible information and the need to conduct a cost-effective and time-efficient study. The resulting estimates of costs are highly uncertain and open to debate.

In this report, EPRI will only address the aggregate cost of the Smart Grid. A separate study has been launched to thoroughly assess the benefits. However, a preliminary update of benefits is included in Section 4.

Why Did the Smart Grid Cost Estimates Change?

There are a number of reasons these estimates changed so dramatically since EPRI last estimated the potential costs of the Smart Gird. First, these changes are due in part to inflation and increasing component costs. Second, they are due to a considerable expansion in the functionality now envisioned in tomorrow's Smart Grid. Table 2-2 highlights these changes.

Table 2-2
Major Elements of Functionality Added to the Smart Grid

Element	Previously Included	Added Benefits
Demand Response	None	Reduced need for generation capacity. Reduced demand for electricity.
Facilitating Renewables	None	Reduced environmental impact of electricity generation.
Plug-In Electric Vehicles or Plug-In Hybrid Electric Vehicles (PEVs)	None	Reduced environment impact from displaced fossil fuels. Grid support (increased system flexibility/ancillary services).
Energy Efficiency	Cost reduction. Reduced need for T&D. Reduced environmental impact.	Reduced need for generation capacity.
Enhanced Energy Efficiency* (additional energy efficiency)	None	Reduced costs. Reduced environmental impact.
AMI	None	AMI-related cost reductions.
Distributed Generation	None	Reduced need for central generating capacity.
Storage	None	Capacity. Reliability and power quality. O&M. Congestion management.

*Enhanced Energy Efficiency includes:
- Continuous Commissioning of Large Commercial Buildings
- Direct Feedback on Energy Usage
- Energy Savings Corresponding to Peak Load Management
- Energy Savings Corresponding to Enhanced M&V Capability

Section 3: Approach

The project team separated the power delivery system into distinct functional areas, and made a number of assumptions about technology development, deployment, and cost over the study period (2010-2030). These assumptions are covered at a high level in this section, and then covered in greater detail for each of the 25 cost components of the four main technology sections of the report – transmission, substations, distribution and customers.

What Constitutes the Power Delivery System?

The power delivery system includes the busbar located at the generating plant (where the power delivery system begins) and extends to the energy-consuming device or appliance at the end user. This means that the power delivery system encompasses generation step-up transformers; the generation switchyard; transmission substations, lines, and equipment; distribution substations, lines, and equipment; intelligent electronic devices; communications; distributed energy resources located at end users; power quality mitigation devices and uninterruptible power supplies; sensors; energy storage devices; and other equipment.

Inadequacies in the power delivery system are manifested in the form of poor reliability, excessive occurrences of degraded power quality, vulnerability to mischief or terrorist attack, the inability to integrate renewables, and the inability to provide enhanced services to consumers.

What Differentiates Smart Grid Enhancement?

Meeting the energy requirements of society will require the application across the entire power delivery system of a combination of current and advanced technologies, including but not limited to the following:

- Automation: the heart of a "smart power delivery system."

- Communication architecture: the foundation of the power delivery system of the future and the enabler of Smart Grid integration.

- Distributed energy resources and storage development and integration.

- Power electronics-based controllers and widely dispersed sensors throughout the delivery system.

- An advanced metering infrastructure.

- A consumer portal that connects consumers and their equipment with energy services and communications entities.

- Power market tools – information systems which enable fluid wholesale power markets.

- Technology innovation in electricity use.

- Appliances and devices which are demand-response ready.

Developing an optimal combination of these technologies will require a significant, sustained RD&D investment. Making such an investment in a critical industry like the U.S. electric power industry is not unprecedented.

Study Steps

To conduct a preliminary quantitative estimate of the level of investment needed over the next 20 years, the project team first separated the core technologies into four broad areas: transmission, substations, distribution and the customer interface. Next, the team subdivided the estimating process into the following segments:

- **Meeting load growth and correcting deficiencies** via equipment installation, upgrading, and replacement to accommodate new customers (new connects), to meet the increasing energy needs of existing customers as their load grows, and to correct deficiencies (e.g., correct power flow bottlenecks and limit high-fault currents that damage critical grid equipment).

- **The Smart Grid:** The project team estimated the investment needed to develop and deploy advanced technologies needed to enhance the functionality of the power delivery system to achieve the level of a Smart Grid.

The first segment represents investments required to maintain adequate capacity and functioning of the existing power delivery system, while the second segment is the additional cost to elevate this system to that of a Smart Grid.

Key Assumptions

The cost estimate was built upon a number of key assumptions:

- Incorporate technologies that not only make the electricity delivery system smarter, but also stronger, more resilient, adaptive, and self-healing.

- Include every reasonable and cost-effective enhancement to accommodate regulatory mandates:

  - Consistent with the functionality requirements of the Energy Independence and Security Act (EISA) of 2007.

  - Meets reasonable cost-benefit assessment.

- Meets North American Electric Reliability Corporation (NERC) reliability standards, maintaining or enhancing today's reliability levels (1 day in 10 years loss of load probability or LOLP).

- Meets System Average Duration Interruption and System Average Interruption Frequency (SADI/SAFI) state guidelines normally suggest 100 minutes SADI/SAFI and power quality (PQ) events to remain at today's levels or to improve.

- Meets performance rate-making targets.

- Meets requirements of future renewable portfolio standards (RPS).

■ Incorporate technology and policies that enhance Smart Grid functionality while meeting load growth, expanding and modernizing the power delivery system.

- Enable a fully functional power delivery system

- Enable consumer connectivity and service enhancement

- Enable integration of distributed energy resources

■ Accommodate expansion of renewable energy resources consistent with PRISM and other EPRI scenarios, and affords the possibility of meeting DOE targets for wind.

- EPRI Prism estimates 135GW of renewables by 2030

- DOE's aggressive target for wind–20% by 2030 seems increasingly plausible.

■ The Energy Information Agency's Annual Energy Outlook 2010 projects that the annual growth rate in electricity for the period 2008 to 2035 is projected to be 1.0%. This is as a result of "structural changes in economy – higher prices – standards – improved efficiency" (EIA, 2009). EPRI estimates that the programs and activities which are part of the Smart Grid as envisioned in this report have the potential to reduce this growth rate to 0.68% per year. In addition, EPRI estimates that peak demand's growth rate will be 0.53% per year (EPRI 1016987). These growth rates were used in assumptions about the increasing needs for assets to serve consumers.

■ Assume simultaneous deployment of Smart Grid functionality. While deployments will realistically be made along parallel paths and in discrete steps, this study assumes they will occur simultaneously and continuously.

■ Assume steady rate of deployment. Deployments are assumed to begin in 2010. Deployment of most technologies will be made at a rate of 1/20th of the maximum assumed penetration each year over the 20-year period. Enhancement and modernization will continue after 2030.

- The Smart Grid will never be finished. It will continue to evolve organically, not as a step function and not as a "revolution," but as new technology becomes available, practicable and reliable.

- The investigators recognize that investments in Smart Grid will not be made linearly over 20 years, or necessarily even within 20, and will not be uniformly distributed around the country.

■ Total power delivery investment costs will exceed Smart Grid investments. They will include investments to meet load growth and to maintain reliability.

■ Technology costs are likely to decrease while performance levels increase in unforeseen, and possibly dramatic, ways over the next 20 years. Reasonable estimates have been made, but they are likely to prove conservative given the rapid pace of technological advances. Historically, massive technology advances such as implied by the Smart Grid are invariably driven by a single breakthrough innovation. Smart Grids don't have just one; instead they have a wide range of ideas – some at the pilot or even experimental stage. In some sense, communications, control, and computational ability is one such set of breakthroughs. However, other advances in storage, power electronics, and sensors are still needed to complete the mosaic this report paints.

■ Appropriately consider operating and maintenance (O&M) costs associated with running utilities which deploy Smart Grid technologies. O&M expenses are a substantial part of total costs and are built into rates at estimated levels. The IT and technology O&M aspects of the Smart Grid need to be included in cost estimates.

■ As the smart grid evolves communications networks will become more ubiquitous and multi-purpose. Utilities may use commercial carriers to provide these networks or they may build out their own networks using dedicated spectrum, share spectrum dedicated to public safety or use unlicensed spectrum. Cost varies significantly for each approach. For this study we have assessed the overall cost for communications networks and have allocated it to the various domains and smart grid applications.

Smart Grid Costs are Particularly Hard to Estimate

Smart Grids are by their nature difficult to estimate for several reasons:

■ They frequently involve the integration of digital technology – Sometimes virtually embedded transmission and distribution assets have different failure rates and life expectancy than the majority of today's grid technologies. These failures and resultant replacement rates must be estimated. Utilizing a reliable component, like a substation transformer, with a 40-year design life and incorporating an information technology with 10, 15 or 20 life forces careful consideration of the costs to upgrade the embedded components.

■ The obsolescence of digital technology is rapid. Increasingly complex and expanding communications and computational ability makes it possible to render Smart Grid components obsolete or inoperable with respect to the rest of the information and communications technology (ICT) system well before the end of their life. Therefore, reasonable replacement costs must be estimated.

- The improvement in Smart Grid technologies and projected decreases in their costs will occur at a greater rate than "conventional" technology.

- Uncertainty in performance – Many Smart Grid technologies are relatively new and unproven. If their performance is marginal or degrades unexpectedly over time, the entire business plan for the technology could be undermined.

- Smart Grid component costs are declining rapidly. As these technologies mature and as production volumes increase, the marginal costs of Smart Grid technologies have the potential to decline rapidly.

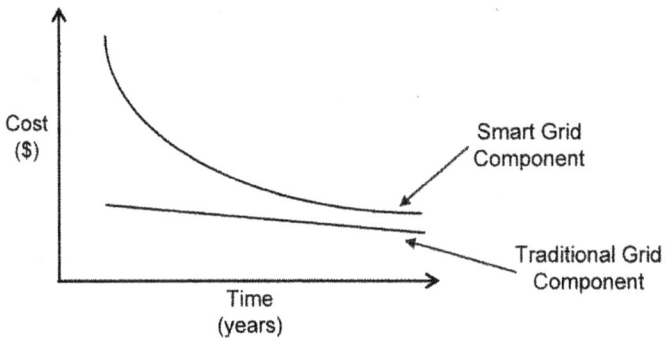

Figure 3-1
Grid Component Costs (Illustrative)

Technology Assessment: What's In and What's Not In?

Table 3-1 summarizes what has been included in the cost analysis and what has not been included. For the most part, T&D line expansion to meet load growth has been excluded.

Figure 3-2 illustrates the scope of the cost estimates included in this report. Investments traditionally made by customers, such as appliances and hybrid vehicles, have been excluded. However, infrastructure integration costs are included.

Table 3-1
Technologies Included and Excluded in the Cost Estimates

Technologies	Costs Included in This Analysis	Costs Excluded
1. Increased use of digital information and controls technology to improve reliability, security, and efficiency of the electric grid.	Sensors, communications and computational ability	None
2. Dynamic optimization of grid operations and resources, with full cyber-security.	All grid-related	None
3. Deployment and integration of distributed resources: storage and generation, including renewable resources.	All integration costs	New transmission lines including those to integrate renewables; and costs of renewable power generation technology*
4. Development and incorporation of demand response, demand-side resources, and energy-efficiency resources	All integration costs	The cost of energy-efficient devices
6. Integration of "smart" appliances and consumer devices.	All integration costs (see above)	Consumer appliances and devices
7. Deployment and integration of advanced electricity storage and peak-shaving technologies, including plug-in and hybrid-electric vehicles, and thermal-storage air conditioning.	Bulk power storage devices and high-value distributed storage – such as bulk storage for wind penetration; distributed storage for grid support; customer-side-of-the-meter storage for end-use energy management	Low-value distributed storage
8. Provision to consumers of timely information and control options to enable consumer engagement.	All enabling costs including cost of consumer display devices	None

Table 3-1(continued)
Technologies Included and Excluded in the Cost Estimates

Technologies	Costs Included in This Analysis	Costs Excluded
10. Identification and lowering of unreasonable or unnecessary barriers to adoption of Smart Grid technologies, practices, and services.	All	None
11. Costs to implement NERC's Critical Infrastructure Protection (CIP) Standards	None	CIP requirements applied to the Distribution System SCADA would incur substantial costs and are excluded here

*It should be noted that some reviewers felt that the cost of new transmission needed to integrate renewables should ultimately be included in estimating the cost of tomorrow's power delivery system.

One additional cost which could be imposed on some utilities involves conversion to the International Electrotechnology Commission's (IEC) standard for substations communication called IEC 61850. If conversion to IEC 61850 were mandated, legacy systems now utilized by some for both distribution SCADA and communications on the power system would become obsolete. These utilities used a form of Multi-Agent Systems or MAS as a simple format for the exchange of digital information on their power system. For some of these utilities, a mandate to convert to implement IEC 61850 would necessitate replacing their MAS infrastructure, all remote MAS radios and Remote Terminal Units (RTUs) in order to provide the increased bandwidth necessary to support IEC 61850.

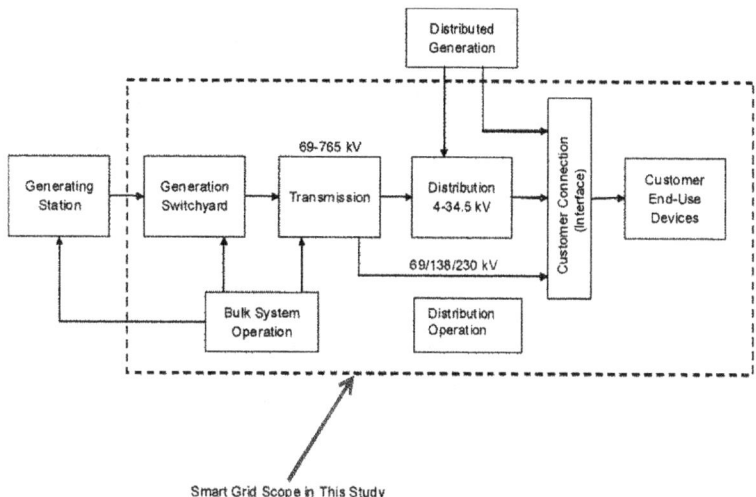

Figure 3-2
Smart Grid Diagram

To conduct a preliminary quantitative estimate of the level of investment needed over the next 20 years to enable the envisioned power delivery system, the project team first decided to treat transmission, distribution, and customer-related costs separately. This is due to fundamental differences in the nature of the transmission and distribution portions of the power delivery system; and uncertainty whether the costs categorized as consumer costs would be borne by utilities, consumers, or third-party service providers. It should be noted that there are, however, substantial areas in which distribution technology enhancements will greatly affect the operation and potentially the configuration of the transmission system. These interactions were not considered in this evaluation. The team also decided to further subdivide the estimating process for transmission and distribution into the following two segments:

- **Load Growth.** Via equipment installation, upgrading, and replacement, transmission and distribution system owners invest in the power delivery system to accommodate new customers (so-called "new connects") and to meet the increasing energy needs of existing customers as their load grows.

- **Power Delivery System of the Future ("Future PDS").** The project team estimated the investment needed to develop and deploy advanced technologies needed to realize the vision of the power delivery system (both transmission and distribution) described above.

Modernizing an Aging Infrastructure

All components of any infrastructure have limited lives regardless if they are roads, bridges, natural gas transmission, water pipelines or telecommunications. From the moment any given infrastructure is installed or renovated, aging begins.

164

Accelerated aging resulting in premature or unexpected failure is important to avoid in any of the infrastructures which provide society essential services Electricity is no exception as failures of components can lead to poor power quality, interruptions or wide-scale blackouts.

In the case of electric power delivery systems, substantial efforts are made to undertake the investments necessary to maintain reliability. Those investments are not included in estimating the cost of the Smart Grid in this study. Transmission and distribution utilities often spend an amount equal to 1 to 2% of their depreciated plant in service on refurbishment so as to maintain reliability. In conducting business as usual, necessary expenditures to accommodate load growth and to maintain reliability will naturally be made with equipment that is compatible with the power delivery system of the future.

Figure 3-3 illustrates how these three cost elements might combine to build the power delivery system of the future. The figure highlights the fact that as utilities make investments to maintain reliability sufficient to accommodate load growth, they are making investments which help build part of the power delivery system of the future.

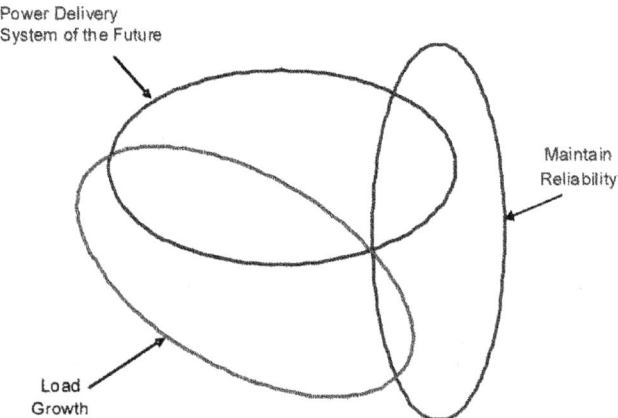

Figure 3-3
Illustration of Synergies in the Three Categories of Needed Transmission and Distribution Investment

Evolving the Smart Grid will mitigate declining reliability caused by aging components on the power delivery system. As Smart Grid components are added to the power delivery system, highlighted as Power Delivery System of the Future in Figure 3-3, and as Load Growth is accommodated as illustrated, the power delivery infrastructure will be strengthened and reliability enhanced. By its nature, the enhanced functionality which the Smart Grid's sensors, communications and computational ability enable will improve O&M, increase reliability and assure that the investments made to maintain reliability are appropriately targeted at infrastructure components which have the greatest risk of failure. In this study, the project team separated the expenditures necessary to accommodate load growth from the expenditures directly related to the power delivery system of the future, in order to elucidate the true cost of the Future PDS.

Section 4: Power Delivery System of the Future: Benefits (The Benefits of the Smart Grid)

Previous EPRI Study

There have been a number of studies which have estimated some of the benefits of a Smart Grid. Each varies somewhat in their approach and the attributes of the Smart Grid they include. None provides a comprehensive and rigorous analysis of the possible benefits of a fully functional Smart Grid. EPRI intends to conduct such a study, but it is outside the scope of the effort presented in this report.

In 2004, EPRI undertook a study to estimate the cost and value (benefits) of the power delivery system of the future. To do so, it developed a flexible framework.

The fundamental approach that was used in the 2004 study involved the identification of *attributes* of the power system (e.g., cost of energy, capacity, security, quality, reliability, environment, safety, quality of life, and productivity). EPRI then developed the framework to quantitatively estimate the dollar value of improving each of these attributes by a defined amount (i.e., percentage improvement).

Existing, documented data sources were used for this estimation process for each attribute. These sources included the U.S. Energy Information Administration, the U.S. Department of Energy's Policy Office Electricity Modeling System, the Federal Energy Regulatory Commission's transmission constraint study, the U.S. Labor Department's Bureau of Labor Statistics, and many more.

Attributes

Table 4-1 shows the various types of improvements that correspond to each of the attribute types used in the root study. A key aspect of the value estimation process in general is its consideration of improvements to the power delivery system (see the left column of Table 4-1), as well as improvements that consumers directly realize (see the right column of Table 4-1). This was done to ensure that emerging and foreseen benefits to consumers in the form of a broad range of value-added services addressed in the estimation of value.

Table 4-1

Attributes and Types of Improvements Assumed in the Value Estimation of the Future Power Delivery System (Left: Power Delivery System Improvements; Right: Improvements That Consumers Realize)

Power Delivery (Improvements/ Benefits)	Attributes	Consumer (Improvements/ Benefits)
O&M Cost Capital Cost of Asset T&D Losses	Cost of Energy (Net delivered life-cycle cost of energy service)	End Use Energy Efficiency Capital cost, end user infrastructure O&M, End User Infrastructure Control/Manage Use
Increased Power Flow New Infrastructure Demand Responsive Load	Capacity	Improved power factor. Lower End User Infrastructure cost through economies of scale and system streamlining, expand opportunity for growth
Enhanced Security Self Healing Grid for Quick Recovery	Security	Enhanced Security and ability to continue conducting business and every day functions
Improve Power Quality and enhance equipment operating window	Quality	Improve Power Quality and enhance equipment operating window
Reduce frequency and duration of outages	Reliability & Availability	Enhanced Security Self Healing Grid for Quick Recovery Availability included
EMF Management Reduction in SF6 (sulfur hexafluoride) emissions Reduction in cleanup costs Reduction in power plant emissions	Environment	Improved Esthetic Value Reduced EMF Industrial Ecology
Safer work environment for utility employees	Safety	Safer work environment for end-use electrical facilities
Value added electric related services	Quality of Life	Comfort Convenience Accessibility
Increase productivity due to efficient operation of the power delivery infrastructure Real GDP	Productivity	Improved consumer productivity Real GDP

The "cost of energy" attribute is the total cost to deliver electricity to customers, including capital costs, O&M costs, and the cost of line losses on the system. Therefore the value of this attribute derives from any system improvement that lowers the direct cost of supplying this electricity. "SQRA" is the sum of the power security, quality, and reliability attributes, because the availability part of SQRA is embedded in the power quality and reliability attributes. The quality of life attribute refers to the integration of access to multiple services, including electricity, the Internet, telephone, cable, and natural gas. This involves integration of the power delivery and knowledge networks into a single intelligent electric power/communications system, which sets the stage for a growing variety of products and services designed around energy and communications.

To quantify the benefit of these improvements for various attributes, the project team developed various "benefit calculator tools." Figure 4-1 shows the relationship of these benefit calculator tools, the attributes, and the overall value. Note that science and technology drivers feed generally into the process.

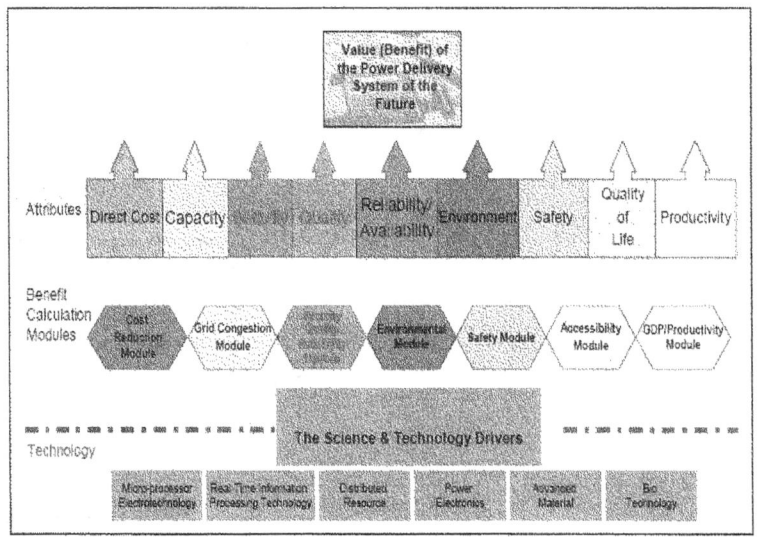

Figure 4-1
Relationship Between Value, Attributes, and Benefit Calculator Tools in the Value Estimation Process

Table 4-2 and Figure 4-1 illustrate the attributes considered in the original estimates of value published in 2004. Table 4-3 summarizes these estimates and escalates them using a chained GDP sequence to 2010 dollars. Table 4-4 lists the major attributes and benefits not included in the original EPRI study.

The increase of benefits using the chained GDP sequence masks the fact that the majority of the overall increase in benefits is from a change in scope of the Smart Grid from a system that can "almost" instantaneously balance supply and demand when the predictability of supply is decreased with the addition of increased amounts of variable renewable resources and the predictability of demand is aggravated with the addition of Plug-in Electric Vehicles, distributed photovoltaics and storage. To rectify this EPRI team analyzed the benefits from Demand Response, PEVs, AMI, Distributed Generation and Storage as shown in Table 2-2. Including these elements has increased the focus of the Smart Grid from operational efficiencies so as to include economic, societal, and energy policy benefits.

The electric power industry is the last industry in the western world to modernize itself through the use of sensors, communications, and computational ability. The combination of these functions allow for a truly interactive power system which can integrate consumer demand with supply interactively.

In addition to the obvious benefits achieved by this enhanced functionality –
namely, improved asset utilization, reduced electricity cost, and improved
reliability, it is now obvious that other substantive benefits will accrue once the
Smart Grid is implemented. Chief among these is the reduction of peak demand,
the adoption of electric vehicles, the use of storage, and the increased use of
renewable power production.

Table 4-2
Summary of Benefit Calculations Included in Original EPRI Study

Benefit Calculations	Attributes
T&D Cost Reduction	Energy efficiency and T&D losses impact on: • Capital cost • O&M Cost • Administrative and general cost
Congestion	• Transmission congestion cost
Security	• Self-healing infrastructure • Mitigating major outages
Power Quality & Reliability/ Availability	• Reliability • Power Quality
Environmental Impact	• SO_2 • CO_2 • NOx
Safety	• Accidental electrocutions • Building fires caused by electrical infrastructure
Quality of Life	• Access to competing suppliers
GDP/Productivity	• Increase GDP from reduced electricity cost

Table 4-3
Benefit Estimates in 2004 EPRI Study Escalated to 2010 Values

Attribute	Net Present Worth (2004) $B		Net Present Worth (2010) $B	
	Low	High	Low	High
Productivity	1	1	1.14	1.14
Safety	11	11	12.54	12.54
Environment	48	48	54.72	54.72
Capacity	49	49	55.86	55.86
Cost	50	50	57	57
Quality	35	57	41.04	64.98
Quality of Life	65	65	74.1	74.1
Security	133	133	151.62	151.62
Reliability	247	390	281.58	444.6
Total	**640**	**804**	**729.6**	**916.56**

Table 4-4 lists the major attributes and benefits not included in the original EPRI study. Hence, the value of the Smart Grid, even with escalation applied, is substantially understated in Table 4-3.

In order to provide a preliminary estimate of at least the major benefits of a fully functional power delivery system, EPRI has attempted to provide estimates for most of the remaining benefits. These are depicted in Table 4-5 using a framework developed by the U.S. Department of Energy (USDOE) and EPRI (EPRI 1020342). The table includes the attributes and benefits explicitly included in the DOE/EPRI framework as well as other attributes not included.

As summarized in the table, the total benefit of all attributes for the Smart Grid is estimated to be between $1,294 billion and $2,028 billion for the period 2010 to 2030. EPRI believes that once all of the attributes and benefits of a Smart Grid are identified and analyzed, estimates of the total benefit will increase even more.

Table 4-4
Major Attributes and Benefits Not Included in Original EPRI Study

Attributes	Benefits
Demand Response	• Reduced need for generation capacity • Reduced demand for electricity
Facilitating Renewables	• Reduced environmental impact of electricity generation
PEVs	• Reduced environmental impact from displaced fossil fuels • Increased system flexibility/ancillary services
Work Force	• Improved utilization of work force
Energy Efficiency	• Generation Capacity deferrals
Enhanced Energy Efficiency (additional energy efficiency)	• Reduced Environmental Impacts
AMI	• AMI-related cost reductions
Distributed Generation	• Facilitating distributed generation
Value-Added Electricity Services	• Comfort and convenience
Synergistic Effects Between Elements	• Compounding between multiple attributes
Storage (various benefits)	• Capacity • O&M • Congestion
Safety	• Personal safety
Transmission O&M	• Ancillary services

Table 4-5
List of Smart Grid Benefits: Based on EPRI/DOE Framework (EPRI 1020342)

Benefit Category	Benefit Sub-Category	Benefit	Included in Original Estimate?	Estimated Value 2010-2030 $Billion Low	Estimated Value 2010-2030 $Billion High	Reference
		Optimized Generator Operation		–	–	not included
		Deferred Generation Capacity Investments	X			Appendix A
		Reduced Ancillary Service Cost				included below
		Distributed Generation		–	–	not included
	Improved Asset Utilization	Storage		48	89	Appendix A
		PEVs as Storage & Load Control		11	11	Appendix A
		Energy Efficiency	X			included below
		Demand Response		–	–	not included
		Enhanced Energy Efficiency*		–	–	not included
Economic		Reduced Ancillary Service Cost		–	–	not included
		Reduced Congestion Cost	X			included below
		Distributed Generation		27	27	Appendix A
		Storage		23	65	Appendix A
		Demand Response		192	242	not included
	T&D Capital Savings	Energy Efficiency	X			included below
		Enhanced Energy Efficiency*		1	3	Appendix A
		Deferred Transmission Capacity Investment	X			included below
		Deferred Distribution Capacity Investment	X			included below
		Reduced Equipment Failures	X			included below

❮ 47 ❯

Table 4-5 (continued)
List of Smart Grid Benefits: Based on EPRI/DOE Framework (EPRI 1020342)

Benefit Category	Benefit Sub-Category	Benefit	Included in Original Estimate?	Estimated Value 2010-2030 $Billion		Reference
				Low	High	
	T&D O&M Savings	More Effective Use of Personnel		--	--	not included
		Economic Benefit of Added Personnel		--	--	not included
		Operations Savings from AMI		4	4	Appendix A
		T&D Efficiency	X	↑		included below
		Reduced Distribution Equipment Maintenance Cost	X	↑		included below
		Reduced Distribution Operations Cost	X	↑		included below
	Theft Reduction	Reduced Electricity Theft		--	--	Not included
Economic	Energy Efficiency	Enhanced Energy Efficiency*		0	2	Appendix A
		Electrification (Net Reduced Energy Use)		--	--	Appendix A
		Reduced Electricity Losses	X	↑		included below
		Productivity Increase	X	↑		included below
		Reduced Electricity Cost	X	↑		Included below
	Electricity Cost Savings	Automatic Meter Reading		91	91	Appendix A
		Customer Service Costs (Call Center)		2	2	Appendix A
		Storage		115	199	Appendix A
		Enhanced National Productivity	X	↑		included below
		Reduced Restoration Cost	X	↑		included below
		Speed of Restoration		--	--	not included

◄ 4-8 ►

Table 4-5 (continued)
List of Smart Grid Benefits: Based on EPRI/DOE Framework (EPRI 1020342)

Benefit Category	Benefit Sub-Category	Benefit	Included in Original Estimate?	Estimated Value 2010-2030 $Billion		Reference
				Low	High	
		Storage		2	20	Appendix A
		Reduced Sustained Outages	X	↑		included below
		Reduced Major Outages	X	↑		included below
		Accessibility	X	↑		included below
		Reduced Momentary Outages	X	↑		included below
	Power Quality	Reduced Sags and Swells	X	↑		included below
		Storage		1	21	Appendix A
		Electrification		21	21	Appendix A
		PEVs		5	123	Appendix A
		Enhanced Energy Efficiency*		1	4	Appendix A
Environmental	Air Emissions	Storage		10	15	Appendix A
		Facilitate Renewables		10	172	Appendix A
		Reduced CO$_2$ Emissions	X	↑		included below
		Reduced SOx, NOx and PM-10 Emissions	X	↑		included below

Table 4-5 (continued)
List of Smart Grid Benefits: Based on EPRI/DOE Framework (EPRI 1020342)

Benefit Category	Benefit Sub-Category	Benefit	Included in Original Estimate?	Estimated Value 2010-2030 $Billion		Reference
				Low	High	
Security	Energy Security	Reduced Imported Oil Usage		—	—	not included
		Personal Security		—	—	not included
		National Security		—	—	not included
		Reduced Wide-Scale Blackouts	X	→		included below
	Safety		X	→		included below
Previous EPRI Estimates – All included in original estimate			X	730	917	
Not included in original estimate				564	1,111	
Total				1,294	2,028	

*Enhanced Energy Efficiency includes:
Continuous Commissioning of Large Commercial Buildings
Direct Feedback on Energy Usage
Energy Savings Corresponding to Peak Load Management
Energy Savings Corresponding to Enhanced M&V Capability

Section 5: Transmission Systems and Substations

The high-voltage transmission system is the "backbone" of the power delivery system. It transmits very large amounts of electric energy between regions and sub-regions. Transmission system equipment fails and causes power outages much less frequently than distribution equipment. But when transmission equipment fails, many more customers are affected, and outage costs can be much higher, compared to the impact of a distribution equipment-related outage. This fact, combined with the high cost per mile or per piece of transmission equipment, has historically led to greater attention to transmission system reliability. However, in the last several decades, a variety of factors has led to a significant decrease in investment in transmission system expansion.

Introduction

To estimate the investment needed in the transmission system, a top-down approach was used for the load-growth and correct-deficiencies segments of investment, while a bottom-up approach was used for estimating the elements needed to create a Smart Grid. In the U.S., according to EEI, there are now more than 200,000 miles of high-voltage transmission lines greater than 230 kV. An earlier study by DOE, entitled the *National Transmission Grid Study, 2002*, showed a total of 187,000 miles, broken down by the voltage levels shown in Table 5-1.

The total cost for enhancing transmission system and substation performance to the level of a Smart Grid is estimated between $56 and $64 billion, as summarized in this section. The cost includes several categories of technology whose functionality overlaps significantly between the transmission system and substations as well as some elements of the distribution system described later, as well as enterprise level functions, such as cyber security and back office systems.

Table 5-1
Transmission Line Miles

Voltage (kV)	Miles
230 AC	85,048
345 AC	59,767
500 AC	32,870
765 AC	4,715
250-500 DC	3,307
Total Miles	**184,707**

In general, monitoring of transmission assets is more cost-effective and beneficial than any other asset class (EPRI 1016055). Although transmission lines are one of the critical core backbone elements of the power grid, thousands of miles are unattended and not monitored in any way. Transmission lines have seasonal ratings that need to be considered by operations and planning. For the most part, there is little if any real-time monitoring other than at substations that provide operators with loading information.

Transmission investment trended downward for more than two decades, declining from $4.8 billion in 1975 to $2.25 billion in 1997, then leveled off before beginning to climb again. It reached roughly $5 billion in 2000 and is expected to reach nearly $11 billion in 2010. Smart Grid functionality should help to increase the value of future transmission investment over and the expansion needed to meet load growth.

The number of substations is one of the basic metrics upon which investment costs were determined. There are an estimated 70,000 substations in the U.S. that reduce voltage between the bulk transmission system and the distribution feeder system, and serve as critical hubs in the control and protection of the electricity grid. This figure was derived from FERC data that shows investor-owned utilities (IOUs) operate a total of 40,619 substations at voltage levels ranging from just above 1 kV to 765 kV. Since IOUs represent roughly 70% of all U.S. customers, the number of existing substations was thus calculated to be 58,027 (40,619/.7 = 58,027).

As elucidated in Chapter 3, load was estimated in the study to grow at a rate of 0.68%/year. Compounded over the 20-year period of the study (2010-2030), this would imply an additional 8,423 substations will be required by 2030. Accordingly, the base figure used throughout this report for substations is 58,027 which could potentially be upgraded and 8,423 which will be new. In addition to this base, another 700 substations will be required by 2030 to handle renewable generation.

Other key benchmarks used in the analysis include an estimate of 8 feeders/substation serving lower-voltage customers downstream. Thus, there are 464,216 feeders that are eligible to be upgraded with intelligent electronic devices for a fully functioning Smart Grid. In addition, there are 67,384 new feeders to be added to accommodate load growth. Segments of the feeders that can be isolated electrically in case of faults and/or reconfiguration are called "pods," and for purposes of the Smart Grid, the analysis team used roughly 4 pods/feeder, yielding more than 2,260,000 isolatable pods for purposes of monitoring and control.

Table 5-2 lists these assumptions.

Table 5-2
Number of Substations and Feeders

Substations	Number
Existing substations	58,027
New substations to accommodate load growth (2030)	8,423
New substations to accommodate renewables (2030)	700
Distribution	
Number of existing feeders	464,216
New feeders to accommodate load growth (2030)	67,384

An underlying assumption in the report is that the digital devices to be deployed in the Smart Grid will comply with the International Electrotechnology Commission (IEC) Standard 61850 (IEC 61850). That standard applies to substation automation and protection, distribution automation, distributed energy resources, hydro generation, SCADA to field devices, and applies to protective relays, SCADA Master, DER, PQ meters, fault recorders and other applications.

Cost Components for the Smart Grid: Transmission Systems and Substations

The core components of cost for the transmission and substation portion of the Smart Grid are as follow:

- Transmission line sensors including dynamic thermal circuit rating
- Storage for bulk transmission wholesale services
- FACTS devices and HVDC terminals
- Short circuit current limiters
- Communications infrastructure to support transmission lines and substations
- Core substation infrastructure for IT
- Cyber-security
- Intelligent electronic devices (IEDs)
- Phasor measurement technology for wide area monitoring

- Enterprise back-office system, including GIS, outage management and distribution management
- Other system improvements assumed to evolve naturally include:
  - Faster than real-time simulation
  - Improved load modeling and forecasting tools
  - Probabilistic vulnerability assessment
  - Enhanced visualization

Substation upgrades will enable a number of new functions including, but not limited to:

- Improved emergency operations
- Substation automation
- Reliability-centered and predictive maintenance

Dynamic Thermal Circuit Rating (DTCR)

Dynamic rating and real-time monitoring of transmission lines are becoming important tools to maintain system reliability while optimizing power flows. Dynamic ratings can be considered a low-cost alternative for increased transmission capacity. Dynamic ratings are typically 5 to 15% higher than conventional static ratings. Application of dynamic ratings can benefit system operation in several ways, in particular by increasing power flow through the existing transmission corridors with minimal investments.

Dynamic rating increases the functionality of the Smart Grid because it involves the monitoring of real-time system data that can be used in various applications:

- Real-time monitors yield a continuous flow of data to system operations – line sag, tension or both, wind speed, conductor temperature, etc. – traditionally not available to operators.
- Monitored data can be processed to spot trends and patterns.
- Real-time monitored data may be turned into useful operator predictive intelligence (e.g., critical temperature and percent load reduction needed in real time).

The New York Power Authority (NYPA) has engaged with EPRI in a demonstration project that will evaluate the instrumentation and dynamic thermal ratings for overhead transmission lines. An area of possible application for DTCR is the growing penetration of wind generation; when turbines are operating, one expects higher dynamic ratings because of increased wind speed. The project will use EPRI's DTCR software, which uses real-time or historical weather and electrical load data to calculate dynamic ratings for overhead lines in real time based on actual load and weather conditions that generally are accessed through the utility's SCADA/EMS system.

The study team assumed that AC transmission lines rated 115 kV to 230 kV in the U.S. are most susceptible to being thermally limited. They considered lines rated 345 kV and above are more likely voltage-limited. Albeit, there may be a few lines rated at these higher voltages that are thermally limited – they would be an exception and not the rule. Furthermore, of the lines which are potentially thermally limited, only 50% actually are. There are 85,048 miles of lines at 230 kV. To dynamically rate all 85,048 miles would require one unit per 7.5 miles or 11,340 units. The cost of DTCR deployed in quantity is estimated to be $20,000 initially, declining to $10,000.

It should be noted that the above calculation does not include any transmission lines at voltages lower than 230 kV, e.g., 115, 138 and 161 kV. It is expected that these transmission lines will benefit from DTCR.

Table 5-3
Cost of Dynamic Thermal Circuit Rating

Technology	Total Units	Units	% Sat	Cost/ Unit Low $	Cost/ Unit High $	Total Cost Low– High $M
Dynamic- Thermal Circuit Rating	11,340	Units/7.5 miles DTCR line	100	10,000	20,000	113.4- 226.8

Sensors and Intelligent Electronic Devices

The Smart Grid will require a more diverse and wider array of sensors and other Intelligent Electronic Devices (IEDs) throughout the power system to monitor conditions in real time. In particular, sensors in transmission corridors and in substations can address multiple applications:

- **Safety:** The application of sensors for transmission line or substation components will allow for the monitoring and communication of equipment conditions. Information that a transmission line or substation component is in imminent risk of failure will enable actions to be taken to address the safety of utility personnel.

- **Workforce Deployment:** If the condition of a component or system is known to be at risk, personnel can be deployed to prevent an outage.

- **Condition-Based Maintenance:** Knowledge of component condition enables maintenance actions to be initiated at appropriate times rather than relying on interval-based maintenance.

- **Asset Management:** Improved knowledge of the condition of equipment and stresses that they have been subjected to will allow managers to better manage the assets. Sensor data used together with historic performance information, failure databases and operational data allows better allocation of resources.

- **Increased Asset Utilization:** The rating of transmission components is influenced by a range of factors such as ambient weather conditions, loading history and component configuration. In order to address this complexity, static ratings are usually based on conservative assumptions of these factors. Higher dynamic ratings can be achieved with more precise, real-time knowledge of the asset's condition.

- **Forensic and Diagnostic Analysis:** After an event occurs, there is limited information to understand the root cause. Sensors allow the capture of pertinent information in real time for a more rigorous analysis.

- **Probabilistic Risk Assessment:** Increased utilization of the grid is possible if contingency analyses are performed using a probabilistic, rather than deterministic, methods. To use probabilistic methods, knowledge of the condition of components and the risks they pose are needed

The transmission system of the future will utilize a synergistic concept for the instrumentation of electric power utility towers with sensor technology designed to increase the efficiency, reliability, safety, and security of electric power transmission. The system concept is fueled by a list of sensing needs illustrated in Figure 5-1 (EPRI 1016921). This system scope is limited to transmission line applications (i.e., 69 kV and above), not distribution, with the focus on steel lattice and pole structures, not wooden.

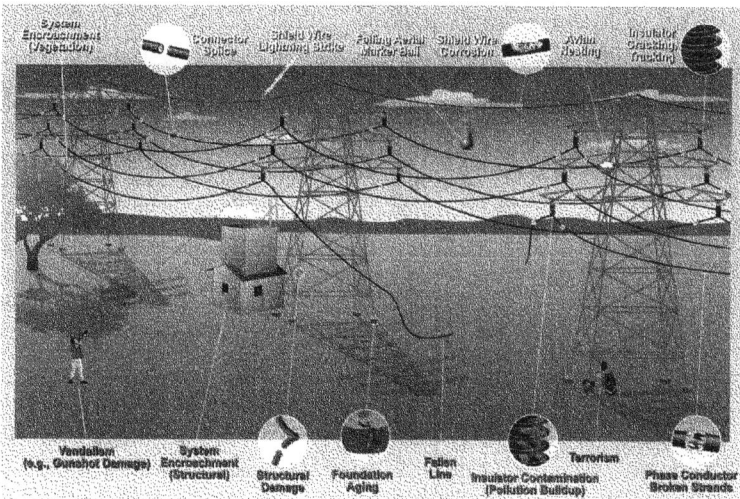

Figure 5-1
Illustration of Sensor Needs for Transmission Lines and Towers (EPRI 1016921)

In this concept, the addition of wiring to interconnect and/or power distributed sensors is not viable because of electromagnetic susceptibility concerns and labor

182

intensive installation. Consequently, sensor concepts will rely on wireless and/or fiber optic technology.

Figure 5-2
Image Showing a Single Structure Illustrating Some of the Concepts (EPRI 1016921)

Figures 5-1 and 5-2 depict some of the high-level concepts that are listed below:

- Sensors distributed on transmission structures and /or conductors.

- Sensors that may or may not communicate with the "hub" installed on the structure – either wireless or wired.

- Sensor information is collected, stored and analyzed in a "central database" which is part of the utility's current data management system. The data is collected /communicated from the sensors /hubs to the central database using one of the following methods:

  - Wirelessly back to the central database from the individual structure hub, e.g. RF directly, via satellite or cell phone network.

  - Collected using a vehicle traveling the length of the line. The data from the collection vehicle is transferred during or after the inspection. The following is a list of possible data collection vehicles:
    - o Unmanned Airborne Vehicle (UAV)
    - o Manned Aerial Vehicle
    - o Line Crawler Robot

  - If the vehicle data collection approach is utilized:
    - o The vehicle may collect the data wirelessly directly from the sensors (possibly excluding the need for a structure "hub").
    - o The vehicle may also have sensors aboard recording data during the collection process, (e.g. video, UV, IR, still images)

These concepts are discussed in detail in the following sections.

There are a number of possible sensors to address each of these applications. Table 5-4 below elucidates the range of sensor needs. The study team assumed that by 2030 one-half of all substations would have installed an advanced sensors package on the transmission system, costing roughly $50,000-100,000/substation. The total Smart Grid investment for sensors approaches $1.5 to 2.9 billion. In addition, 100% of new substations built to accommodate load growth would incorporate a suite of sensors at a cost ranging between $421 to 842 million.

Table 5-4
Sensor Needs

	Item	Cause	Result	Update Interval	Probability	Consequence	Sensing Technologies
1	System Tampering	Terrorism	Tower/line down	Real-time	Low	High	Vibration, Acoustic, E-Field, Optical
2	System Encroachment	Man-made	Safety hazard, Less reliable	3-12 mo	High	Med	Optical, Satellite, Proximity, Vibration, E-Field
3	System Encroachment	Vegetation	Flashover, Fire	3 mo	High	High	Optical, Satellite, LIDAR, Line-of-Sight, Proximity
4	System Encroachment	Avian Nesting, Waste	Flashover	6-12 mo	High	High	Optical, Vibration, Leakage Current, Proximity, E-Field
5	Shield Wire	Corrosion	Flashover, Outage	3-6 years	Med	High	Optical, IR Spectroscopy, Eddy Current, MSS
6	Shield Wire	Lightning	Flashover, Outage	1 year	Med	High	Optical, IR Spectroscopy, Eddy Current, MSS, Lightning Detection, Vibration
7	Insulator (Polymer)	Age, Material Failure	Outage	6 years	Med	High	Optical, Vibration, RFI, UV, IR
8	Insulator (Ceramic)	Age, Material Failure	Outage	12 years	Low	High	Optical, Vibration, RFI, UV, IR
9	Insulator	Contamination	Flashover	3 mo	Med	Med	Optical, RFI, UV, IR, Leakage Current
10	Insulator	Gun Shot	Outage	Real-time, 3 mo	Med	High	Optical, Vibration, RFI, UV, IR, Acoustic

185

Table 5-4 (continued)
Sensor Needs

Item		Cause	Result	Update Interval	Probability	Consequence	Sensing Technologies
11	Phase Conductor	External strands broke	Line Down, Fire	1 year	Low	High	Optical, Vibration, RFI, UV, IR
12	Phase Conductor	Internal strands broke	Line Down, Fire	1 year	Low	High	E-MAT, MSS, Electromagnetic
13	Phase Conductor	Corrosion of steel core	Line Down, Fire	1 year	Low	High	E-MAT, MSS, Electromagnetic, IR Spectroscopy, Optical
14	Connector Splice	Workmanship, thermal cycling, age	Line Down, Fire	1 year	Med	High	Direct Contact Temperature, IR Temperature, Ohmmeter, RFI, E-MAT, MSS
15	Hardware	Age	Line Down, Fire	6 years	Low	High	Optical, IR Spectroscopy
16	Phase Spacer	Age, galloping event	Line Down, Fire	6 years	Low	Med	Optical, UV, RFI
17	Aerial Marker Ball	Vibration Damage, Age	Safety concerns	1 year	Low	Med	Optical, UV, RFI
18	Structure (Steel Lattice)	Corrosion	Reliability Concerns	10 years	Med	Med	Optical, IR Spectroscopy
19	Structure (Steel Lattice)	Bent, damaged members	Reliability Concerns	1 year	Med	Med	Optical, Strain, Position, Tilt
20	Structure (Steel Pole)	Corrosion, age	Reliability Concerns	10 years	Med	Med	Optical, IR Spectroscopy

Table 5-4 (continued)
Sensor Needs

	Item	Cause	Result	Update Interval	Probability	Consequence	Sensing Technologies
21	Structure (Steel Pole)	Internal Deterioration	Reliability Concerns	1 year	Med	Med	Optical, MSS, Ultrasonics
22	Foundation (Grillage)	Age, corrosion	Reliability Concerns	10 years	High	High	Excavation, MSS, Radar, GPR Imaging, Half Cell, Voltage Potential
23	Foundation (Anchor Bolt)	Age, corrosion	Reliability Concerns	10 years	Low	High	Optical, Ultrasonics, E-MAT, Vibration
24	Foundation (Preform)	Age, corrosion	Reliability Concerns	10 years	Med	High	Optical, Ultrasonics, E-MAT, Vibration
25	Foundation (Stub Angles)	Age. Corrosion	Reliability Concerns	10 years	Low	High	Optical, Ultrasonics, E-MAT, Vibration
26	Foundation (Direct Embedment)	Age, corrosion	Reliability Concerns	10 years	High	High	Excavation, MSS, Half Cell, Voltage Potential
27	Foundation (Anchor Rods, Screw-In)	Age, corrosion	Reliability Concerns	10 years	High	High	Excavation, MSS, Half Cell, Voltage Potential, Ultrasonics
28	Grounding	Age, corrosion, tampering	Reliability, Lightning, Safety concerns	6 years	Med	Med	AC impedance, DC resistance, Impulse
29	TLSA (Transmission Line Surge Arrestor)	Lightning Strikes, age	Reliability, Lightning Concerns	1 year	Med	Med	Optical, IR, Leakage Current, Lightning Strike Counter

The sensor system architecture is comprised of sensors that acquire diagnostic data from components of interest and from communications hubs that collect the sensor data and relay it to a central repository. Sensors may be directly attached to the item being monitored, or may be remotely located such as in the case of a camera. Communications hubs may be mounted on or near towers or may be located on a wide variety of mobile platforms, such as manned airplanes or unmanned line crawlers or UAVs. Sensors and hubs may operate and be polled periodically (e.g., at intervals of minutes, hours, days) or continuously monitored (e.g., a real-time alarm) depending on the application. In any case, sensors communicate their results via hubs to a central repository. Figure 5-3 illustrates the architecture and flow of data.

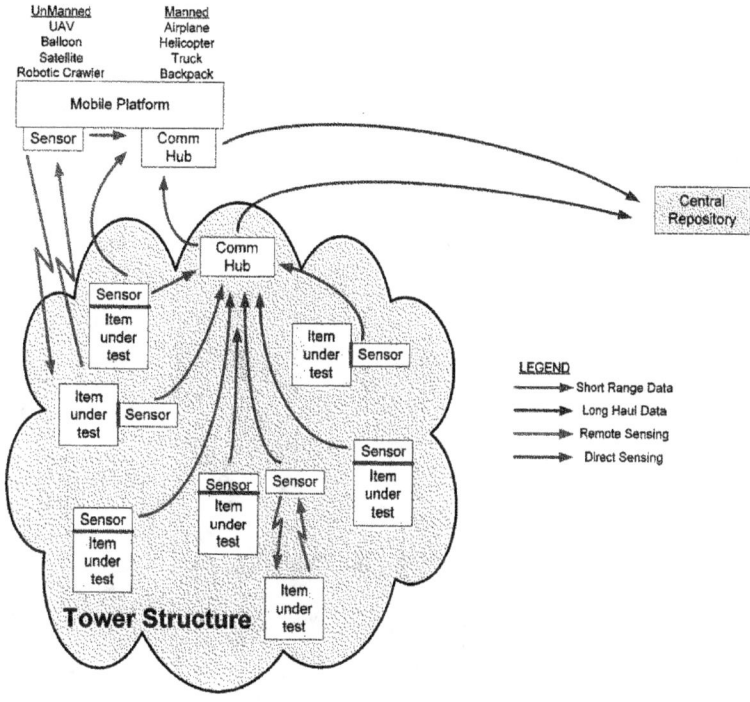

Figure 5-3
Sensor System Architecture (EPRI 1016921)

Table 5-5
Cost of Sensors

Technology	Total Units	Units	% Sat	Cost/ Unit Low $	Cost/ Unit High $	Total Cost Low~ High $M
Transmission Line Sensors	58,027	Number of existing substations	50*	50,000	100,000	1,451- 2,901
Transmission Line Sensors	8,423	Number of new substations	100	50,000	100,000	421- 842

*Assumes 50% of substations will have sensors listed in Table 5-4 partially deployed.

Examples of Transmission Line Sensors

Figure 5-4
RF Conductor temperature and current sensor, offering power harvesting, live working install, and low cost. Cost is an order of magnitude lower than that of other available technologies.

Figure 5-5
Clamp-on RF leakage current sensor for transmission line applications installed on a 115-kV composite insulator.

Figure 5-6
An RF leakage current sensor installed on a post insulator in a substation. It also shows dry band arcing (the discharge activity due to contamination and wetting that causes the leakage currents) captured during a contamination event. The sensor recorded the event

Figure 5-7
Antenna array installed on a portable trailer deployed in a 161-kV substation.

Figure 5-8
A solar-powered tank-top temperature sensor installed on a transformer.

Figure 5-9
An MIS sensor for measuring acetylene gas levels in oil. The use of MIS gas-in-oil
sensors can increase the number of transformers monitored due to their lower cost.

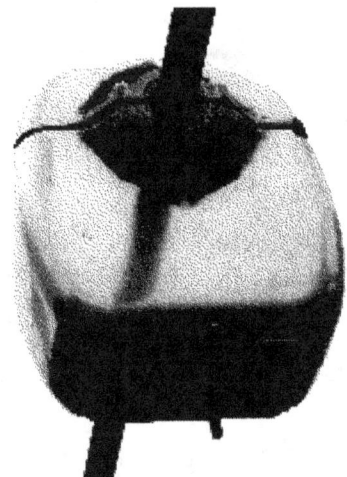

Figure 5-10
Conductor temperature sensors installed on a steel-reinforced aluminum cable
(ACSR) conductor. The sensor communicates using a cell phone modem.

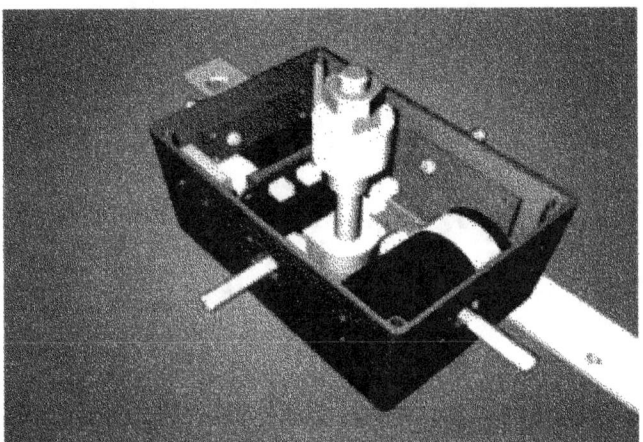

Figure 5-11
Inside the housing of a leakage current monitoring sensor utilized on post-type
insulators. The two lithium polymer batteries utilized to power the sensor for 14
years.

Short-Circuit Current Limiters (SCCL)

The short-circuit current limiter (SCCL) is a technology that can be applied to
utility power delivery systems to address the growing problems associated with
fault currents. The present utility power delivery infrastructure is approaching its

maximum capacity and yet demand continues to grow, leading in turn to increases in generation. The strain to deliver the increased energy demand results in a higher level of fault currents. The power-electronics-based SCCL is designed to work with the present utility system to address this problem. It detects a fault current and acts quickly to insert an impedance into the circuit to limit the fault current to a level acceptable for normal operation of the existing protection systems.

The SCCL incorporates advanced Super GTO (SGTO) devices for a higher-performing and more compact system that incorporates the most advanced control, processing, and communication components. This enables it to function as a key part of the Smart Grid.

The study team expects installation of the SCCL to begin in ten years and to be phased in slowly, beginning in 2020. From 2020, penetration will rise to 2% of transmission substations by 2030. The installation cost will be about $500,000 for SCCL at transmission substations, and about $50,000 at distribution substations. Total installed cost for transmission circuit current limiters through 2030 is estimated at $2.03 billion.

Table 5-6
Cost of Transmission Short Circuit Current Limiters

Technology	Total Units	Units	% Sat	Cost/ Unit Low $	Cost/ Unit High $	Total Cost Low– High $M
Transmission Short-Circuit Current Limiters	58,027	Number of substations	2	500,000	500,000	580.3-580.3

Flexible AC Transmission System (FACTS)

There are a number of flexible AC transmission (FACTS) technologies which are critical to the Smart Grid. These all incorporate power electronics and can be applied to the transmission system. These include both the control and operation of the power system and applications that will extend eventually to transformers themselves.

FACTS devices can be used for power flow control, loop flow control, load sharing among parallel corridors, voltage regulation, enhancement of transient stability, and mitigation of system oscillations. FACTS devices include the thyristor controlled series capacitor (TCSC), thyristor controlled phase angle regulator (TCPAR), static condenser (STATCON), and the unified power flow controller (UPFC). AEP installed the first UPFC at its Inez substation in eastern Kentucky in 1998 (EPRI 1010633).

FACTS is a concept invented by EPRI in the 1980s, for which EPRI developed several patents licensed to Siemens. It involves the injection of a variable-voltage source, which adjusts the power flow across a transmission line, resulting in variable voltage, impedance, and phase angle. Six successful major FACTS installations were demonstrated, each with several key features. For example, one installation at the New York Power Authority's Marcy Substation resulted in the ability to increase power-transfer capacity into New York City by 200 MW, resulting in substantial savings. However, these FACTS devices did not spawn an anticipated revolution in the control of power flow, as was expected. They were plagued by three technical problems, the combination of which made FACTS 20% more costly than "conventional" solutions. These technical problems included the cost and performance of the control systems within the devices; the performance of the systems used to cool the electronics; and the cost and performance of the power-electronic devices themselves. As of now, the first two of these issues have been resolved, and with the successful demonstration of advanced power electronics, there is relative certainty that the industry is poised for a rebirth of FACTS. This is especially important with the introduction of increasing amounts of variable generation – like wind and solar – located far from load pockets.

The advantages of FACTS technology are as follows:

- Increases the amount of power that can be imported over existing transmission lines.

- Provides dynamic reactive power support and voltage control.

- Reduces the need for construction of new transmission lines, capacitors, reactors, etc which mitigate environmental and regulatory concerns, and improves aesthetics by reducing the need for construction of new facilities.

- Improves system stability.

- Controls real and reactive power flow.

- Mitigates potential Sub-Synchronous Resonance problems.

FACTS for HVDC – The backbone of HVDC transmission is a FACTS device which is the converter station that converts AC to DC for transmission and then converts DC back to AC at the other end of the line. Power electronics do the heavy lifting in these applications. In the 400- to 1000-MW range, second-generation IGBT power-electronic devices are used, while in the 1000-MW and above range, older technologies are used (namely, thyristors). As load growth increases and the use of renewable power generation located far from the load pocket becomes the norm, there will be increased demand for DC technology. DC technology may be the only effective means of increasing power flow on an existing corridor that was originally built for AC transmission.

FACTS for Controlling Reactive Power – TCSCs are a derivation of the FACTS technology that uses power electronics to inject capacitance into the power system to improve power flow by controlling reactive power. There are roughly 100 TCSCs installed in the U.S., primarily in the west. Eastern utilities have been reluctant to take advantage of this technology due to concerns about sub-synchronous resonance. Further development of thyristor control can dampen those oscillations.

FACTS for Electronic Transformers – Power-electronic transformers will eventually become part of the transmission system. Existing mechanical switching is accomplished within about 6 AC cycles. This is rapid enough for most applications. However, there is a need for a power-electronic device that could reduce short-circuit currents and then act as power-electronic circuit breaker at the same time. The potential maximum for short-circuit currents is growing so as to exceed the maximum capacity of today's breaker fleet. A distribution device to limit short-circuit currents using power electronics has been demonstrated. Further development could allow this device to scale up to high voltage and, coincidentally, act as a fast-switching power-electronic breaker. This device would have the added benefit of eliminating the use of SF6. Considerable effort has been expended in the development of distribution transformers based on power electronics. Using power electronics in transformers can eliminate the majority of the inductance and, along with it, all of the oil used as coolant, resulting in a substantial reduction in the losses. In addition, it offers a great deal more flexibility in voltage control. The potential to scale up the distribution version of this transformer is promising. Further advancement in power electronics will be needed to realize for this transmission application.

Power Electronic Devices for Mitigating Geomagnetically-Induced Currents – Geomagnetically-induced currents can cause serious problems to high-voltage equipment and promote blackouts. The future application of power electronics to be applied to the grounded neutral on substation transformers could neutralize these currents. Further development is needed to realize this technology in the 2020 to 2030 time frame.

Table 5-7
DC Lines and Terminals Known to be Under Consideration

Renewable Studies	No. Lines	Type
EWITS	10	800 kV DC lines
WWIS	1	600 kV DC lines
HPX	2	500 kV DC lines
Santa Fe	1	500 kV DC lines
Clean Line	1	500 kV DC lines
Total	15	Lines or 30 HVDC Terminals

Table 5-8
Estimated Additional FACTS Devices Needed by 2030

FACTS Devices	Number of Units	Cost/Unit		Total Costs	
		Low $	High $	Low $	High $
STATCON & UPFC	20	35,000,000	45,000,000	700,000,000	900,000,000
TCSC	100	10,000,000	12,000,000	1,000,000,000	1,200,000,000
HVDC Terminals	60	22,500,000	27,500,000	1,350,000,000	1,650,000,000
Power Electronic Transformers	25	40,000,000	40,000,000	1,000,000,000	1,000,000,000
Geomagnetic Controllers	25	5,000,000	7,000,000	125,000,000	175,000,000
Total FACTS Cost				**4,175,000,000**	**4,925,000,000**

◄ 5-21 ►

Storage

Bulk storage is one of the major limitations in today's "just in time" electricity delivery system and one of the great opportunities for Smart Grid development in the future. Only about 2.5% of total electricity in the U.S. is now provided through energy storage, nearly all of it from pumped hydroelectric facilities used for load shifting, frequency control, and spinning reserve. System balancing is provided by PH, combustion turbines, and the cycling of coal power systems. (In contrast, some 10% of the electricity produced in Europe is cycled through a storage facility of some kind, and Japan stores 15% of the electricity it produces.) Deployment and policy have been instrumental in long-term resource planning and management.

Storage is essential for electricity consumers where power quality and reliability is critical, such as at airports, broadcasting operations, hospitals, financial services, data centers, telecommunications, and many finely tuned industrial processes. Such operations frequently install energy storage as part of an uninterruptible power supply. In the future, storage—as both an end user and electric utility energy management resource —will become possible due to a confluence of high TOU rates, dynamic pricing, and lower cost energy storage systems.

Compressed air energy storage (CAES), pumped hydro, and advanced lead-acid batteries are the primary options for utilities pursuing bulk storage for T&D grid support and system and renewables integration. Table 5-9 shows the cost for CAES running from $810-1045/kW, whereas lead-acid batteries typically exceeds $2000/kW (EPRI 1017813).

Table 5-9
Highest Value Electrical Storage Technologies

Highest Value Storage Market	kW	Best Technology Fit	Cost/kW
Wholesale Services Without Regulation	5,800,000	CAES	$810-1045
Wholesale Services With Regulation	2,800,000	CAES	$810-1045
Home Backup	2.8	Lead-Acid	$2200
Industrial Power Quality & Reliability	1.8	Advanced Lead-Acid	$2300-2400
Transportable Storage Systems (for Distribution Deferral)	1.7	Advanced Lead-Acid	$2180-2900

Note: Li-ion systems may, in the long run, be a potentially low-cost option for grid support. This maybe driven by the large-scale manufacturing underway for the automotive maker. These may be the most compelling for energy durations under four hours with one to three hours being the sweet spot.

The study team estimated the installation of 5,800,000 kW of CAES storage capacity for wholesale services in areas without regulation, and 2,800,000 kW of capacity for wholesale services with regulation. The total investment cost over the next 20 years is estimated at $4.7 billion to $6.1 billion, as shown in Table 5-10.

Table 5-10
Cost of Storage Technology

Technology	Total Units	Units	% Sat	Cost/ Unit Low $	Cost/ Unit High $	Total Cost Low– High $M
Storage for Bulk Transmission Wholesale Services Without Regulation	5,800,000	kW	100	810	1045	4,698-6,061
Storage for Bulk Transmission Wholesale Services With Regulation	2,800,000	kW	100	810	1045	2,268-2,926

Recently, it was announced that several Li-ion systems are going in to provide fast regulation services. It is speculated that their costs may be as low as $1,200 per kW. These technologies could provide frequency regulation at lower costs and less lumpy investments than CAES. However, CAES is needed to avoid wind curtailment under high penetration wind scenarios.

Communications and IT Infrastructure for Transmission and Substations

Smart substations require new infrastructure capable of supporting the higher level of information monitoring, analysis, and control required for Smart Grid operations, as well as the communication infrastructure to support full integration of upstream and downstream operations.

The substation of the future will require a wide-area network interface to receive and respond to data from an extensive array of transmission line sensors, dynamic-thermal circuit ratings, and strategically placed phasor measurement units. The smart substation must be able to integrate variable power flows from renewable energy systems in real time, and maintain a historical record or have access to a historical record of equipment performance. Combined with real-time monitoring of equipment, the smart substation will facilitate reliability-centered and predictive maintenance.

The core and distributed IT infrastructure will be able to coordinate the flow of intelligence from critical equipment, such as self-diagnosing transformers, with downstream operations, and be able to differentiate normal faults from security breaches. It will be able to distill and convey critical performance data and maintenance issues to back office systems.

The smart substation will build upon the existing platform. There is already a significant installed base of sensors at substations, but there is still limited bandwidth connecting the substation to the enterprise. Historically, the communications channel to the substation was justified as part of the installation of the energy management system (EMS) and supervisory control and data acquisition (SCADA) systems. A key consideration for the future is that these legacy systems have limited bandwidth.

The study team estimates a cost of $50,000-$75,000 per substation to achieve the optimal performance level required for the Smart Grid. Substation upgrades will phase in slowly over the next 20 years, reaching an 80% penetration level of all existing substations by 2030. This suggests a cumulative investment between $2.9 billion and $4.2 billion by 2030. All new substations will incorporate communications and IT infrastructure at the time of construction.

Table 5-11
Cost of Communications and IT Infrastructure for Transmission and Substations

Technology	Total Units	Units	% Sat	Cost/ Unit Low $	Cost/ Unit High $	Total Cost Low-High $M
Core Substation Infrastructure for IT: Smart Substations	58,027	Number of existing substations	80	50,000	75,000	2,321-3,481
Communications Infrastructure to Support Transmission Lines & Substations	58,027	Number of existing substations	80	14,400	14,400	668.5-668.5
Total IT & Communications Infrastructure for Existing Substations						2,989.5-4,149.5
Core Substation Infrastructure for IT: Smart Substations	8,423	Number of new substations	100	50,000	75,000	421-632

Table 5-11 (continued)
Cost of Communications and IT Infrastructure for Transmission and Substations

Technology	Total Units	Units	% Sat	Cost/ Unit Low $	Cost/ Unit High $	Total Cost Low– High $M
Communications Infrastructure to Support Transmission Lines & Substations	8,423	Number of new substations	100	14,400	14,400	121-121
Total IT & Communications Infrastructure for New Substations						542-753

Intelligent Electronic Devices (IEDs)

Intelligent Electronic Devices (IEDs) encompass a wide array of microprocessor-based controllers of power system equipment, such as circuit breakers, transformers, and capacitor banks. IEDs receive data from sensors and power equipment, and can issue control commands, such as tripping circuit breakers if they sense voltage, current, or frequency anomalies, or raise/lower voltage levels in order to maintain the desired level. Common types of IEDs include protective relaying devices, load tap changer controllers, circuit breaker controllers, capacitor bank switches, recloser controllers, voltage regulators, network protectors, relays etc.

With available microprocessor technology, a single IED unit can now perform multiple protective and control functions, whereas before microprocessors a unit could only perform one protective function. A typical IED today can perform 5 to 12 protection functions and 5 to 8 control functions, including controls for separate devices, an auto-reclose function, self-monitoring function, and communication functions etc. It can do this without compromising security of protection – the primary function of IEDs.

The study team estimated the cost of incorporating IEDs to monitor and control critical functions at substations at an average cost of $110,000/substation, and assumed approximately 80% of the substations would be brought up to Smart Grid levels by 2030 and 100% of new substations would incorporate them.

Table 5-12
Cost of Intelligent Electronic Devices

Technology	Total Units	Units	% Sat	Cost/ Unit Low $	Cost/ Unit High $	Total Cost Low– High $M
Intelligent Electronics & Sensors	58,027	Number of existing substations	80	110,000	110,000	5,106-5,106
Intelligent Electronics Devices (IED)– Relays & Sensors	8,423	Number of new substations	100	110,000	110,000	927-927

Phasor Measurement Technology

Phasor measurement units (PMUs) or synchrophasors provide real-time information about the power system's dynamic performance. Specifically, they take measurements of electrical waves (voltage and current) at strategic points in the transmission system 30 times/second. These measurements are time stamped with signals from global positioning system satellites, which enable PMU data from different utilities to be time-synchronized and combined to create a comprehensive view of the broader electrical system. Widespread installation of PMUs will enhance the nation's ability to monitor and manage the reliability and security of the grid over large areas.

Synchrophasor technology has demonstrated the potential to enhance grid planning and operations processes. Recent industry R&D efforts have focused on developing a variety of applications including situational awareness, small signal stability behavior, event analysis, model validation, state-estimation enhancement, and on-line voltage stability assessment. Currently, about 150 PMUs have been installed in North America (as shown in Figure 5-12), and over 850 additional PMUs will be installed during the next 3 to 5 years across the U.S., as part of the DOE Smart Grid Investment Grant. While the industry continues to explore the use of PMU data in real time and off-line environments, the lack of killer applications has impeded the widespread use of synchrophasor technology. A concerted industry R&D effort is warranted among the research community, end users (grid operators and planners) and EMS vendors to produce production-grade PMU data applications for the users. To that end, EPRI is collaborating with the industry by forming an executive team to help accelerate the deployment of advanced control room applications.

PMUs provide system operators with feedback about the state of the power system with much higher accuracy than the conventional SCADA systems which typically take observations every four seconds. Because PMUs provide more

202

precise data at a much faster rate, they provide a much more accurate assessment of operating conditions and limits in real time.

The ultimate link between PMUs and other Smart Grid technologies is only now beginning to be revealed. PMUs are themselves an enabling technology that may make investments in advanced communication infrastructures and IEDs more desirable. The full potential benefits of PMUs will not materialize by simply installing PMUs. Wide area measurement systems (WAMS) or wide area control systems (WQACS), which include PMUs, communications infrastructure, other control devices and software application algorithms, will be required to fully realize the potential for PMUs. These other costs are included under separate headings.

The study team expects utilities to install approximately 1,250 PMUs throughout the grid over the next 20 years at a total cost of $26-39 million.

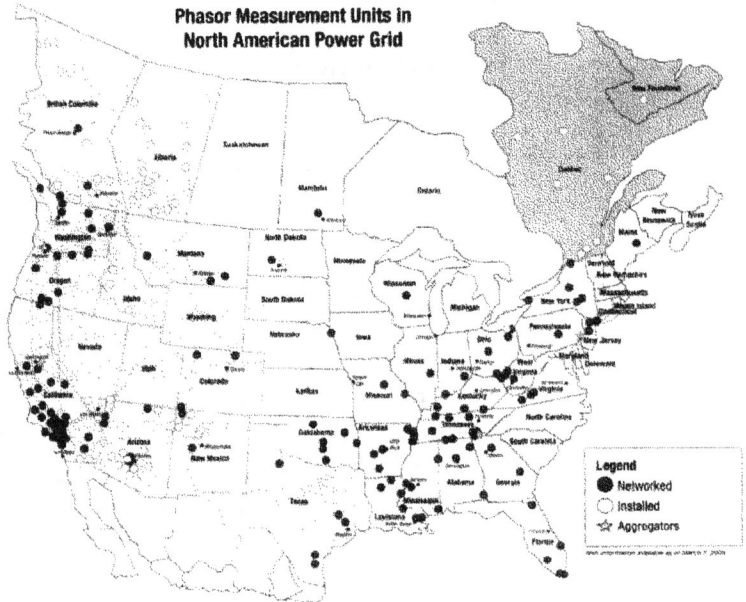

Figure 5-12
Phasor Measurement Units (PMUs) Installed in North America as of September 2009

Table 5-13
Cost of Phasor Measurement Units

Technology	Total Units	Units	% Sat	Cost/ Unit Low $	Cost/ Unit High $	Total Cost Low~ High $M
Phasor Measurement Units (PMU)	1,250	Numbers of	100	125,000	125,000	156-156

Cyber Security

Electric utilities have been incorporating cyber security features into their operations since the early 2000s. In recent years as the Smart Grid became increasingly popular, cyber security concerns have increased significantly. While there have to date been few reliable reports of cyber attacks on power systems, there is a great deal of urban lore which suggests alleged attempts to disrupt the reliability of U.S. electricity supply.

Cyber security is an essential element of the Smart Grid. It is the protection needed to ensure the confidentiality and integrity of the digital overlay which is part of the Smart Grid.

The North American Electric Reliability Corporation (NERC) has created eight Critical Infrastructure (CIP) Standards. These include standards for Critical Cyber Asset Identification (CIP002) and Security Management Controls (CIP003) as well as others. Meeting these standards are part of Smart Grid costs.

At present, utilities are considering cyber security as part of information technology (IT) projects for:

- Advanced metering infrastructure
- Plug-in electric vehicle (PEV) management systems
- Distribution automation
- Substation automation
- Transmission upgrades

Interviews with industry suppliers by the EPRI team indicate that as a percentage of IT project costs, cyber security costs range from 10 to 15% of SCADA and distribution automation and approximately 20% for AMI.

Cyber security costs vary by the size of the utility. Urban utilities are likely to be more aggressive in Smart Grid deployment than suburban or rural utilities. For purposes of estimating cyber security costs, three utility sizes were used – small, medium, and large – corresponding to rural, suburban, and urban.

Mere compliance with cyber security standards will not assure security. It is assumed large utilities will use applicable industry standards and best practices, including emerging security standards like NIST's Smart Grid Interoperability Standards Framework and AMI-SEC System Security Requirements, for end-to-end security of the Smart Grid. Most will implement intrusion detection and prevention services (IDS/IPS) as well as security information event management (SIEM). They will likely use a system-of-systems approach to cyber security by deploying International Organization for Standardization and International Electrotechnical Commission (ISO/IEC), National Security Agency InfoSec Assessment Methodology (NSA IAM), Information Systems Audit and Control Association (ISACA), and International Information Systems Security Certification Consortium (ISC2).

Many utility AMI systems will likely use a certificate-based solution for identifying and authenticating trusted devices, authorizing commands and encrypting communications between user entities (people, programs, devices). This solution applies industry standard cryptography to privatize all data transmissions and ensures that communications between authorized entities are confidential, trusted and legitimate.

There is very little information available as to what actual cyber security costs are. To make estimates, the project team interviewed IT suppliers who specialize in cyber security. As a result, these estimates were developed:

Table 5-14
Estimates of Cyber Costs by Utility Size

	Initial Cyber Costs $K/Utility*	Ongoing Cyber Costs $K/Year
Small	100	10
Medium	400	40
Large	2,200	200

*To be renewed every 10 years

To categorize utilities into small, medium and large, the team first identified investor-owned utilities (IOU), rural electric cooperatives (Co-ops), municipal utilities (Munis), and Power agencies. To estimate the number of utilities which would need to make provisions to engage in investments to secure their Smart Grid-related cyber information technology activities, the project team used Platt's 2010 Directory of Electric Power Producers and Distributors (Platts, 2009).

Platts estimates that there are a total of 342 investor-owned utilities (IOUs). This includes 60 holding companies, 29 transmission companies, and a number of other service and wholesale generation companies. Platts estimated 893 rural electric cooperatives (Co-ops), including both distribution and generation and transmission (G&T) entities. Platts' data for municipal utilities (munis) is divided into two categories: One is municipal and local government utilities; the second is the array of federal, state and district government utilities in the U.S.

Table 5-15 summarizes the Platts' estimates. The team subsequently estimated the breakdown of size between each of these utility types.

Table 5-15
Breakdown of Utility Types

Type	Total #	Small		Medium		Large	
		%	#	%	#	%	#
IOU	342	0	–	25	86	75	256
Co-op	893	50	446	40	357	10	90
Muni	2,118	50	1,059	45	953	5	106
Total	**3,353**		**1,505**		**1,396**		**452**

Table 5-16
Cyber Cost Estimates

Size	Number	Cyber Investment $M 2010-2030		Ongoing Cyber Cost $M 2010-2030	
		Per Utility Each 10 Years	Total	Per Utility Per Year	Total
Small	1,505	.10	301.5	.01	30.1
Medium	1,396	.40	1,116.8	.14	111.7
Large	452	2.20	1,988.8	.20	108.8
Total	**3,353**		**3407.1**		**322.6**

Enterprise Back-Office Systems

All large utilities already have enterprise back-office systems which include geographic information systems (GIS), outage management, and distribution management systems (DMS). To enable the Smart Grid, additional features will be required, including an historic data function in conjunction with analytic tools to take in data streams, compare and contrast with historical patterns and look for anomalies in the data.

Enterprise systems will be needed to be upgraded by virtually all utilities. Medium and large utilities will need complete systems of their own. Small utilities may aggregate their needs or use service providers. Table 5-17 summarizes the project teams estimates for enterprise back-office systems.

206

Table 5-17
Cost of Enterprise Back Office Systems

| Utility Size | Number | Back-Office Investment $M 2010-2030 Each 10 Years | |
		Per Utility	Total
Small	1,505	1,000	3,010
Medium	1,396	4,000	11,168
Large	452	20,000	18,080
Total	**3,353**		**32,258**

Incremental Ongoing System Maintenance

The project team estimated that the PMUs and sensors installed on transmission lines and substations would cause additional incremental maintenance of $50,000 per substation.

Table 5-18
Smart Grid Incremental Maintenance

Technology	Total Units	Units	% Sat	Cost/ Unit Low $	Cost/ Unit High $	Total Cost Low–High $M
Incremental Ongoing Maintenance	58,027	Number of existing substations	50	50,000 /yr	50,000 /yr	15,232- 15,232
Incremental Ongoing Maintenance	8,423	Number of new substations to meet load growth	100	50,000 /yr	50,000 /yr	4,422- 4,422
Incremental Ongoing Maintenance	700	Number of new sub-stations to accommodate renewables	100	50,000 /yr	50,000 /yr	368- 368

Impacts on System Operators

Independent system operators (ISOs), transmission system operators (TSOs), and other independent operators (referred to as ISOs here) are making investments in an increasingly robust communications infrastructure as well as an enhanced analytical and forecasting capability. These investments are being made

◄ 5-31 ►

in response to requirements for ISOs to incorporate increasing functionality in order to maintain reliability, meet load growth, and to comply to new regulations which are increasing grid compliance with FERC rules, increasing the use of distributed resources, demand response and energy efficiency. At the same time, market operations are becoming increasingly more complex, the threat of cyber security is increasing, and pressures is mounting to maintain costs and improve the use of assets.

All ISOs initiated these investments as part of sustaining core capabilities even before the nation began to evolve the concept of a "Smart Grid." For example, the development of techniques for real-time simulation and enhanced visualization have been under development since the 1990s. Today, they are considered part of the Smart Grid, but would have simply been viewed as necessary improvements a decade ago.

The project team identified three ISO functions which are considered part of the Smart Grid in this study. They include:

1. Enhancing the visibility of the grid (transparency), increasing reliability, and energy efficiency.

2. Enable the effective integration of increasing amounts of distributed resources including renewables, energy storage, and demand response.

3. Enable effective response to increasingly sophisticated cyber and physical security threats including natural events.

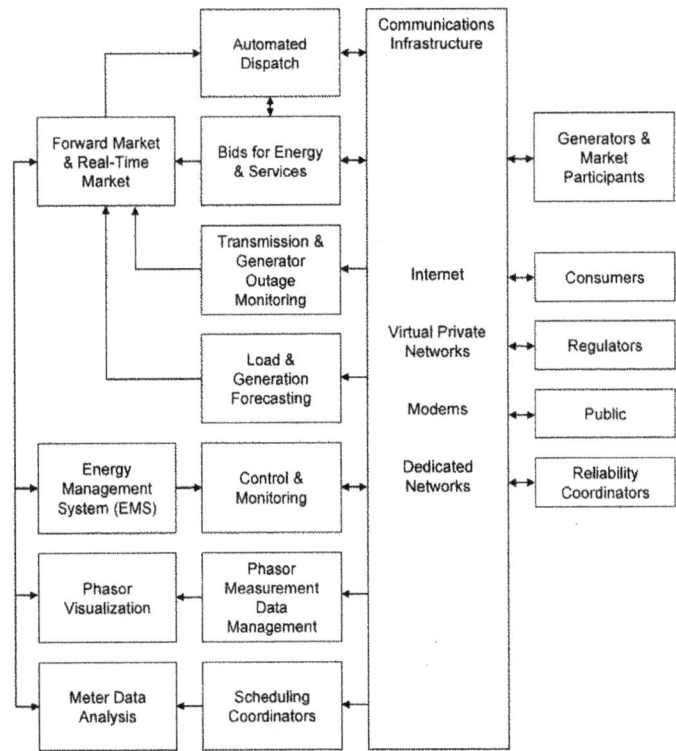

Figure 5-13
Key Components of an ISO Infrastructure

Figure 5-13 illustrates the key applications that form an ISO. The functional enhancements which the Smart Grid is expected to yield will require changes in these applications in the following areas:

- Communications – The increasing number of external interfaces will necessitate that telemetry data is compatible with IEC 61850. This will require retrofits and upgrades. In addition, the Smart Grid will contain many more transmission line sensors and sensors from wind and solar installations.

- Prices-To-Devices – Truly enabling consumers to respond to variations in prices and system constraints will require the ability to broadcast that information to consumers both through the Internet and through secure, automated area networks.

- Upgrade to NIST-Proposed Standards – The National Institute for Standards and Technology (NIST) has established a Smart Grid Interoperability Panel (SGIP), and the panel is developing Priority Action Plans (PAP) to encourage adoption of standard business practices and interface specifications. It is likely that these interfaces will require widespread adoption of the Common Information Model (CIM).

- Forecasting – Developing far more accurate data to forecast renewable energy production (particularly wind and solar) including very short-term forecasts and ramp rates. In addition, it will be necessary to forecast demand response participation.

- Cyber Security – The North American Electric Reliability Corporation is developing Critical Infrastructure Protection Standards (CIP) to meet Federal Energy Regulatory Commission (FERC) and Congressional requests.

- Integrate Synchrophasors – Integrating synchrophasor data sources with visualization and to enable advanced-state estimation and dynamic stability analysis incorporating the data from numerous new distributed generation sources.

- Evolve New Markets – The advent of widespread consumer connectivity coupled with increasing participation of third parties allows for new markets to evolve. The ISO's system will need to be sufficiently flexible so as to allow the evolution of new markets.

- Other Applications – Individual ISOs may add other functionality like the ability to quickly add new market participants, to improve outage management data and other enhancements.

Each of these and other features will require an ongoing investment in the ISO's applications.

The costs of the enhancements to an ISO's system to respond to changes in the applications are very difficult to estimate. Each ISO is in a different phase of development. Each varies as to the projected nature and penetration of distributed resources and as to the extent of other market participants.

The project team interviewed executives from several ISOs including PJM, CAISO, NEISO, and NYISO in order to establish estimates for the costs which an ISO will occur in order to accommodate the functionality required to achieve Smart Grid goals. Their informal estimates varied but generally included five or six full-time staff ($2.4 million per year) and at least several million dollars annually in software. For one ISO, this totaled 10% of its annual budget or $12 million per year.

The following are the ten regional power markets in the U.S.

- California (CAISO)
- Midwest (MISO)
- New England (ISO-NE)
- New York (NYISO)
- Northwest
- PJM
- Southeast
- Southwest

- SPP
- Texas (ERCOT)

Table 5-19 summarizes the estimated cost of alignment of ISOs with Smart Grid requirements.

Table 5-19
Cost to Align ISOs with Smart Grids

Technology	Total Units	Units	% Sat	Cost/ Unit Low $	Cost/ Unit High $	Total Cost Low– High $M
ISO Smart Grid	10	Regional markets	100	12/yr	12/yr	2,400- 2,400

Summary of Transmission and Substations Costs

The cumulative cost for bringing the nation's transmission and substations system up to the performance levels required for Smart Grid operation is estimated to cost between $6,312 and $7,280 million by 2030, as shown in the summary Table 5-20, below. Smart Grid related investment in the transmission system will continue well beyond 2030. This does not include related investments to meet load growth which are estimated to cost between $56,350 and $63,702 million by 2030.

Table 5-20
Smart Grid Transmission and Substation Costs

Technology	Total Units	Units	% Sat	Cost/ Unit Low $	Cost/ Unit High $	Total Cost Low-High $M
Dynamic-Thermal Circuit Rating	11,340	Number of substations with one DTCR unit/7.5 miles of line	100	10,000	20,000	113.4-226.8
Substation & Transmission Line Sensors	58,027	Number of existing substations	50	50,000	100,000	1,451-2,901
Transmission Short-Circuit Current Limiters	58,017	Number of substations	2	500,000	500,000	580.3-580.3
Storage for Bulk Transmission Wholesale Services w/o Regulation	5,800,000	kW	100	810	1045	4,698-6,061
Storage for Bulk Transmission Wholesale Services with Regulation	2,800,000	kW	100	810	1045	2,268-2,926
FACTS Devices	330	Numbers of	100	Various	Various	4,175-4,925
Communications: Core Infrastructure for Smart Substations	58,027	Number of substations	80	50,000	75,000	2,321-3,481
Communications to Substations	58,027	Number of substations	80	$1,200/mo	$1,200/mo	668.5-668.5
Phasor Measurement Units (PMU)	1,250	Numbers of	100	125,000	125,000	156-156

Table 5-20 (continued)
Smart Grid Transmission and Substation Costs

Technology	Total Units	Units	% Sat	Cost/ Unit Low $	Cost/ Unit High $	Total Cost Low–High $M
Intelligent Electronics Devices (IED) – Relays & Sensors	58,027	Number of substations	80	110,000	150,000	5,106-6,963
Cyber Security – Enterprise-Wide	1,454	Number of utilities	100	100,000*	2,200,000*	3,729.2-3,729.2**
Enterprise Back Office System – GIS, Outage management, Distribution Management	1,454	Number of utilities	100	1,000,000*	20,000,000*	32,258-32,258
ISO Smart Grid	10	Regional markets	100	12,000,000/yr	12,000,000/yr	2,400-2,400
Incremental Ongoing System Maintenance***	58,027	Numbers of substations	50	50,000/yr	50,000/yr	15,232-15,232
Total						**75,157-82,509**

SCCL will begin in 2020 and reach a 2% penetration rising to 7% by 2030.
*Varies by size.
**Includes annual cost.
***Additional ongoing maintenance costs are included in individual cost components.

◄ 5-37 ►

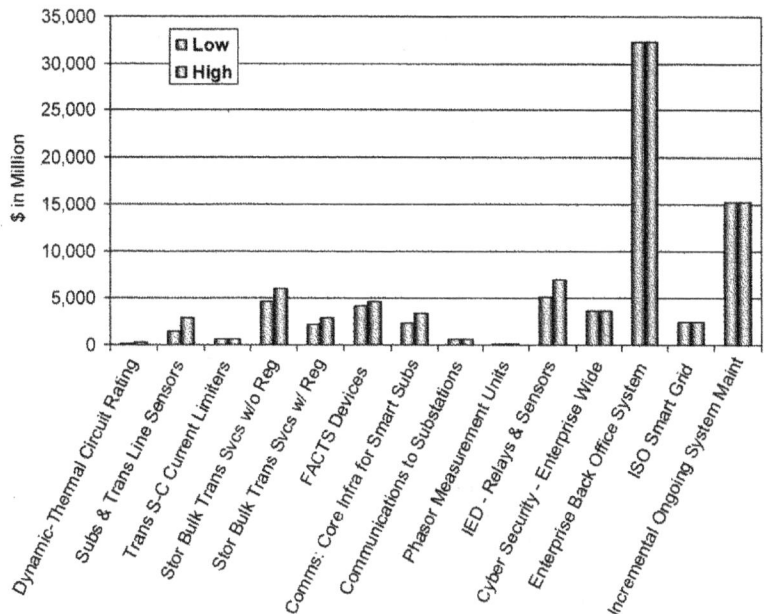

Figure 5-14
Smart Grid Transmission and Substation Costs

Table 5-21
Smart Grid Transmission and Substation Cost to Meet Load Growth

Technology	Total Units	Units	% Sat	Cost/ Unit Low $	Cost/ Unit High $	Total Cost Low– High $M
Substation & Transmission Line Sensors	8,423	Number of new substations	100	50,000	100,000	421-631.7
Communications: Core Infrastructure for Smart Substations	8,423	Number of substations	100	50,000	75,000	421-632
Transmission Systems & Communications to Substations	8,423	Number of substations	100	14,400	14,400	121-121
Intelligent Electronics Devices (IED) – Relays & Sensors	58,027	Number of substations	80	110,000	150,000	927-1,264
Incremental Ongoing System Maintenance	8,423	Number of substations	100	50,000	50,000	4,422-4422
Total						**6,312-7,280**

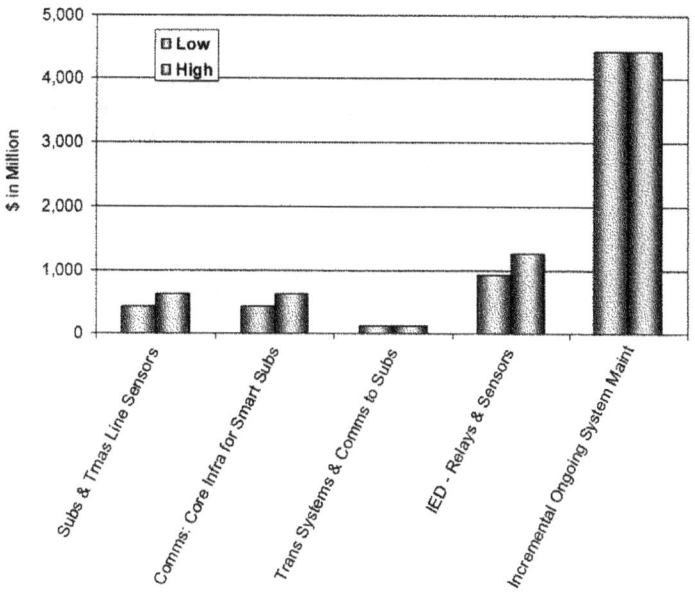

Figure 5-15
Smart Grid Transmission and Substation Cost to Meet Load Growth

Table 5-22
Smart Grid Transmission and Substation Costs to Meet Renewables

Technology	Total Units	Units	% Sat	Cost/ Unit Low $	Cost/ Unit High $	Total Cost Low– High $M
Communications: Core Infrastructure for Smart Substations	700	Number of substations	100	50,000	75,000	35-53
Transmission Systems & Communications to Substations	700	Number of substations	100	14,400	14,400	10-10
Phasor Measurement Units (PMU)	700	Number of substations	100	125,000	125,000	88-88
Intelligent Electronics Devices (IED) – Relays & Sensors	700	Number of substations	100	110,000	150,000	77-105
Incremental Ongoing System Maintenance	700	Number of substations	100	50,000/yr	50,000/yr	368-368
Total						**577-623**

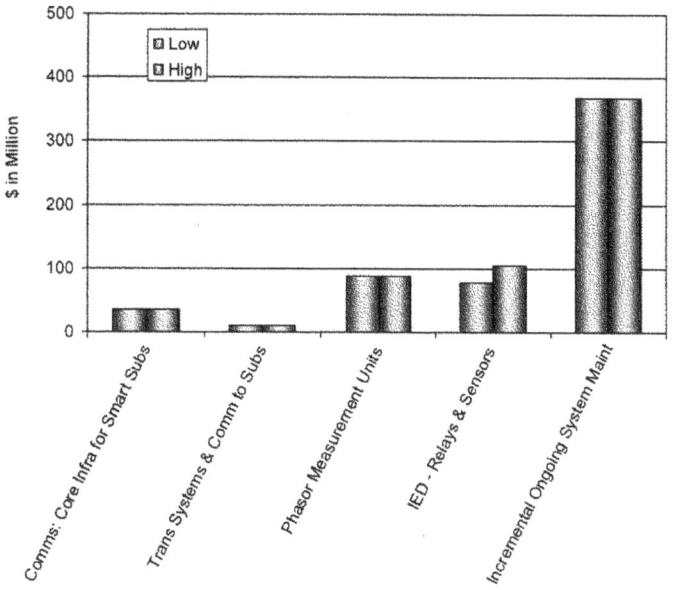

Figure 5-16
Smart Grid Transmission and Substation Costs to Meet Renewables

Table 5-23
Total Smart Grid Transmission and Substation Costs

Costs to Upgrade the Existing System ($M)		
	Low	**High**
Transmission and substations	75,157	82,509
Costs to Embed Smart Grid Functionality While Accommodating Load Growth ($M)		
	Low	**High**
Transmission and substations	6,312	7,281
Costs to Embed Smart Grid Functionality While Accommodating Large-Scale Renewables ($M)		
	Low	**High**
Transmission and substations	577	623
Total	**82,046**	**90,413**

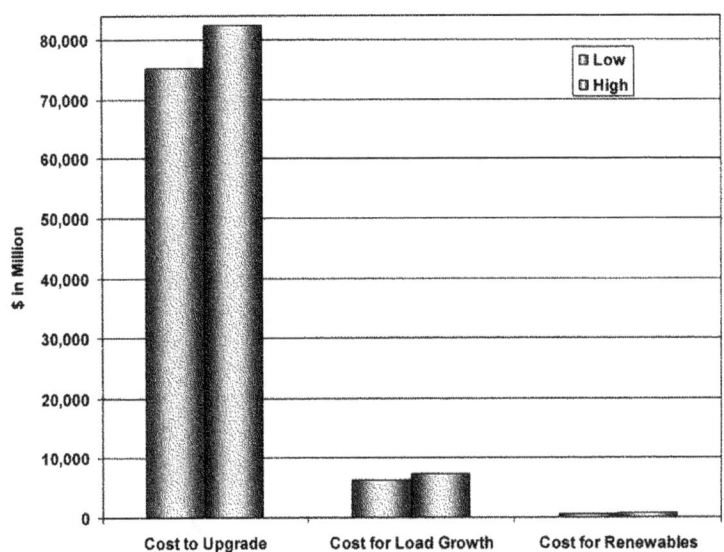

Figure 5-17
Total Smart Grid Transmission and Substation Costs

Section 6: Distribution

While a small percentage of electricity customers are served directly from the transmission system, the vast majority of the 165 million customers in the U.S. are served by the distribution system, which is comprised of a complex network of substations, lines, poles, metering, billing and related systems to support the retail side of electricity delivery.

The study team estimated the cost of Smart Grid distribution investment between $309 to $403 billion over the next 20 years.

Introduction

Utility distribution systems are generally challenged by an aging infrastructure, conventional designs, and increased demands for digital-quality power. There are a few exceptions where distribution utilities have implemented a reasonably smart grid and are working to make it smarter – but these are the exception. Compared to the transmission system, the greater complexity, exposure, and geographic reach of the distribution system results in inherently lower reliability, reduced power quality, and greater vulnerability to disruptions of any kind. Using a reliability measure of average total duration of the interruptions experienced by a customer in a year, over 90% of the minutes lost by consumers are attributable to distribution events. In 2004, EPRI estimated that a fully automated distribution system could improve reliability levels by 40%. Advances in Smart Grid technologies are not a substitute for good maintenance practices, inspection, and vegetation management.

Investment in the distribution system has averaged $12 to $14 billion per year for last few decades, primarily to meet load growth, which includes both new connects and upgrades for existing customers. An urban utility may have less than 50 feet of distribution circuit per customer, while a rural utility can have more than 300 feet of primary distribution circuit per customer. Assuming a rough average 100 feet of line for each of the 165 million U.S. customers indicates the U.S. has an installed base of more than 3 million miles of distribution line. Upgrading a system this extensive to the level of performance required in a Smart Grid will require a substantial investment.

Estimates of the cost of individual distribution system components in this analysis were based largely upon utility experience in deploying the first wave of Smart Grid investment. Estimates of AMI and distribution automation investment were drawn, for example, from FirstEnergy, SCE, SDG&E, Dayton Power & Light, FortisAlberta, Inc, and Idaho Power, among others. Representative costs are shown in the two tables below (FirstEnergy, 2009).

Table 6-1
Distribution Cost Estimates Per Feeder (FirstEnergy, 2009)

Function	No. of Feeders	Total Cost	Cost/Feeder
Distribution Automation	59	$18.2 M	$308,000
Volt/VAR Control	33	$8.5 M	$258,000

Table 6-2
Distribution Cost Estimates Per Customer (FirstEnergy, 2009)

Function	No. of Customers	Total Cost	Cost/Customer
Direct Load Control	34,000	$24.6 M	$728
AMI with DR	44,000	$41.2 M	$940

Cost Components for the Smart Grid: Distribution

Smart Grid investments in the distribution system entail wider high bandwidth communications to all substations, intelligent electronic devices (IED) that provide adaptable control and protection systems, complete distribution system monitoring that is integrated with larger asset management systems, collaborative distributed intelligence, including dynamic sharing of computational resources of all intelligent electronic devices and distributed command and control to mitigate power quality events and improve reliability and system performance. The key cost components for the distribution portion of the Smart Grid are as follows:

- Communications between all digital devices on the distribution system including to feeders for AMI and distributed smart circuits

- Distribution automation

- Distribution feeder circuit automation

    - Intelligent reclosers and relays at the head end and along feeders

    - Power electronics, including distribution short circuit current limiters

    - Voltage and VAR control on feeders

- Intelligent universal transformers

- Advanced metering infrastructure (AMI)

- Local controllers in buildings, on microgrids, or on distribution systems for local area networks

Communications

Communications constitute the critical backbone for integrating customer demand with utility operations. Detailed, real-time information is key to effectively managing a system as large and dynamic as the distribution power grid. Each smart meter in the advanced metering infrastructure (AMI), described later in this section, must be able to communicate with a wide range of user control systems, as well as reliably and securely communicating performance data, price signals, and customer information to and from an electric utility's back-haul system.

No single technology is optimal for all applications. Among the communications media now being used for AMI applications are cellular networks, licensed and unlicensed radio, and power line communications. In addition to the media, the type of network is also an important part of communications design. Networks used for Smart Grid applications include fixed wireless, mesh networks, and a combination of the two. Several other network configurations, including Wi-Fi and Internet networks are also under investigation.

Interoperability remains one of the most critical success factors for Smart Grid communications. There is growing interest in the potential use of the Internet Protocol Suite (TCP/IP) as a networking protocol that could run over many different communication technologies. In June, 2009, EPRI submitted a *Report to NIST on the Smart Grid Interoperability Standards Roadmap* that became the starting point for NIST's own roadmap, released in September, 2009 (www.nist.gov/smartgrid).

Communication architectures remain diverse for integrating residential devices with the grid. Some utilities envision using the meter as a gateway to the home for price and feedback information, whereas others envision using the Internet or other communication channels. Radio frequency (RF) networks communicating in both licensed and unlicensed radio bands and have been chosen by the majority of Smart Grid deployments in the U.S. Mesh networks incorporate multi-hop technology where each node in the network can communicate with any other node. Star networks utilize a central tower that can communicate with a large number of end devices over a wide area. Each type has certain advantages and disadvantages, and is selected based on the unique needs and circumstances of the utility. Power line carrier networks where communications are carried via electric power lines are also used by a large number of utilities both in the U.S. and abroad.

Communications to feeders for AMI and distribution smart circuits were estimated to cost about $20,000 per feeder, and to be fully installed on 80% of existing feeders and 100% of new feeders by 2030, for a total cost of nearly $9 billion.

Table 6-3
Cost of Communication to Feeders for AMI

Technology	Total Units	Units	% Sat	Cost/ Unit Low $	Cost/ Unit High $	Total Cost $M
Communication to Existing Feeders for AMI & Distribution Smart Circuits	464,216	Number of feeders	80	20,000	20,000	7,427
Communication to New Feeders for AMI & Distribution Smart Circuits	67,384	Number of feeders	100	20,000	20,000	1,348

Distribution Automation

Distribution automation (DA) involves the integration of SCADA systems, advanced distribution sensors, advanced IED's and advanced two-way communication systems to optimize system performance. In a dense urban network it will also include network transformers and network protectors. The SCADA system collects and reports voltage levels, current demand, MVA, VAR flow, equipment state, operational state, and event logging, among others, allowing operators to remotely control capacitor banks, breakers and voltage regulation. Substation automation, when combined with automated switches, reclosers, and capacitors, will enable full Smart Grid functionality.

This includes not only building intelligence into the distribution substations and into the metering infrastructure but also into the distribution feeder circuits and components that link these two essential parts of the grid. This means automating switches on the distribution system to allow automatic reconfiguration, automating protection systems and adapting them to facilitate reconfiguration and integration of DER, integrating power-electronic based controllers and other technologies to improve reliability and system performance, and optimizing system performance through voltage and VAR control to reduce losses, improve power quality and facilitate the integration of renewable resources.

- **Intelligent head-end feeder reclosers and relays.** Replacing electromechanical protection systems with microprocessor-based, intelligent relays and reclosers are an integral part of Smart Grid operation. Advantages include multiple functionality, including both instantaneous and time-overcurrent protection, greater sensitivity, better coordination with other devices, and the ability for self diagnosis. Approximately 70% of all feeders will include intelligent reclosers and relays by 2030, at an estimated unit cost of $50,000.

- **Intelligent reclosers.** The use of intelligent switching and protection devices on feeders (referred to as "mid-point or tie-reclosers") to allow isolation of segments of feeders to enhance reliability. Approximately 25% of all feeders will have intelligent reclosers and relays by 2030 at an estimated cost of $100,000 to $150,000.

- **Remotely controlled switches.** Remotely controlled switches contain distributed intelligence and use peer-to-peer communications to take actions without the need for central control intervention in order to isolate faults and restore power quickly in the event of an outage. As a result, distribution system operators will no longer be the only ones that can perform that function. It is estimated that 5% of all feeders will use one remotely controlled switch at a cost of $50,000 to $75,000 by 2030.

- **Power electronics, including distribution short circuit current limiters.** Advances in power electronics allow not only greater fault protection but flexible conversion between different frequencies, phasing, and voltages while still producing a proper ac voltage to the end user. Power electronics will be deployed on about 5% of 57,000 substations by 2030 at an average cost $80,000/package.

- **Voltage and VAR control on feeders.** Voltage/VAR controls are a basic requirement for all electric distribution feeders to maintain acceptable voltage at all points along the feeder and to maintain a high power factor. Recent efforts by distribution utilities to improve efficiency, reduce demand, and achieve better asset utilization, have indicated the importance of voltage/VAR control and optimization. Utilities continue to face system losses from reactive load, such as washing machines, air conditioners. By optimizing voltage/VAR control great efficiencies can be realized. An estimated 55% of the 566,000 distribution feeders will include voltage/VAR control by 2030, at an average cost of $258,000/feeder.

Smart Grid software will be able to combine information flowing from the automated substations with SCADA data points throughout the distribution system to analyze and recommend re-configuration of the distribution system for optimum performance. Circuit optimization will minimize line loses and integrate customer data from AMI to regulate voltage while still maintaining acceptable levels for customers. This functionality will help support conservation voltage reduction (CVR) strategies to achieve energy savings. As the Smart Grid evolves, this one dimension of optimization will expand to include optimization of reliability, power quality and asset management – among others.

By 2030, an estimated 55% of all existing distribution feeders will be integrated with advanced distribution automation systems at a cost of $308,000 per feeder, and 100% of all new feeders will be equipped by 2030. In this analysis, there are additional costs assumed for automation of the feeders themselves. The Smart Grid investment of distribution automation is estimated at nearly $96 billion, as shown in Table 6-4.

Table 6-4
Cost of Distribution Automation

Technology	Total Units	Units	% Sat	Cost/Unit Low $	Cost/Unit High$	Total Cost $mill
Distribution Automation	464,216	Number of existing feeders	55	Varies	Varies	124,134-177,008
Distribution Automation	67,384	Number of new feeders	100	308,000	308,000	20,754-20,754

As shown in Table 6-5, the total investment required for distribution feeder automation exceeds $92 billion from 2010 to 2030.

Table 6-5
Cost of Distribution Feeder Automation for Existing Systems

Technology	Total Units	Units	% Sat	Cost/Unit Low $	Cost/Unit High $	Total Cost Low-High $M
Communications to Feeders for AMI & Distribution Smart Circuits	464,216	Number of feeders	80	20,000	20,000	4,428-4,428
Head End of Feeders – Intelligent Reclosers & Relays	464,216	Number of feeders	70	50,000*	50,000*	16,248-16,248
Power Electronics, Include Distribution Short-Circuit Current Limiters	58,027	Number of substations	5	80,000	80,000	2,321-2,321
Smart Switches, Reclosers, Monitored Capacitor Banks, Regulators & Circuit Improvement	464,216	Number of feeders	55	308,000	308,000	78,638-78,638
Voltage &VAR Control on Feeders	464,216	Number of feeders	55	60,000	258,000	15,319-65,873
Intelligent Reclosers	464,216	Number of feeders	25	100,000	150,000	11,605-17,408
Remotely Controlled Switches	464,216	Number of feeders	5	50,000	75,000	1,161-1,741
Direct Load Control (not integrated with AMI)	123,949,916	Number of customers	5 declining to 0	100	100	1,859-1,859
ElectriNet Controllers	464,216	Number of feeders	10	50,000	100,000	2,321-4,642
Incremental Ongoing System Maintenance**						
Total Distribution Feeder Cost						**124,134-177,008**

*Incremental cost – i.e., excludes the cost of the "switch" itself.

**Incremental Ongoing System Maintenance expenses are included in individual cost components.

Table 6-6
Cost of Distribution Feeder Automation for New Feeders

Technology	Total Units	Units	% Sat	Cost/ Unit Low $	Cost/ Unit High $	Total Cost Low-High $M
Communications to Feeders for AMI & Distribution Smart Circuits	67,384	Number of feeders	100	20,000	20,000	1,348-1,348
Head End of Feeders – Intelligent Reclosers & Relays	67,384	Number of feeders	100	50,000*	50,000*	3,369-3,369
Power Electronics, Include Distribution Short-Circuit Current Limiters	8,423	Number of substations	100	80,000	80,000	6,738-6,738
Smart Switches, Reclosers, Monitored Capacitor Banks, Regulators & Circuit Improvement	67,384	Number of feeders		308,000	308,000	20,754-20,754
Voltage &VAR Control on Feeders	67,384	Number of feeders	100	258,000	258,000	4,043-17,385
Intelligent Reclosers	67,384	Number of feeders	25	100,000	150,000	1,685-2,527
Remotely Controlled Switches	67,384	Number of feeders	5	50,000	75,000	169-253
ElectriNet Controllers	67,384	Number of feeders	25	50,000	100,000	842-1,684
Incremental Ongoing System Maintenance**						
Total Distribution Feeder Cost						**38,948-54,059**

*Incremental cost – i.e., excludes the cost of the "switch" itself.

**Incremental Ongoing System Maintenance expenses are included in individual cost components.

Intelligent Universal Transformers

Conventional transformers suffer from poor energy conversion efficiency at partial loads, use liquid dielectrics that can result in costly spill cleanups, and provide only one function—stepping voltage. These transformers do not provide real-time voltage regulation nor monitoring capabilities, and do not incorporate a communication link. At the same time, they require costly spare inventories for multiple unit ratings, do not allow supply of three-phase power from a single-phase circuit, and are not parts-wise repairable. Future distribution transformers will also need to be an interface point for distributed resources, from storage to plug-in hybrid electric vehicles.

The intelligent universal transformer (IUT) is a first-generation, power-electronic replacement of conventional distribution transformers. EPRI has developed an IUT which can serve as a "Renewable Energy Grid Interface" (REGI). The new concept includes a bi-directional power interface that provides direct integration of photovoltaic systems, storage systems, and electric vehicle charging. It will also incorporate command and control functions for system integration, local management, and islanding.

REGI will become a key enabler in the overall Smart Grid development strategy. It plays a transformational role by combining the traditional functions of a power transformer with new interface capabilities. It can seamlessly integrate widespread renewable energy technologies, including energy storage, electric vehicles, and demand response, while also providing an architecture that allows the operation of reliable local energy networks. The controller will interface with distribution management systems, energy management systems, and demand response systems to optimize overall grid performance and improve reliability.

The cost of the IUT, as shown in Table 6-7, are expected to decline dramatically over the next 20 years, from \$1.50 to 2.00/W today to \$0.20/W in 2030. As a result, deployment is expected to grow rapidly, from for example, 10,000 25-kW units in 2015 to 1 million in 2030.

Table 6-7
Declining Costs and Growing Deployment of the IUT

IUT	2010	2015	2020	2025	2030
$/Watt With Storage Integration Option	$1.50-$2.00	$.75	$.50	$.35	$.20
25 kW No. Units	Demo	10 x 10³	50 x 10³	200 x 10³	1 x 10⁶
50 kW No. Units	Demo	5 x 10³	25 x 10³	100 x 10³	500 x 10³
$/watt With PV Inverter	$3.00-$1.00	$.75	$.50	$.35	$.20
25 kW No. Units	Demo	10 x 10³	50 x 10³	200 x 10³	1 x 10⁶
50 kW No. Units	Demo	5 x 10³	25 x 10³	100 x 10³	500 x 10³

Total Smart Grid investment in IUTs through 2030, based upon the expected deployment of three million units, ranging in cost from $7,500 to $100,000, is $76 to $131 billion, as shown in Table 6-8.

Table 6-8
Cost of Intelligent Universal Transformers*

Technology	Total Units	Units	% Sat	Cost/ Unit Low $	Cost/ Unit High $	Total Cost Low–High $M
Intelligent Universal Transformer With Storage	1,500,000	Number of	various	37,500	100,000	12,563-12,688
Intelligent Universal Transformer With PV Inverter	1,500,000	Number of	various	7,500	50,000	12,437-12,937
Total IUT Cost	**3,000,000**	**Number of**	**various**			**25,000-25,625**

Note: It is assumed that 50% of IUTs will be installed on existing feeders and 50% will be installed on new feeders added for load growth.

---

* IUT may include some part of energy storage costs. Energy storage cost would be $500/ kWh of storage installed.

Advanced Metering Infrastructure (AMI)

An advanced metering infrastructure (AMI) involves two-way communications with smart meters, customer and operational data bases, and various energy management systems. AMI, along with new rate designs, will provide consumers with the ability to reduce electricity bills by using electricity more efficiently, to participate in Demand Response Programs and to individualize service, and provide utilities with the ability to operate the electricity system more robustly.

Smart meters are the main component of AMI and generally the first technology deployed by an electric utility in a Smart Grid program. Although smart meters have been used by commercial and industrial customers for decades, only recently have they become economical for widespread use in residential settings. The broader AMI system in which smart meters operate involves the two-way communication network to exchange energy usage, price and curtailment signals, and operational control signals. Integral to AMI is a common enterprise bus network architecture linking all key enterprise systems including meter data management, customer care, auto-demand response system, and energy management. The goal is to provide a highly secure, resilient and flexible technology upgrade to the core business of electric utilities and to integrate electricity usage into Smart Grid dynamics. Three basic functions are involved:

- **Smart meters** capable of two-way communication with the utility, remotely programmable firmware, and, optionally, a remotely manageable service disconnect switch. In addition to consumption measurements, smart meter functionality includes: voltage measurement and alarms that can be integrated with distribution automation projects to maximize CVR benefits, and interval data to support dynamic pricing and demand response programs.

- **Communications** system that is highly secure (encrypted), redundant and self-healing, and related hardware and software systems to communicate between smart meters, substation and distribution automation equipment, customer energy management systems, and head-end software applications / meter data management systems.

- **Meter data management system** capable of storing and organizing data, allowing for advanced analysis and processing, and interfacing AMI head-ends with a range of other enterprise software applications.

AMI Cost Assumptions

- **Residential meter** costs are based more on volume than other factors

  - Meter + AMI $40-80/unit
  - Meter + AMI + Disconnect $70-130/meter
  - Meter + AMI + Disconnect+ HAN $80-140/meter

- **Commercial and Industrial meter** costs are based more on features selected than other factors.

  - Meter + communications $120-150/meter

- - GT&D
  - Meter $1500-5000
- Installation costs
  - Residential $7-10/meter
  - Commercial and industrial $20-65/meter
  - AMI network and backhaul equipment $3-11/endpoint
  - Head end software and integration $4-10/endpoint
  - System initiation and management $2-4/endpoint
- Ongoing maintenance $3-11/year/endpoint

AMI Costs for the Smart Grid

Based upon these unit costs and the assumption of an average of 83% saturation, the total costs for the AMI portion of Smart Grid investment from 2010 to 2030 ranges from $15 to $42 billion, as shown in Table 6-9.

Table 6-9
Cost of Advanced Metering Infrastructure (AMI) for Existing Customers

Technology	Total Units	Units	% Sat	Cost/Unit Low$	Cost/Unit High$	Total Cost Low-High $M
Advanced Meter Infrastructure (AMI) Residential Meters	123,949,9166	Number of customers	80	70	140	7,437-13,387
Installation of Residential Meters	123,949,9166	Number of customers	80	7	15	694-1,487
Advanced Meter Infrastructure (AMI) Commercial & Industrial Meters	18,170,986	Number of customers	100	120	500	2,240-9,284
Installation of Commercial & Industrial Meters	18,170,886	Number of customers	100	20	65	364-1,184
Other AMI Costs	142,121,652	Number of customers	83	Various	Various	1,062-2,949
Ongoing System Maintenance	142,121,652	Number of customers	83	3/year	11/year	3,716-13,624
Total AMI Costs	**164,982,450**	**Number of customers**	**83**			**15,513-41,915**

Table 6-10
Cost of Advanced Metering Infrastructure (AMI) for New Customers

Technology	Total Units	Units	% Sat	Cost/Unit Low$	Cost/Unit High$	Total Cost Low-High $M
Advanced Meter Infrastructure (AMI) Residential Meters	19,978,760	Number of customers	80	70	140	1,498-2,697
Installation of Residential Meters	19,978,760	Number of customers	80	7	15	140-300
Advanced Meter Infrastructure (AMI) Commercial & Industrial Meters	2,800,932	Number of customers	100	120	500	360-1,493
Installation of Commercial & Industrial Meters	2,800,932	Number of customers	100	20	65	59-190
Other AMI Costs	22,907,634	Number of customers	100	Various	Various	586-1,523
Ongoing System Maintenance	22,907,634	Number of customers	100	3/year	11/year	722-2,646
Total AMI Costs	**22,907,634**	**Number of customers**	**100**			**3,365-8,850**

Controllers for Local Energy Network

Local energy networks are means by which consumers can get involved in managing electricity by reducing the time and effort required to change how they use electricity. If usage decisions can be categorized so they are implemented based on current information, and that information can be readily collected and processed, then consumers will purchase and operate such a system, install and operate a home area network (HAN.) A HAN is an electronic information network, connected to a central or "master" control which acts as an energy management system (EMS.) The HAN accommodates the flow of information to and from network nodes. Each node is associated with a device or element of the household's electric system. Nodes can be hard-wired devices that account for substantial portions of electricity used like the HVAC, a pool pump, lighting circuits, or smaller plug loads like TVs, entertainment centers, and a multitude of chargers. Communication among devices and the EMS is accomplished through wireless, wired, or power line carrier media that define and make operational the HAN.

An EMS is a decision processor which controls energy use within the building, organizes response to Demand Response participation, controls distributed generation, electric vehicle charging and storage and interfaces with retail electricity markets. An EMS is an intelligent device that acts as the coordinator for the devices that comprise the home area network. It maintains certain user-defined rules for interior temperature settings as well as when appliances and other household loads should not ever turn off. These rules can be based on the price of electricity at a particular instance of time (e.g., when it exceeds some threshold), on current conditions (e.g., the time of day a household service is typically expected to run), or in response to a command to do so from an external agent (e.g., a curtailment order from a curtailment service provider).

The EMS is the controller, making decisions based on exigent conditions viewed in light of a predefined instruction set, and the HAN is the neural system that conveys information about the state of the nodes and delivers commands and verifies their receipt and enactment. The EMS is an electronic device whose purpose is to manage household electricity consumption better than the household can do so in its absence. Achieving that result requires the very difficult task of understanding how the household members use and value electricity, establishing ways for them to negotiate differences in value systems, and establishing a holistic household utility function that establishes the relative value under different system states and executes pre-established operational decisions.

Architectures are evolving for marrying the Smart Grid with low-carbon central generation, local energy networks (LEN) and electric transportation. LEN includes a combination of end-use energy service devices, distributed generation, local energy storage, and integrated demand-response functions at the building, neighborhood, campus or community level. These architectures to facilitate a highly interactive network based upon a distributed, hierarchical control structure that defines the interactions of LEN, distribution systems and the bulk power system (Gellings, 2010).

These architectures facilitate the inclusion of multiple centralized generation sources linked through high-voltage networks. The design implies full flexibility to transport power over long distances to optimize generation resources and to deliver the power to load centers in the most efficient manner possible. In particular, these architectures enable the inclusion of inherently less controllable variable resources such as wind, solar and certain kinetic energy sources by offering a variety of balancing resources. To enable integration of these elements, these architectures must address the key transformative technical challenges shown in Table 6-11.

Table 6-11
Key Technical Challenges for Tomorrow's Distribution Architecture

Operational Area	Modeling, Simulation, and Control	Monitoring, Data Management, and Visualization	Advanced Control Architecture	Control Structure
System Operations	Physical and distributed models, real time simulation, local and global system constraints	Widespread sensor integration with simulation tools and expert systems. visualization tools for decision making	Distributed intelligence and control architecture – local optimization integrated with system management integration with simulation tools and expert systems. visualization tools for decision making	Autonomous Control Devices
Market Operations	Aggregate resource models, local vs. global optimization	Market and participant awareness	Market structures to support distributed control architecture	Autonomous Control Devices

The concept of these distribution architectures is to optimize performance locally without complete dependence on the bulk power system infrastructure by taking advantage of the overall infrastructure to optimize energy efficiency and energy use. A key transformative element will be the development of distributed, intelligent control devices that will be able to constantly balance generation and load and more at the device, home, neighborhood, city, area and regional levels. To achieve this vision, specific controllers will need to be designed and prototyped, tested and demonstrated in field applications to verify their interactions.

One of the key advantages of the new architecture is the efficiency that can be achieved in terms of energy savings and tons of avoided emissions. Estimated energy savings by 2030 are between 56 to 203 billion kWh, with a corresponding reduction in annual carbon emissions of 60 to 211 million metric tons of CO_2. On this basis, the environmental value to the U.S. is equivalent to converting 14 to 50 million cars into zero-emission vehicles each year (EPRI 1016905).

As shown in Table 6-12, the estimated cost of local energy network (LEN) controllers by 2030 is roughly $3 to $6 billion.

Table 6-12
Cost of Controllers to Enable Local Energy Networks

Technology	Total Units	Units	% Sat	Cost/Unit Low $	Cost/Unit High $	Total Cost Low–High $M
EMS Controllers for Local Area Networks (LEN) on Existing System	464,216	Number of feeders	10	50,000	100,000	2,321-4,642
EMS Controllers for Local Area Networks (LEN) on New Feeders	67,384	Number of feeders	25	50,000	100,000	842-1,685

Summary of Distribution Costs

The cumulative cost for bringing the electrical distribution system up to the technology levels required for the Smart Grid is estimated at $167 to $249 billion by 2030. Smart distribution investment will continue well beyond 2030 and will be influenced by the increasing functionality and lower costs of future technology as well as the changing needs of the full array of consumers.

Table 6-13
Smart Grid Costs for Upgrading the Existing Distribution System

Technology Group	Total Cost $M	
	Low	High
Distribution Automation	124,134	177,008
Intelligent Universal Transformers	25,000	25,625
Advanced Metering Infrastructure	15,513	41,915
LEN Controllers	2,321	4,642
Total	**166,968**	**249,190**

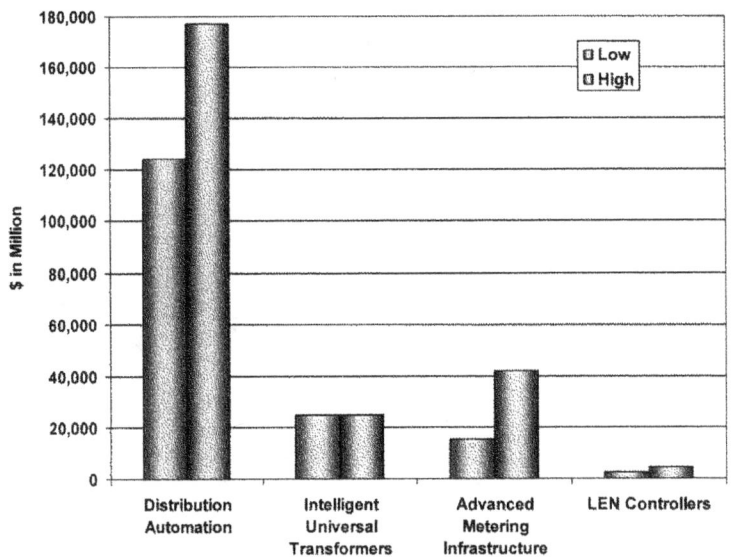

Figure 6-1
Smart Grid Costs for Upgrading the Existing Distribution System

Table 6-14
Smart Grid Costs for Distribution System to Meet Load Growth

Technology Group	Total Cost $M	
	Low	High
Distribution Automation	38,948	54,059
Intelligent Universal Transformers	25,000	25,625
Advanced Metering Infrastructure	3,365	8,850
LEN Controllers	842	1,685
Total	68,155	90,219

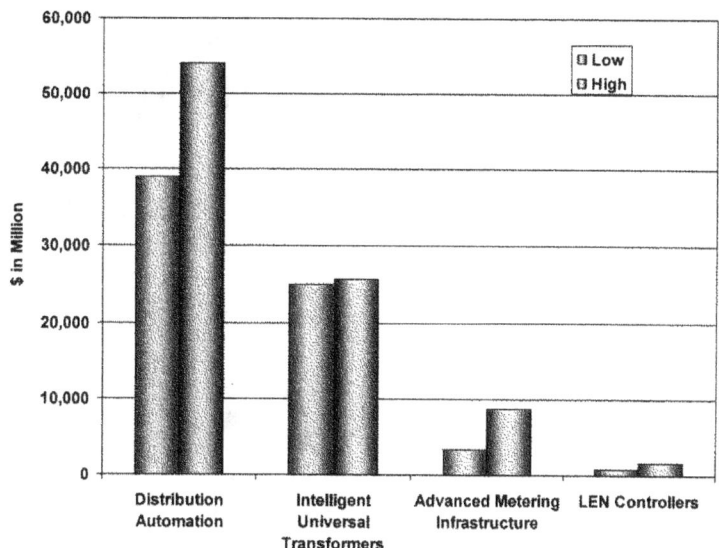

Figure 6-2
Smart Grid Costs for Distribution System to Meet Load Growth

Table 6-15
Total Smart Grid Distribution Costs

Costs to Upgrade the Existing System ($M)		
	Low	High
Distribution	164,647	249,190
Costs to Embed Smart Grid Functionality While Accommodating Load Growth ($M)		
	Low	High
Distribution	67,313	90,219
Total	**231,960**	**339,409**

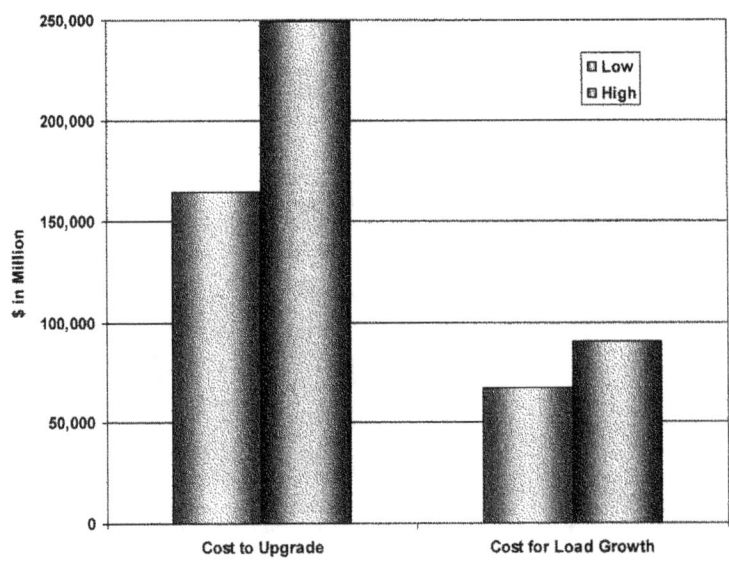

Figure 6-3
Total Smart Grid Distribution Costs

Section 7: Customers

The Smart Grid allows the electricity customer to become fully integrated with the traditional electricity supply system. Such integration began decades ago with commercial and industrial accounts, but with more advanced and lower cost IT and communications technology, it is beginning to gain traction with residential customers. In 2008, FERC estimated 4.7% of the U.S. customers had advanced meters, and that 8% of U.S. customers were engaged in some form of demand response (DR) program. These programs increased the nation's demand-response potential to 5.8% of peak demand by 2008, or more than 40,000MW. DR potential is expected to grow rapidly over the next 20 years as AMI fully saturates the market.

Introduction

There are more than 142 million customers in the U.S., of which 13% represent commercial and industrial accounts. The customer base is expected to grow 16% over the next 20 years to more than 165 million as shown in Table 7-1. However, since most consumer appliances will by then be DR-ready, the actual number of individual communication-connected end nodes will more than double.

Table 7-1
U.S. Electricity Customers

Number of Electric Customers	2007	2030	Load Growth
Residential	123,949,916	143,928,676	19,978,760
Commercial	17,377,219	20,178,151	2,800,932
Industrial	793,767	921,709	127,942
Transportation	750	750	0
Total	**142,121,652**	**165,029,286**	**22,907,634**

Over the next 20 years, integrating the customer into the Smart Grid will enable new functionality to take hold:

- **Increase demand response capabilities** and enable time varying rates to reduce costs, improve load factors, and optimize the economic performance of the grid.

- **Facilitate integration of DER**, including a variety of customer-owned systems, such as rooftop photovoltaic (PV) systems.

- **Integrate the transportation system into the grid** via electric vehicles (EVs) and plug-in hybrid electric vehicles (PHEVs), which can provide the distributed means for large scale electricity storage for the grid, balance daily load cycles, and substantially reduce the nation's oil dependency.

- **Improve energy management** in the home and commercial buildings, reducing peak demand and improving energy efficiency via intelligent agents.

- **Optimize grid performance** by using the demand-side as a resource for stabilizing the grid, for shaving peak demand, and for reducing the capital investment requirements for peaking generation.

Many of the experts who are studying the Smart Grid are increasingly adopting the view that a truly Smart Grid should require as little consumer participation as possible. The Smart Grid does not require consumer participation to succeed. To date, the cost components of the Smart Grid related to the electricity customer have been limited to the costs of grid integration, and exclude costs borne by the consumer to purchase appliances and devices or in enabling the intelligent devices necessary to minimize their direct participation. While, this study does include the costs of the engineering development costs of DR-ready appliances, it does not include the consumer's purchase of energy efficient or DR ready appliances such as PEVs, air conditioners, washing machines, refrigerators, energy efficient devices, and low-value distributed storage. These costs are excluded for the following reasons:

1. Increasingly the performance criteria for energy efficient devices and appliances are driven by Federal Appliance Efficiency Standards and, to some extent energy efficiency provisions of individual State building codes and standards and not as a result of Smart Grid activities.

2. The hypothesis used in this report is that appliances manufactures will be able to include DR ready capability in appliances for little or no marginal cost except for some initial expenses included here. The development of advanced appliances is enabling increased on board processing such that DR ready features will evolve with no marginal cost to consumers. Therefore there is no additional cost which should be attributed to the Smart Grid.

3. Decisions to purchase Plug-in Electric Vehicles are completely independent of Smart Grid investments and are not be included in estimating Smart Grid costs.

4. Albeit impossible to predict, there is growing belief that the enabling technologies to engage with consumers and their end use appliances and devices will originate from entities outside the traditional electric utility industry as part of a service bundle. These providers may include entities like internet search firms, software companies, consumer electronics or IT manufacturers and communications providers. As the capital costs associated with these technologies will be minimal and unknown, they are not included in the Smart Grid estimates in this report.

This distinction is shown in greater detail in the Approach section, "What's In and What's Out" in Table 3-1. The total cost of the consumer portion of the Smart Grid, as shown in this section, is estimated at $32-56 billion by 2030.

Cost Components of the Smart Grid: Consumer/Customer Technologies

The key components for the customer portion of Smart Grid costs are listed below:

- Integrated inverter for PV adoption
- Consumer EMS portal and panel
- In-home displays
- Grid-ready appliances and devices
- Vehicle-to-grid two-way power converters
- Residential storage for back-up
- Industrial and commercial storage for power quality
- Commercial building automation

Who Will Bear These Costs?

Costs in this section are labeled as "customer" costs. However, it is not the intent of the authors to imply that the consumer directly bears these costs. Nor are they necessarily costs that will be borne by the utility to integrate them, or a combination of customer and utility costs. These are costs which must be borne by society and paid directly, bundled with other goods and services or otherwise included by the utility in its cost of service. Like the other technologies, these are critical to achieving the vision of a Smart Grid.

PV Inverters

Inverters are microprocessor-based units used to transform dc to ac power that can be used to connect a photovoltaic (PV) system with the public grid. The inverter is the single most sophisticated electronic device used in a PV system, and after the PV module itself, represents the second highest cost. It is also considered the weakest link. Whereas, solar panels are very robust and carry 25-year warranties, inverter warranties have traditionally been in the 5 to 10 year range. Inverter reliability, however, has been trending up.

There are many types of inverters. Some are stand-alone units isolated from the grid and used to support a stand-alone rooftop system; others are grid-tied, in which case the microprocessor circuits are more elaborate and require additional functionality, including lightning protection. Central inverters are used in large applications. Many times they can be connected according to "master-slave" criteria, where the succeeding inverter switches on only when enough solar radiation is available. Module inverters are used in small photovoltaic systems, such as household rooftops.

A new generation of micro-inverters holds promise to increase PV performance. With current PV design, all solar panels are connected in series, so that if any panel in the series is shaded, it brings down the performance of the entire system. Moreover, for a series module to work, all panels have to have the same orientation and tilt, which limits roof top configuration. The micro-inverter scheme, on the other hand, allows each panel to be connected to its own micro-inverter, increasing overall system performance and providing flexibility for the staggered roof designs of many modern homes. Austin Energy, among others, is testing new micro-inverter designs.

The study team estimated the aggregate cost of inverter integration with 10MW of PV capacity by 2030 at a unit price of $800-1000/kW at $8-10 billion.

Table 7-2
Cost of PV Inverters

Technology	Total Units	Units	% Sat	Cost/ Unit Low $	Cost/ Unit High $	Total Cost Low–High $M
Integrated PV Inverter	10,000	kW of distributed PV	100	800	1000	800–1,000

Residential Energy Management System (EMS)

A residential EMS is a system dedicated (at least in part) to managing systems such as building components or products and devices. Residential EMS systems are not typically called "portals" in today's parlance. Portal is a term commonly applied to a web portal. This subdivides into several components including resident EMS and intelligent home devices (IHD). In addition the system may handle customer preferences and occupancy via a schedule, on-demand, or occupancy sensing automation. The line between a residential management system that handles lighting, family calendars, shopping or replenishment, and an EMS has become fuzzy. While proponents of a dedicated device propose that a homeowner will eventually purchase such a device, we are seeing parallel development of other approaches where the core of the system is a software application bundled on a server located at a third-party data center.

Online energy management portals offer customers insight into their energy usage and automatic management of energy efficiency. Through a central view on a web page, for example, customers can access current energy usage statistics, historical usage patterns, and the amount of carbon dioxide emissions avoided by utilizing a renewable energy source. The portal can also display price signals and tie a customer's energy consumption and production patterns into their utility's rate schedule. Current standards developments may also enable effective aggregation and third party information sharing that will impact the adoption of a residential EMS. As of the writing of this document, preliminary information regarding consumer purchases of advanced residential EMS systems show promise but adoption to date has been low.

These aspects make it difficult to pin the price tag onto the residential EMS. Many components have a dual purpose and exist under separate financial justifications. Consumer reluctance to purchase an EMS may be driven by on-line options that could replace key parts of the functionality of an EMS. These issues could either imply that the cost per customer is low, or the penetration rate is low. However, the end result should be similar regardless of which way we apply this observation.

Based on the plurality of the residential EMS architecture paths, the study team held the residential EMS estimate to 10% of the customer base by 2030 at an average cost $150 to $300, yielding a total cost of $2.2 to $4.3 billion.

Table 7-3
Cost of EMS Portals

Technology	Total Units	Units	% Sat	Cost/ Unit Low $	Cost/ Unit High $	Total Cost Low–High $M
Customer EMS Portal	143,928,676	Number of	10	150	300	2,159–4,318

In-Home Displays and Access to Energy Information

Providing real-time feedback on energy consumption holds significant promise to reduce electricity demand. Several studies over the past 30 years have evaluated the effectiveness of energy savings from home energy displays of varying sophistication. Most of these studies verified savings between 5% and 15% with a longer-term sustained impact toward the lower end of this scale. Other studies have found that information alone does not appear to be sufficient to achieve appreciable reductions. People need a strong motivation to change, such as compensation, confidence they can change, and feedback that changes they do make are having an impact. In addition this feedback must be easy and trustworthy. As such, most successful approaches provide more frequent feedback, as well as feedback on specific behaviors.

As the Smart Grid unfolds, various methods to provide energy, cost, and environmental information are beginning to emerge. A specific class of stand-alone devices has been utilized extensively and is referred to as the in-home display (IHD). Typically, IHDs present basic information, such as real-time and projected hourly electricity cost and electricity consumption (kWh). Some can display additional information, such as electricity cost and consumption over the last 24 hours, the current month (and/or prior month) consumption and cost, projected usage, monthly peak demand, greenhouse gas emissions, and outdoor temperature. A similar approach is a component of a prepayment system, also known as a pay-as-you-go system, since these also have a display. The very nature of the pay-as-you-go billing encourages consumers to keep an eye on the display to monitor their usage and know when they will need to replenish their energy account.

In contrast, there are simpler approaches that may not be quantitative. That is, they do not include feedback on electricity consumption or electric demand. Such devices do not require direct attention, but effectively communicate information peripherally. For example, a small glowing ball has been used to indicate a higher electric price or energy demand by changing colors. Other implementations have developed a simple plug-in device with red, green, and yellow lights as simple indicators of energy demand or price.

Small-scale demonstrations have utilized more sophisticated home energy displays. Typically, they provide much greater detail about electricity consumption broken down by different end uses, or circuits, and use richer display graphics. In most cases, these advanced displays are part of a more comprehensive system that may have many features beyond energy management. This makes allocation of the Smart Grid component cost more difficult to pin down since energy management may not be the driving force behind the decision to interact with the device.

Alternative methods to provide energy information to the consumer continue to emerge. As the standards development processes move forward, additional innovations in this area will continue to become available. Some of these products offer an alternative to the dedicated IHD device or at least a subset of functionality. Although the consumer must have a physical means to view the information, the means may already exist in some form such as the PC/laptop, cell phone and PDA. Additional developments might use any products with a consumer facing display as a location to display energy information. This includes appliances, security systems and any new consumer product categories that may be on the drawing board.

Standards resulting from the NIST PAP 10 work may allow product manufacturers the option to include energy information on multi-purpose devices and "other-purposed" devices. This will tend to eliminate or hide the cost making it difficult to identify the cost of access to the energy consumer energy information. This could be accommodated in the cost estimates by indicating that the cost per customer becomes lower over time. This could also be accommodated by indicating that the penetration rate of the stand-alone single-purpose IHD will not ramp up over time due to the alternative methods of information access. With this understanding in mind, the study team estimated 20% of utility residential customers would have an in-home display by 2030. The average cost was estimated at $20 to $50 per unit leaving the Total Smart Grid cost estimate at $1.4 to 2.9 billion for this item.

Table 7-4
Cost of In-Home Displays

Technology	Total Units	Units	% Sat	Cost/ Unit Low $	Cost/ Unit High $	Total Cost Low– High $M
In-Home Displays	143,928,676	28,785,735	20	20	50	575-1,439

Grid-Ready Appliances and Devices

Grid-ready appliances do not require truck rolls to retrofit with remote communications and control capabilities. Grid-ready appliances and devices, which are often referred to as "DR-ready," are manufactured with demand-response (DR) capabilities already built in. The universal entry of grid-ready devices into the marketplace, which is fully anticipated to take shape in the next several years, will lead to ubiquitous demand-response capability.

The average American home has 4.67 appliances per home, with the refrigerator being the most universal (99.8% of U.S. households have a refrigerator based on U.S. Census data). The number of households in the U.S. is projected to reach 143,928,676 by 2030. The study team assumed that the first grid-ready appliances will start to appear in 2011. The penetration of DR-ready appliances is expected to approach 40% over the next 20 years. To account for homes with electric water heating and air conditioning "appliances," the team increased the average to 5.67 appliances per home. The reader could argue that many water heaters are electric and not every home has AC. There is a balancing argument that the number of appliances per home may also grow, leaving the team comfortable with the 5.67 appliances per home as being a reasonable estimated average.

By using data from AHAM (Association of Home Appliance Manufacturers) for the life expectancy of major appliances, the team used the average life expectancy of 13.91 years for each appliance. By dividing this into the 20-year span of the study, the average appliance would be replaced 1.44 times during this time.

Sales figures indicate that a single year of appliances sales has trended toward 10% of the current installed base. This number includes both new construction and replacement sales. Since we are using the projected number of households for the year 2030 which includes new construction, the replacement from appliance life-expectancy was used against the year 2030 households projection to avoid double counting. The additional cost to incorporate grid-ready functionality into future appliances is estimated at $10 to $20 per unit for the first generation, but declining to zero within 10 years as the grid-ready design becomes standard. Rather than accounting for engineering cost separately, this was included with the component costs per appliance. The final number should indicate the cost to the consumer which includes other costs calculated from the materials and production costs.

The team estimated that in the year 2011, the total consumer cost per appliance would be $40. This appears as higher costs during the first several years. Both the engineering cost and the component cost are assumed to reduce over time as is the norm in a product development cycle. In the future, the grid-ready design is expected to become part of a standard appliance design. Furthermore, the team assumed that the appliances will have communication technology built in that is justified for other non-energy usages and additional consumer benefits. This should make this cost become negligible after 10 years.

Penetration by 2030 may be limited by consumer model selection and the fact that certain appliance products (such as cooking and refrigeration) may not be nearly as appropriate as others for grid messaging or demand management. Therefore, the study shows the penetration starting to level off as it approaches 40%. The assumptions used by the study team are summarized in Table 7-5, and indicate the expected non-linear penetration and costs as smart-grid appliances enter the market.

Table 7-5
Assumptions of Grid-Ready Appliance Costs

Grid-Ready Appliance Costs	Value	Notations
Total U.S. Households	143,928,676	Includes projected growth to 2030 (1)
Total appliances per household	4.67	Including electric HW and AC (2)
Penetration by 2030	38%	See chart for ramp up (3)
Appliance Life Expectancy	13.91 yrs	Averaged from cooking, cleaning, food preservation (4)
Rate of replacements	1.44	In 20 years each appliance will be purchased 1.44 times (5)
Estimated Total Appliances Purchases	1,173,442,029	58,672,101 per year average (6)
Total Cost	$230,531,663-$412,354,482	Assumes penetration ramp up starting between 0.2% and .5% in 2011 (7)

(1) U.S. Census information
(2) Extrapolated from U.S. Census information
(3) Penetration estimated. Note that Smart Grid appliances and other products will ramp up. Estimated from averaging a number of informal sources (that may be changing daily).
(4) By using data from AHAM (Association of Home Appliance Manufacturers) for the life expectancy of major appliances, the team used the average life expectancy of 13.91 years for each appliance.
(5) Simple application of the average life expectancy of 13.91 years for each appliance and dividing this into the 20-year span of the study, the average appliance would be replaced 1.44 times during this time.
(6) Simply the application of the other numbers and averages to determine the number of appliances purchased (conversation with AHAM staff).
(7) The ramp-up rate of products containing Smart Grid enablements. As in (3) above, the ramp rate is arguable and could shift widely over the time period estimated. Currently, the ramp-up happens at an initial pace that is defensible at this point in time. Most manufacturers are not willing to share a lot of sales and projected sales of new and unannounced product models.

Variables used to select the cost range are largely dependant on two factors. The first is the cost of the engineering and components added to the appliances. The second significant factor is the rate of market penetration of the Smart-Grid enabled appliances. Since the component cost is projected as being higher in the early years, as diagrammed in Figure 7-1, faster deployment can push the total cost of Smart Grid appliances upward. However, an earlier drop in components cost due to higher volume might counteract this to some degree bringing the cost of common appliance Smart Grid components down earlier in the cycle. The range of cost of grid-read appliances was estimated between $230 to $412 million.

Table 7-6
Cost of Grid-Ready Appliances

Technology	Total Units	Units	% Sat	Cost/ Unit Low $	Cost/ Unit High $	Total Cost Low– High $M
Grid-Ready Appliances	143,928,676	Number of	38%	230	412	230-412

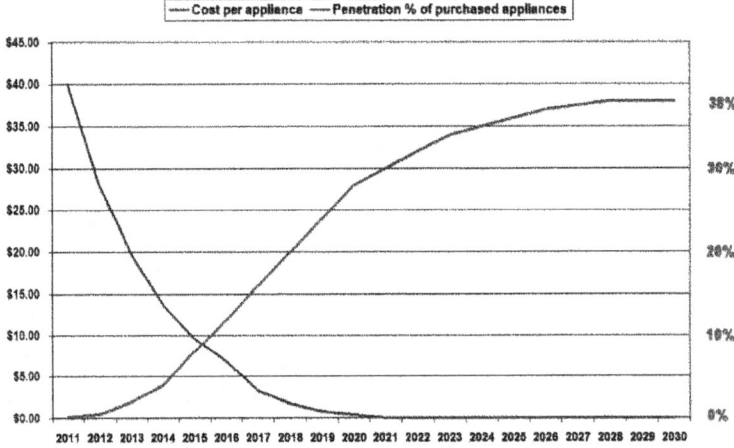

Figure 7-1
Appliance Cost and Penetration

Plug-in Electric Vehicle Charging Infrastructure and On-Vehicle Smart Grid Communications Technologies

Plug-in electric vehicles (PEVs) are defined as any hybrid vehicle with the ability to recharge its batteries from the grid, providing some or all of its driving through electric-only means. Almost all of the major automotive manufacturers have announced demonstration or production programs in the 2010-2014 timeframe, and their announced vehicles feature all-electric, plug-in hybrid-electric and extended-range electric vehicle configurations. Notable and earliest introductions among these are globally targeted production vehicles from General Motors (Chevrolet Volt, extended-range EV) and Nissan (Leaf, battery-only EV). It should also be noted that Tesla remains the only manufacturer today producing a two-seater roadster and continues to work on their lower-priced, family (Model S) sedan.

EPRI is deeply involved in developing and demonstrating on- and off-vehicle PEV to Smart Grid connectivity technologies. Automotive manufacturers are bringing their early generations of PEVs designed for grid rechargability with uni-directional electric power flow capability (from grid to vehicle). These first generation of vehicles will be relying on the consumer inputs via in-vehicle interactive touch panel display or through cellular/Telematics connectivity, to program vehicle charging, input pricing or other information related to on-board energy management. Given this state of technology on board the PEVs, the initial impetus for any type of demand response and load control as well as critical peak pricing program implementation has been on the off-board "charging" station equipment. The technical term is electric-vehicle supply equipment (or EVSE), given that the off-board equipment is just a glorified 240 Volt outlet while the actual AC/DC power conversion and energy management actually occurs on-board the PEVs. The most predominant costs associated with the Smart Grid infrastructure capability will, therefore, be in developing Smart Grid to PEV and Smart Grid to "charging" station (aka, electric vehicle supply equipment or EVSE) connectivity and communications.

All applications which may provide reverse flow power capability such as vehicle-to-grid (V2G) are unproven. Their impact on battery durability, utility/automotive/consumer acceptance, and economics are yet to be demonstrated. In addition, it is uncertain what services they will enable and whether policies and incentives will be needed to bring them to reality. V2G, therefore, remains an R&D agenda item of several automotive manufacturers, EPRI, some ISOs/RTOs, and several R&D institutions. It is, however, too early to predict the direction and magnitude of this technology's installed base in the near future.

In terms of load management or time shifting of the load due to PEV charging, both utilities and automotive manufacturers have agreed to jointly pursue standardization activity that will enable the PEVs to act as just another appliance on the AMI or HAN. Two activities within SAE, under J2836 (use cases) and J2847 (data specification) with extensive automotive and utility participation, are in the process of defining the requirements for PEV to Smart Grid communications, which will enable the PEVs to be utility-controllable distributed resources for load shifting, demand response, and pricing-signaling purposes. ZigBee Alliance and HomePlug Alliance have created the Smart Energy Initiative, which is crafting the Smart Energy 2.0 (SE2.0) specification for AMI and HAN applicability. SAEJ2836/J2847 and SE2.0 teams are working together to coordinate the data exchange requirements between Smart Grid and PEVs. A draft Marketing Requirements Document (MRD) and Technical Requirements Document (TRD) are currently being refined for late 2010 ratification. SE2.0 and J2847 are expected to define a consistent set of data specifications for the charging load of the PEVs to be controllable by the same utility load management systems that determine demand response and load control signals for other loads, such as air conditioners, smart thermostats, or other smart appliances. The SAEJ2836 use cases include enabling PEV owners to enroll into utility demand response, load control and special incentive pricing programs and then program their vehicles to accept or reject utility requests for participating in demand response, load control and critical peak pricing-related events.

For the first generation of PEVs, the technology options for integrating PEVs with the Smart Grid will reside off-board, in the form of the closed, proprietary networks of charging station operators such as Coulomb Technologies, ECOtality, and Silver Spring Networks. Whereas significant public funding to the tune of $300M through stimulus awards from federal government and state and local authorities has been directed towards focused regional charging infrastructure build-out, the focus has been on enabling PEV technology adoption in early adopter markets, rather than on scalability and cost competitiveness of these technologies longer term.

EPRI's collaborative research with the automotive industry indicates that for PEVs to be widely deployed, the infrastructure overhead for them would need to be reduced to "minimal to none,", with each PEV carrying its own required technology on-board that can connect to the nearest Smart Grid node It would use either AMI/HAN to connect to the Smart Grid through the "front-end" or the on-board Telematics-based technology to connect through the "back end" to the utility back office systems, and to meter data management systems through standardized server- to-server communications.

To this end, EPRI envisions the PEV manufacturers to quickly integrate the standards-driven communication technologies on-board the PEV. The only significant costs for "Smart Grid-enabled" PEVs will, therefore, be the cost-plus for incorporating the communications hardware necessary to send/receive data from the utility based on applicable standards. The automotive and utility industries have agreed for PLC- (power line carrier-) based wired interface to be the physical interface between the PEV and the AMI/HAN, with the PLC(HomePlug AV or IEEEP1901 are the currently adopted technologies) transceiver chipset and associated Smart Grid communications "application layer" software with requirements defined by SAEJ2836/J2847 and SE2.0, residing on-board. That would include a PLCto X bridge residing off-board, with X being the transport layer of the AM I/HAN network, which also implements SE2.0-based messaging as the application layer.

The per-vehicle cost overhead for PLC transceiver is about $20 per vehicle in the near term, reducing to $10 per vehicle longer term, as PLC is already a very widely deployed technology. On the PLC to X bridge aspect, the X in most cases is ZigBee, but WiFi (802.11x-based) is also rapidly emerging as the HAN contender. The per-unit PLC/ZigBee or PLC/WiFi chipset prices vary between $10 and 20 per unit as well. Given that there are likely to be 1.2 charging stations long term for every PEV sold, the per-PEV PLC to X bridge costs will run to

$12 to $24. Therefore, the per-PEV infrastructure costs will run to between $25 and 50 for long-term and short-term volumes respectively. Assuming 2030 PEV installed base volume to be about 10 million vehicles, the cost of deploying Smart Grid infrastructure will approach $250 million ($25 per unit times 10 million vehicles) in 2030.

Table 7-7
Cost of Vehicle to Grid Converter

Technology	Total Units	Units	% Sat	Cost/ Unit Low $	Cost/ Unit High $	Total Cost Low– High $M
Vehicle to Grid Power Converter	30,000,000	Number of vehicles	50	300	500	4,500- 7,500

Communication Upgrades for Building Automation

Today, over one-third of the conditioned and institutional buildings in the U.S. have some form of energy management and control systems installed (EPRI 101883). Automated demand response (ADR) can be accomplished by communicating to advanced building energy management systems using an Internet-communicated signal or some other form of direct link. Legacy systems deployed today lack this capability. Open automated demand-response (Open-ADR) involves a machine-to-machine communication standard that provides electronic, Internet-based price and reliability signals linked directly to the end-use control systems or related building and automated control systems (EPRI 1016082). The building automation system is pre-programmed to reduce load according to the messages it receives, and it may also provide real-time energy consumption information back to the utility or service provider.

Employing Open-ADR presumes the building has an advanced EMS system. There are two cost components that enable the building to respond to DR signals. The first is enable the building's EMS to receive the DR signals. In some cases, this might mean upgrading the software, and in other cases, this might mean installing a "simple client" whose only purpose is to receive the DR signals and pass them on to the EMS system. One of the features of Open-ADR is to allow very simple and inexpensive clients to be built that can interface to existing EMS systems via dry-relay contacts. Dry relay contacts seem to be the near-universal interface mechanism for EMS systems.

The second and perhaps largest cost component is the programming of load control strategies in the EMS. The cost is primarily one of manpower that involves audits of loads in the facility and specialized knowledge of how to convert the EMS to implement load control strategies. Auditing building use and load characteristic is not a trivial exercise. In this regard, the simple response levels sent as part of an Open-ADR signal can be used. In many cases, it is more convenient for the facility manager to think in terms of "normal, moderate, and high" response levels instead of prices or specific dispatch commands. Also, it is not insignificant that if the engineers set up their load control strategies based upon simple levels, then they can more easily move between different programs without the need to reprogram their EMS system.

The study team estimated that by 2030 some 5% of the 20,178,151 commercial buildings would be upgraded to the level of complete energy automation at a cost of $5,000 to $20,000 per building. The total Smart Grid cost is estimated between $5–20 billion.

Table 7-8
Cost of Communication Upgrades for Building Automation

Technology	Total Units	Units	% Sat	Cost/ Unit Low $	Cost/ Unit High $	Total Cost Low– High $M
Communication Upgrades for Building Automation	20,178,151	Number of buildings	5	5,000	20,000	5,045-20,180

Electric Energy Storage

Advanced lead-acid batteries represent the most prevalent form of electric energy storage for residential, commercial and industrial customers wanting to maintain an uninterruptible power supply (UPS) system. In the future stationary lithium-ion batteries may also be deployed for use in consumer premises.

As shown in Table 7-9, commercial and industrial systems can supply power for up to 8 hours at 75% efficiency, and maintain performance through more than 5000 cycles. Residential versions typically involve two hour duration at 75% efficiency and 5000 cycle performance.

Table 7-9
Electric Energy Storage Options for Customers

Application	Technology option	Capacity (MWh)	Duration (hours)	Efficiency %	Total cycles	Cost $/kW
Residential	Advanced lead-acid	0.8	8	75%	5000	2300-2400
Commercial & Industrial	Advanced lead-acid	10	2	75%	5000	2200-2400

Both standby and online UPS technologies are available. The online UPS is ideal for environments where electrical isolation is necessary or for equipment that is very sensitive to power fluctuations. Although once previously reserved for very large installations of 10 kW or more, advances in technology have permitted it to now be available as a common consumer device, supplying 500 watts or less. The online UPS is generally more expensive but may be necessary when the power environment is "noisy" such as in industrial settings, or for larger equipment loads like data centers, or when operation from an extended-run backup generator is necessary.

In an online UPS, the batteries are always connected to the inverter, so that no power transfer switches are necessary. When power loss occurs, the rectifier simply drops out of the circuit and the batteries keep the power steady and unchanged. When power is restored, the rectifier resumes carrying most of the load and begins charging the batteries, though the charging current may be limited to prevent the high-power rectifier from overheating the batteries and boiling off the electrolyte.

The main advantage to the on-line UPS is its ability to provide an electrical firewall between the incoming utility power and sensitive electronic equipment. While the standby and Line-Interactive UPS merely filter the input utility power, the Double-Conversion UPS provides a layer of insulation from power quality problems. It allows control of output voltage and frequency regardless of input voltage and frequency.

The study team estimated that by 2030 roughly 1.8 GW of on-site back-up storage will be installed in commercial and industrial facilities at a unit cost of $2300 to 2400/kW. An additional 2.8 GW of battery storage for residential backup applications will be installed at an average unit cost of $2200 to 2400 kW.

Table 7-10
Cost of Electric Energy Storage

Technology	Total Units	Units	% Sat	Cost/ Unit Low $	Cost/ Unit High $	Total Cost Low-High $M
Integrated PV Inverter	10,000	kW of distributed PV	100	800	1000	8.0-10.0
Consumer Energy Management System	143,928,676	Number of	10	150	300	2,159–4,318
In Home Displays	143,928,676	Number of	20	50	100	1,439–2,878

Summary of Customer Costs

The cost to bring the customer interface of the electric infrastructure up to Smart Grid performance levels so that it can support a broad array of customer services—ranging from DR-ready appliances to V2G charging—is estimated at $24 to $44 billion, as shown in Table 7-11. This cost does not include the sizeable investment that will be made by customer in appliances, PHEVs, HVAC equipment, and the like.

Table 7-11
Smart Grid Costs for Customers

Technology	Total Units	Units	% Sat	Cost/ Unit Low $	Cost/ Unit High $	Total Cost Low– High $M
Integrated PV Inverter	10,000	kW of distributed PV	100	800	1000	8.0-10.0
Consumer Energy Management System	143,928,676	Number of	10	150	300	2,159– 4,318
In Home Displays	143,928,676	Number of	20	50	100	1,439– 2,878
Grid-Ready Appliances	143,928,676	Number of	40%	10	20	222–443
Vehicle to Grid Power Converter	30,000,000	Number of vehicles	50	300	500	4,500– 7,500
Communication Upgrades for Building Automation	20, 178,151	Number of buildings	5	5,000	20,000	5,045– 20,180
Industrial & Commercial Storage for Backup	1,800,000	kW	100	2,300	2,400	4,140– 4,534

255

Table 7-11 (continued)
Smart Grid Costs for Customers

Technology	Total Units	Units	% Sat	Cost/ Unit Low $	Cost/ Unit High $	Total Cost Low– High $M
Residential Storage for Backup	2,800,000	kW	100	2,200	2,400	6,160– 6,720
Ongoing System Maintenance						
Total Cost Customer						**23,672– 46,368**
Allocated to Existing Customers						**20,386– 39,932**
Allocated to New Customers						**3,286– 6,436**

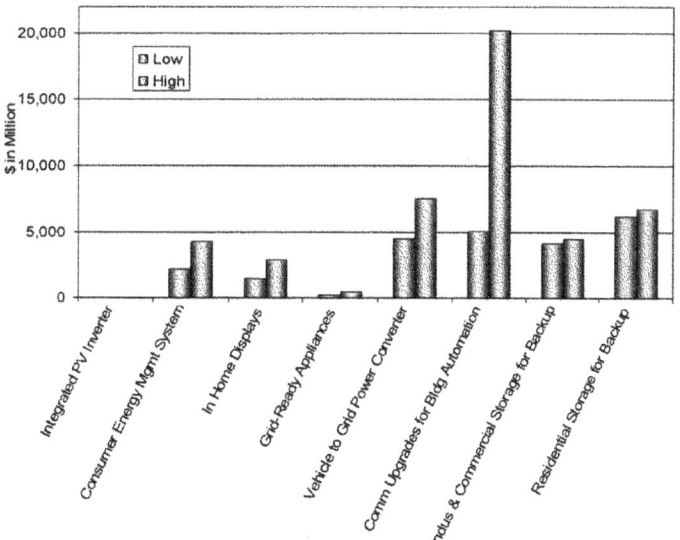

Figure 7-2
Consumer Costs for a Smart Grid

256

Section 8: References

"Power Delivery System of the Future: A Preliminary Estimate of Costs and Benefits," EPRI, Palo Alto, CA: 2004. 1011001.

"Characterizing and Quantifying the Societal Benefits Attributable to Smart Metering Investments," EPRI, Palo Alto, CA: 2008. 1017006.

"The Green Grid: Energy Savings and Carbon Emissions Reductions Enabled by a Smart Grid," EPRI, Palo Alto, CA: 2008. 1016905.

"Transforming America's Power Industry: The Investment Challenge 2010-2030," prepared by The Brattle Group for The Edison Foundation, November 2008.

"Methodological Approach for Estimating the Benefits and Costs of Smart Grid Demonstration Projects," EPRI, Palo Alto, CA: January 2010. 1020342.

"Power Delivery System of the Future: A Preliminary Estimate of Costs and Benefits," EPRI, Palo Alto, CA: 2004. 1011001.

"Transforming America's Power Industry: The Investment Challenge 2010-2030," prepared by The Brattle Group for The Edison Foundation: November 2008.

"Energy Storage Market Opportunities: Application Value Analysis and Technology Gap Assessment," EPRI, Palo Alto, CA: 2009. 1017813.

"Sensor Technologies for a Smart Transmission System," EPRI, Palo Alto, CA: 2009. 1020619.

"Technical Architecture for Transmission Operations and Protection: Envisioning the Transmission Smart Grid," EPRI, Palo Alto, CA: 2008. 1016055.

"Assessment of Commercial Building Automation and Energy Management Systems for Demand Response Applications," EPRI, Palo Alto, CA: 2009. 1017883.

"Automated Demand Response Tests: An Open ADR Demonstration Project," EPRI, Palo Alto, CA: 2008. 1016082.

"Energy Storage Market Opportunities: Application Value Analysis and Technology Gap Assessment," D. Rastler, EPRI Technical Update 1017813, December 2009.

"Development of a 15 kV Class Solid-State Current Limiter," EPRI, Palo Alto, CA: 2009, 066335.

"Assessment of National EHV Transmission Grid overlay Proposals: Cost-Benefit Methodologies and Claims," Christensen Associates Energy Consulting, LLC, February 22, 2010.

"FirstEnergy Smart Grid Modernization Initiative," their report to the U.S. Department of Energy, 2009.

"National Electric Transmission Congestion Report," U.S. Department of Energy (US DOE), October 2007.

"The ElectriNet," C. Gellings and P. Zhang, *ELECTRA*, CIGRÉ, June 2010.

"Energy Independence and Security Act of 2007," http://www.eia.doe.gov/oiaf/aeo/otheranalysis/aeo_2008analysispapers/eisa.html.

"Energy Outlook 2010," Energy Information Agency, www.eia.doe.gov, December 2009.

"Final Report on the August 14, 2003 Blackout in the United States and Canada: Causes and Recommendations," North American Electric Reliability Corporation for the U.S.-Canada Power System Outage Task Force, April 2004.

"The Smart Grid: An Estimation of the Energy and CO_2 Benefits, Rev. 1," Pacific Northwest National Laboratory (PNNL) for the U.S. Department of Energy, January 2010.

"The Aging Workforce: Electricity Industry Challenges and Solutions," L.B. Lave, M. Ashworth and C.W. Gellings, *Electricity Journal*, January 2007.

"8th FACTS Users Group Meeting: Presented Material at the 8th FACTS Users Group Meeting," EPRI, Palo Alto, CA: 2005. 1010633.

"The Power to Reduce CO_2 Emissions: The Full Portfolio: 2009 Technical Report," EPRI, Palo Alto, CA: 2009. 1020389.

"Assessment of Achievable Potential from Energy Efficiency and Demand Response Programs in the U.S.: (2010-2030)," EPRI, Palo Alto, CA: 2009. 1016987.

EPRI's IntelliGrid[SM] Initiative, http://intelligrid.epri.com.

GridWise Architecture Council, http://www.gridwiseeac.org.

"The Modern Grid Initiative Version 2.0," conducted by the National Energy Technology Laboratory for the U.S. Department of Energy Office of Electricity Delivery and Energy Reliability, January 2007, http:www.netl.doe.gov/moderngrid/resources.html.

"Report to NIST on the Smart Grid Interoperability Standards Roadmap," (Contract No. SB1341-09-CN-0031), D. Von Dollen, EPRI, June 17, 2009.

LogRhythm and NERC CIP Compliance, www.logrhythm.com, 2009.

"The Cost of Power Disturbances to Industrial and Digital Economy Companies," EPRI, Palo Alto, CA: 2009. 1006274.

"A National Assessment of Demand Response Potential," Federal Energy Regulatory Commission, June 2009.

"National Action Plan on Demand Response," Federal Energy Regulatory Commission, June 2010.

"Smart Grid Strategy and Roadmap," Southern California Edison, 2010.

"Electric Energy Storage Technology Options: A Primer on Applications, Costs and Benefits," EPRI, Palo Alto, CA: 2010, 1020676.

"Methodological Approach for Estimating the Benefits and Costs of Smart Grid Demonstration Projects," EPRI, Palo Alto, CA: 2010. 1020342.

"Sizing Up the Smart Grid," A. Faruqui, R. Heldik and C. Davis, Elster Energy Axis User Conference, February 24, 2009.

"Characterizing and Quantifying the Societal Benefits Attributable to Smart Metering Investments," EPRI, Palo Alto, CA: 2008. 1017006.

"The U.S. Smart Grid Revolution: KEMA's Perspectives for Job Creation," GridWise Alliance, January 13, 2008.

"Smart Grid: Enabling the 21st Century Economy," S. Pullins, Governor's Energy Summit, West Virginia, 2008.

"Estimating the Benefits of the GridWise Initiative," Phase I Report, Pacific Northwest National Laboratory, TR-160-PNNL, Rand Corporation, 2004.

"GridWise: The Benefits of a Transformed Energy System," Pacific Northwest National Laboratory, PNNL-14396, September 2003.

"Energy Storage for the Electricity Grid: Benefits and Market Potential Assessment Guide," prepared by Sandia National Laboratories for the U.S. Department of Energy, SAND2010-0815, February 2010.

"Energy Storage Market Opportunities: Application Value Analysis and Technology Gap Assessment," EPRI, Palo Alto, CA: 2009. 1017813.

"TAG™ Technical Assessment Guide, Vol. 4: Fundamentals and Methods, End Use," C.W. Gellings and P. Hanser, EPRI, Palo Alto, CA: 1987. P-4463-SR.

"The Green Grid: Energy Savings and Carbon Emissions Reductions Enabled by a Smart Grid," EPRI, Palo Alto, CA: 2008. 1016905.

"The Smart Grid: An Estimation of the Energy and CO2 Benefits," Pacific Northwest National Laboratory for the U.S. Department of Energy, PNNL-19112, January 2010.

"Smart Meters and Power Planning," C. King, Utility Automation and Engineering T&D, October 2006.

"Grid Supporting Technologies," J. Wellinghoff presentation to National Electrical Manufacturers Association (NEMA) Energy Storage Council Technical Committee Meeting, October 6, 2008.

"Illinois Statewide Smart Grid Collaborative: Collaborative Report," Illinois Statewide Smart Grid Collaboration (ISSGC), Illinois Commerce Commission, September 30, 2010.

"The Modern Grid Initiative Version 2.0," conducted by the National Energy Technology Laboratory (NETL) for the U.S. Department of Energy Office of Electricity Delivery and Energy Reliability, January 2007, http://www.netl.doe.gov/moderngrid/resources.html.

"Smart Grid VC Investment Surge," greentechgrid, January 21, 2010, www.greentechmedia.com.

"A Framework for Assessing the Net Benefits of Home Area Networks to Enable Demand Response," S. Mullen and B. Neenan, Electric Power Research Institute (EPRI), Working Paper, 2010.

"An Updated Annual Energy Outlook (AEO) 2009 Reference Case Reflecting Provisions of the American Recovery and Reinvestment Act and Recent Changes in the Economic Outlook," EIA, 2009.

"2010 UDI Directory of Electric Power Producers and Distributors," 118th Edition of the Electrical World Director, E. Giles & K.L. Brown, Platts Division of the McGraw-Hill Companies, Inc., New York, NY, 2009.

Appendix A: Notes Pertaining to Table 4-5: List of Smart Grid Benefits

Facilitating Plug-In Electric Vehicles (PEVs)

Hi Band Estimate – EPRI's Prism analysis estimates a potential CO_2 emissions reduction in 2030 of 9.3% as a result of electricity displacing gasoline and diesel to fuel a substantial portion of the vehicle fleet. EPRI bases this estimate on the assumption that plug-in electric vehicles (PEVs) are introduced to the market in 2010, consistent with product plans of many automakers, and the rapid growth of market share to almost half of new vehicle sales within 15 years. Net emissions reduction estimates from the increasing market share of PEVs are based on research by EPRI and others (EPRI/NRCD, 2007), factoring vehicle miles traveled, carbon savings from gasoline not burned, and the trend for the electric system to become "cleaner" – i.e., for an increasing share of power generation to emit less or no CO_2.

EPRI Prism analysis assumptions:

- 100 million PEVs in the fleet by 2030; and

- Fraction of non-road transportation applications (e.g., forklifts) represents three times the current share by 2030.

PEV Low Band – The PEV low band used the results of the EPRI-NRDC study from 2007 and, somewhat arbitrarily, attributed up to 20% of the carbon savings to the presence of a Smart Grid. The reasoning was that a Smart Grid is an enabling factor, but not the sole determining factor, in the market growth of PEVs. PEV-to-Smart Grid interface-related incremental costs, which run about $25 to $50 per vehicle (on- and off-board, $50 short term, $25 long term, and only include the PLC and PLC/X interface chipset BOM costs). So the incremental cost estimate is $250 million.

Facilitating Electrotechnologies

The 2009 analysis estimates a potential CO_2 emissions reduction in 2030 of 6.5% as a result of electric technologies displacing traditional use of primary energy consumption for certain commercial and industrial applications. Electrotechnology research (EPRI/ELEC, 2009) indicates that there are applications through which net reductions in CO_2 emissions can be achieved.

This projection is based on replacing significant use of direct fossil-fueled primary energy with relatively de-carbonized electricity for a range of possible applications, e.g., heat pumps, water heaters, ovens, induction melting, and arc furnaces. It is assumed that 25% of these electro technologies are facilitated by the Smart Grid. A total of 4.5% or primary energy supplied by fossil fuels is replaced by electricity by 2030.

Facilitating Renewable Energy Resources

Hi Band Estimate – EPRI's 2009 analysis estimates a potential CO_2 emissions reduction in 2030 of 13% as a result of substantially increased deployment of renewable generation facilitated by the Smart Grid. This assumes the penetration of diverse renewable generation resources based on consideration of existing and potential state and federal programs, cost and performance improvements, and grid integration challenges. This assumption corresponds to 135 gigawatts (GW) by 2030 consisting of ~100 GW new wind; ~20 GW new biomass; and ~15 GW other technologies including solar. The average new generation over 20 years will be equal to 67.5 GW corresponding to a reduction in 3.41 Billion tons of CO_2 at $50 per ton or $172 billion.

Low Band Estimate – The renewables low band estimate was based on 100 additional GW of renewable capacity. Of that, 50 GW was assumed to be wind power. Assuming a 61% load factor, 267 billion kWh of additional energy would be by wind. The study attributed 50% of the realization of this energy from wind to the resolution of the intermittency challenge of wind, and then further, attributed up to 50% of the credit for resolving the intermittency challenge to the presence of a Smart Grid. The rationale used was that Smart Grid is not the sole criterion for such large-scale wind integration, but it is a critical component. The study then applied the estimated CO2 intensity of generation in 2030 to get the 37 MMtons figure.

Table A-1
Environment Benefits From Renewables, PEVs and Electrotechnologies: High Band Estimates

| | Technology | % Savings | 20-Year Savings (@ ramp 0 to 100%) | | |
			Total	%Attribute to Smart Grid	$ @ $50/Ton
Gross CO_2 in 2030	Additional Renewables	13	3.431E+09	100	1.72E+11
263,900,000 Million	PEV	9.3	2.454E+09	100	1.23E+11
Metric Tons	Electrotech	6.5	1.715E+09	25	2.14E+10

Table A-2
Low-Band Estimates of PEVs and Renewables (EPRI)

	Million Tons Co₂ (2030)		Total (over 20 years)		$B @ $50/Ton	
	Low	**High**	**Low**	**High**	**Low**	**High**
Renewables	19	37	190	270	9.5	18.5
PEVs	10	60	100	600	5.0	30.0

Table A-3
Value of PEVs: High Band

			$B Low	$B High
Value of PEVs as a grid support technology*	50% of 30 million vehicles in 20 years	$1,500 per vehicle	11.3	11.3

*Wellinghoff, 2008

Expanded Energy Efficiency

EPRI provides estimates of benefits of expanded energy efficiency not included in its 2004 report in a subsequent study (EPRI 1016905). This is shown in Table A-4.

Table A-4
Value of Expanded Energy Efficiency

Type	Billion kWh (2030)		Value $ (@ 7¢/kWh)		Co2 Reduction Mill. Metric Ton		Value @ $50/Ton	
	Low	**High**	**Low**	**High**	**Low**	**High**	**Low**	**High**
Continuous commissioning	2	9	140M	630M	1	5	50M	250M
Energy efficiency benefits from demand response	0	4	0	280M	0	2	0	100M
Feedback	40	121	280M	847M	22	68	1100M	3400M
Total	42	134	420M	1757M	23	75	1150M	3750M

Related T&D capital savings can be calculated using the following assumptions

- 25% Load Factor and T&D Savings $ 800/kW = 2030 savings range from $ 1B to $ 3B

AMI Benefits

Table A-5
Edison SmartConnect™ Cost Benefit Information and U.S. Estimate

Benefits	Amount ($M)
Meter Services	$3,909
Billing Operations	187
Call Center	96
Transmission & Distribution Operations	92
Demand Response – Price Response	1,044
Demand Response – Load Control	1,242
Conservation Effect	828
Other	39
Total Benefits	**$7,437**

Southern California Edison (SCE) filings to the California Public Utilities Commission (CPUC) Proceedings: D.08-09-039, A.08-06-001; A.08-07-021 estimated benefits for several AMI attributes over a 20-year period including: Meter Services = $3,909 million; Billing Operations = $1,187 million; and Call Center = $96 million. Estimate made using SCE estimates of $4,874,890 customers and total U.S. estimated customers in 2030 of 142,121,652.

Table A-6
Southern California Edison Company Estimates of AMI Attributes

	$	$/Meter	Potential Benefit	Estimated Benefit
Meter services	3,909,000,000	801.8642	1.13962E+11	91,169,817,193
Billing operations	187,000,000	38.35984	5,451,763,819	4,361,411,055
Call center	96,000,000	19.69275	2,798,766,453	2,239,013,162
Total SCE meters	4,874,890			
Total U.S. meters	142,121,652			

Avoided Generation Investment from EE and DR

The Brattle Group estimates (Brattle, 2008) for the period 2010 to 2030 of avoided generation cost investment due to energy efficiency and demand response to be between $129 billion and $242 billion.

Table A-7
Avoided Generation Investment from Energy Efficiency and Demand Response (Brattle, 2008)

	Low	High
(Avoided) generation investment due to EE/DR	(192)	(242)

Energy Storage Benefits

Table A-8
Storage Benefits by Attribute (20 years) (EPRI 1017813 and Sandia, 2010)

Type	$Million	
	Low	High
Improved Asset Utilization		
Electric Energy Time Shift	4,936	7,367
Electric Supply Capacity	3,239	8,908
Load Following	16,354	32,561
Area Regulation	1,236	1,519
Electric Supply Reserve Capacity	1,915	2,634
Voltage Support	497	1,326
Transmission Support	221	937
Transmission Congestion Relief	19,745	33,743
Total	**48,142**	**88,995**
T&D Capital Savings		
T&D Upgrade Deferred	8,257	21,421
Renewables Capacity Firming	6,483	17,828
Wind Integration – short	958	2,865
Wind Integration – long	7.662	22,911
Total	**23,360**	**65,024**

Table A-8 (continued)
Storage Benefits by Attribute (20 years) (EPRI 1017813 and Sandia, 2010)

Type	$Million	
	Low	High
Electricity Cost Savings		
TOU Energy Cost Management	96,855	139,502
Demand Charge Management	18,245	58,976
Total	**115,100**	**198,478**
Reliability		
Substation On-Site Power	55	791
Reliability	1,731	19,774
Power Quality	700	21,026
Total	**2,485**	**41,591**
Environmental		
Renewables Integration	9,871	14,733
Total	**9,871**	**14,733**
Total All Storage	**198,959**	**408,821**

Table A-9
Distributed Generation Transmission Capacity Assumptions

Distributed Generation	Core Value
Capacity per participating customer (kW)	3.0
Grid connected PV systems	70000
Penetration growth rate	10.0%
Capacity factor	15.0%
Capacity reduction per kW – Transmission	0.45 kW
Total reduction 2010 – 2030	**$27 Billion**

Electrification Energy Benefits

Table A-10
Reduced Net Energy Required by Electrification (EPRI 1014044 and 1018871)

	High Case	Low Case
2030 decrease in quadrillion BTUs	5.32	1.71
2010-2030 decrease in quadrillion BTUs	53.2	17.1
Value @ $6.000 per million BTUs	$319.2M	$102.6M

Table A-11
Electric Sector Carbon Dioxide Emissions (AEO, 2009 Updated)

	Million Metric Tons is 2030
Petroleum	41
Natural gas	365
Coal	2222
Other*	12
Total	**2639**

*Includes emissions from geothermal power
and non-biogenic emissions from municipal waste.

267

The Electric Power Research Institute Inc., (EPRI, www.epri.com) conducts research and development relating to the generation, delivery and use of electricity for the benefit of the public. An independent, nonprofit organization, EPRI brings together its scientists and engineers as well as experts from academia and industry to help address challenges in electricity, including reliability, efficiency, health, safety and the environment. EPRI also provides technology, policy and economic analyses to drive long-range research and development planning, and supports research in emerging technologies. EPRI's members represent more than 90 percent of the electricity generated and delivered in the United States, and international participation extends to 40 countries. EPRI's principal offices and laboratories are located in Palo Alto, Calif.; Charlotte, N.C.; Knoxville, Tenn.; and Lenox, Mass.

Together...Shaping the Future of Electricity

Program:
Smart Grid Demonstrations

1022519

Electric Power Research Institute
3420 Hillview Avenue, Palo Alto, California 94304-1338 • PO Box 10412, Palo Alto, California 94303-0813 USA
800.313.3774 • 650.855.2121 • askepri@epri.com • www.epri.com

Dr. HOWARD. It outlines that the costs over a period of 20 years is at the top end of about $500 billion.

Now capacity, a lot of times we mistake capacity as being generation capacity. But we also have to think about it in terms of capacity on transmission and distribution as well. And so, you know, is the capacity available in the distribution system to provide the energy?

The study that we did in 2011 looks at all of those costs and we estimate that, at that point, about $500 billion is the top end.

The CHAIRMAN. Is there agreement on the panel that this is the number that we're looking at as effectively, $500 billion, top end?

Ms. BARTON. I think generally speaking, yes. One of the key things I think that's important to remember is that if you have market-based solutions you can make sure that these technologies, using your words, Senator Murkowski, pass the do no harm test, but I would add, at a reasonable cost to consumers. And what that cost to consumers is with respect to these integrated technologies really varies in terms of what is that technology is thought to be integrated in order the system needs. If it's a market-based solution then consumers will really choose what's important to them.

We have used, for example, in Presidio, Texas, which is on the border of Mexico. And it would have cost a lot of money for a new transmission line to be put in place there. We basically have two, 2.4 megawatt NaS batteries in place to basically serve to enhance the transmission grid there. So that's just an example of how technology can be used to meet a specific need. And I think it's very important that it be solution-based.

The CHAIRMAN. Senator Cantwell, I have been asking a series of questions and have gone well over my allotted five minutes. But if you're ready to proceed?

Senator CANTWELL. Yes.

The CHAIRMAN. I will turn to you. Thank you.

Senator CANTWELL. Thank you. Thank you, Madam Chair.

I don't know all the questions you asked, but we were discussing earlier about, obviously, how do we make investment and get recovery on these issues. Ms. Barton, your company operates in 13 states and I understand that in Ohio the federal government funded about half of your 150 million smart grid demonstration projects.

Ms. Edgar, your association represents regulators in every state. In your own state, Florida Power and Light's smart meter installation program was funded in part through $200 million in the Recovery Act.

What examples can each of you give of investments and the cost of recovery mechanism?

Ms. BARTON. Sure, I'll start. As you mentioned, we have transmission facilities in 13 states. We have regulated utility operations in 11 states and by virtue of that territory it gives us a unique perspective and information from which to draw from.

So in Indiana, for example, we have launched a solar project, a pilot program, where we're looking to basically have about 16 megawatts worth of solar capacity integrated into the system. That's been very successful.

I just recently talked about a project in Presidio, Texas, and we also had one in West Virginia which uses NAS batteries which is Sodium Sulfur batteries.

These technologies, while they are expensive, in unique grid situations can really amount to be the best solution. We've had tremendous success in a number of our states with bolt bar technology. And what bolt bar technology is is basically controlling through a series of complex communication the voltage on the system through regulators and capacitors. What that does is basically reduce the level of voltage that folks receive in their home and it results in significant energy savings.

So in the jurisdictions that our states have been supportive of that technology, we have gone forward with that technology. We try to stay very close with our states and our regulators to make sure that we're implementing the types of technologies that meet the state's needs.

Ms. EDGAR. Thank you, Senator, for the question. Much of what we've been discussing today is the recognition that the grid is a technology integration network and that distributed generation, although offering many advantages, is often intermittent and that it does require investment in both new transmission and distribution and in modernizing the system, both transmission and distribution that is already there.

So the role of the federal government in supporting research and in bringing that research in new technologies to scale, cost effectively, so that states and communities and consumers can take advantage of those resources is key, I believe, in helping us move forward.

Senator CANTWELL [presiding]. So how well established do you think we are in those kinds of recovery ideas?

Ms. EDGAR. You mean cost recovery?

Senator CANTWELL. Yes.

Ms. EDGAR. The rate making process at the state level is tried and true. The cost causer pays looking at actual cost before putting them into the rate base and then looking at projected cost. The challenge now moving forward is how to continue to attract capital investment so that consumers are getting a good value from the investment that they are making and also trying to look at alternative rate making mechanisms to take advantage of the cost efficiencies that are out there in the future that may not be here right now. And that is something that we are grappling with.

Senator CANTWELL. I remember probably 20 years ago when some of the first ideas were being surfaced on metering and things of that nature and they were turned down by UTC. But now, how many states have policies around this?

Ms. EDGAR. Probably most. I don't have an exact number, but I would say more than half and probably beyond that.

You mentioned a project that I mentioned in my testimony, and yes, that project was partially paid by federal dollars. I do believe it's an important investment in my state to helping us move the grid forward.

But of course, we're still learning how to take advantage of that technology so that consumers really benefit and then all of the other issues with data collection, data management, privacy and security.

Senator CANTWELL. Dr. Taft, could you talk about the Recovery Act funds that were part of the $4.5 billion of money that was spent and what lessons we've learned?

Dr. TAFT. I can talk about some of that. I don't have all the details with me, but a lot of those projects were to improve the basic ability to measure on the grid. So we saw a lot of investment in phasor measurement units. We saw a lot of investment in AMI. We also saw large scale projects such as the one in the Pacific Northwest to take a look at how to integrate large numbers of resources and coordinate them in such a way that they would operate collectively even though many of them were not part of the actual utilities.

And the lessons coming out of those are going to be available shortly. There's going to be a, actually, a report and seminar about that project in particular.

So these projects move things forward because they brought a lot of money off the sidelines and helped establish things that the utilities needed to understand. When it comes to things like storage there are quite a few projects that are now finishing up this year that will have lessons learned, and there are a number of states working on determining the cost justifications based on those lessons. So, very valuable in helping people understand how the technology can work, but also how the economics of those technologies will work out going forward.

And moving forward, fundamentally, the basic measurement in observability/capabilities of the grid so that we begin to get the data that we'll need for the next stages also has been a crucial aspect of that very valuable for the utilities.

Senator CANTWELL. Well I know we, in the region, talk so much about some of the things we were able to do out on the Olympic Peninsula, but maybe it's worth reminding people of some of the efficiencies that were delivered through those analysis and how we now take that data and try to scale it.

Dr. TAFT. So I haven't brought all the numbers with me. I'd be happy to supply them.

[The information referred to follows:]

Pacific Northwest Smart Grid
Demonstration Project

A COMPILATION OF
SUCCESS STORIES

272

To learn more visit:
www.bpa.gov/goto/smartgrid

Or contact:
TechnologyInnovation@bpa.gov

273

THE BONNEVILLE POWER ADMINISTRATION INVESTS in the Northwest's energy future. Through its · Technology Innovation program, BPA funds an annual portfolio of projects that advance technologies and enable breakthroughs and operational improvements in support of BPA's mission of providing low-cost, reliable electric power to the Pacific Northwest. BPA's research goals, as outlined in its strategic direction and technology roadmaps, focus on key areas such as advancing energy efficiency technologies, preserving and enhancing generation and transmission system assets, and expanding balancing capabilities and resources. Since 2005, Technology Innovation's disciplined research management approach has led to unprecedented levels of success.

BPA works closely with the U.S. Department of Energy and Electric Power Research Institute, and continues to strategically expand its partnerships with electric utilities, universities, researchers and technology developers. Technology Innovation has led the collaborative development of roadmaps that pinpoint the technology needs for the electric power industry for the next five to 20 years. To date, BPA has partnered with industry experts, researchers and others to develop technology roadmaps for energy efficiency, transmission and demand response. These roadmaps now serve as a resource for BPA and others to prioritize their technology investments and identify partnership opportunities.

BPA's R&D investments deliver savings of millions of dollars in avoided costs and increased efficiencies, and result in a smarter, more dynamic, efficient and reliable Northwest electric power system. Today, BPA is investing about $17 million annually in R&D, which is nearly five times the U.S. industry average. The largest project in the TI portfolio is the $178 million Pacific Northwest Smart Grid Demonstration Project. The nation's largest smart grid demonstration project involves 60,000 metered customers across five states, 11 public and private utilities, two universities and six infrastructure partners. Through BPA's $10 million cost-share contribution, the project has deployed $80 million in new smart grid technologies in the Northwest, setting the stage for regional smart grid growth.

Terry Oliver
BPA CHIEF TECHNOLOGY INNOVATION OFFICER

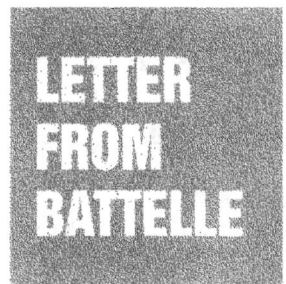

WHEN WE EMBARKED UPON THIS AMBITIOUS PROJECT in 2010, we began an exciting, and highly challenging, journey. It was not an entirely new endeavor — we were building on the earlier GridWise® Olympic Peninsula demonstration. But no project had tackled the breadth and scope of implementing a key new smart grid technology called transactive control, for coordinating demand response from 11 utilities across a five-state region. In addition, participants identified their own individual smart grid technology objectives. In all, the project will have evaluated about 80 different technology test cases. Like any project of this size and nature, we worked our way through perplexing technical issues and challenges we didn't anticipate, but we also experienced rewarding accomplishments.

This booklet chronicles the Pacific Northwest Smart Grid Demonstration Project's successes. It's exciting to read about Lower Valley Energy's great experience with water heater demand response, Portland General Electric's creation of the Salem Smart Power Center, the University of Washington's broader understanding of electricity consumption in campus buildings, and Avista's visionary work in establishing a smart grid city. There are a number of successes to report among the participating utilities, and I hope you'll take time to peruse the articles and learn more about other examples of progress. It's also illuminating to read about aspects of the project that didn't turn out as planned — essentially "lessons learned" that, along with successes, will be important in informing the industry for future smart grid technology deployment.

Battelle and the other participants in this demonstration are pleased and proud to have been a part of a monumental project that reflects the unique grid-related capabilities — and ingenuity — of the Pacific Northwest. I have no doubt that this project and the knowledge that has been gained will successfully prepare the region for a bold energy future that strengthens our economy, protects our environment and enhances our quality of life.

Ron Melton
BATTELLE PROJECT DIRECTOR

Pacific Northwest Smart Grid Demonstration Project
SUCCESS STORIES

AVISTA

AVISTA®

Creating a smart city by focusing on grid efficiencies

Before Washington was granted statehood, the utility known as Avista had already built the world's longest transmission line and would later go on to create the country's first electric stove. Today, Avista's rich history of innovation is being applied to one of the greatest challenges facing the energy industry — integrating new technologies. Avista's vision for modernizing its grid resulted in the region's first smart grid city as part of the Pacific Northwest Smart Grid Demonstration Project.

With an investment of $19 million, matched by funds from the Department of Energy, Avista has the momentum to deploy a system-of-systems architecture model. Washington State University partnered with Avista as part of its project.

As Avista's funding accelerated, so did the pace of the upgrades. Yet the approach to planning one of the region's first smart cities remained strategic and forwarding-looking.

The case for a system of systems

"You have to look at the business case with respect to the current reality and potential new realities," said Curtis Kirkeby, Avista's principal investigator for the smart grid demonstration. "The economics that we have used forever in utilities may not be the economics that are actually valid anymore."

Instead of looking at one particular system, Avista's business case for smart grid

AVISTA CORPORATION
Spokane, Washington

- Investor-owned utility since 1889
- 125-year history of innovation
- 30,000 square-mile service territory
- Serves population of 1.5 million in eastern Washington, northern Idaho and eastern Oregon
- Electricity and gas service
- 359,000 electric customers

INVESTMENT
$19 million

HIGHLIGHTS

- Voltage optimization
- Capacitor bank controls
- Smart transformers
- Customer thermostats and apps
- WSU air handlers, chillers and generators

FOR MORE INFORMATION
Curtis Kirkeby (509) 495-4763
www.avistautilities.com

concentrated on a much broader set of objectives, with a deliberate focus on interoperability and the ability to share information across multiple systems.

'It made for strong solutions as we move forward," said Kirkeby. "You don't want to limit yourself or short the vision."

After all, to create a smart city, scale is important.

An Avista dispatcher is able to see, in real-time, the current state of portions of Avista's distribution network.

'Our goal was to do everything possible that would result in better operation of our system. At the same time, we wanted to maximize reliability, system efficiency and the customer experience. We knew that's where we would find the greatest value, when we could achieve all three of these objectives," said Heather Rosentrater, Avista's director of engineering.

Smart circuits set the stage

First, Avista upgraded electrical facilities and automated the electrical distribution system in Spokane and Pullman. A distribution management system was put in place to serve as the brains for the

smart city, along with intelligent devices and a communications system to benefit more than 110,000 electric customers.

Smart circuits reduce energy losses, lower system costs and improve reliability and efficiency in the electricity distribution system.

The system efficiencies will save about 42,000 megawatt-hours per year, enough to power 3,500 homes, and prevent

14,000 tons of carbon from being released into the atmosphere from power generation.

Problem areas on the system are instantly identified by the advanced distribution management system, which was deployed for one-third of the customer base. A whole new level of information is displayed, which can be operated manually or fully automated around the clock — not a typical installation. The distribution management system features predictive applications and auto-restoration technology.

The new, advanced distribution management system revolutionizes how the system is designed, built, tuned and operated. Real-time power management systems require an accurately maintained

calculation model so that the distributed resources can be managed regardless of location.

Smart transformers

Every home or business uses different amounts of energy that's distributed through a transformer on a pole to multiple sites. Smart transformers installed in the Pullman area gather information about how much energy each transformer supplies, so Avista can determine the appropriate size transformer to meet customers' energy needs. The "right sizing" of transformers makes the distribution system more efficient as energy is delivered to customers.

Biggest bang for your buck

Efficiency equals managing voltage and power factors.

Using voltage optimization, the utility can lower the voltage on the feeder — the line from a substation to the home or business — to minimize the loss of electricity.

"As we scoped the project, we realized this is where the biggest bang for the buck is," said Kirkeby. "This is where the real dollars are. We estimated 1.86 percent savings by applying this technology to the feeders in Pullman on WSU's campus based on a regional study, but we're actually seeing 2.5 percent."

There's also still room to grow this efficiency savings. Fine tuning continues.

"Be part of the study that may change everything"

The two-way communication foundation requires installation of advanced meters at the customer's location. All of Pullman and Albion — a total of 13,600 customers — now have advanced meters. The digital meters operate via a secure wireless network, allowing two-way, real-time communication between the customer's meter and Avista.

"We didn't have any opposition to putting advanced meters there, which may not be typical across the country," Kirkeby said.

Building awareness and understanding among customers was critical to successfully deploying new technology and engaging people. Inspirational key messages were disseminated through focus groups, targeted email and direct mail, print advertising, town meetings and board meetings. Cutting through the clutter of busy lives was challenging. Customers responded best to the in-person communication.

"They had the feeling that they were part of a project of national importance. The research aspect of our work resonated in a college community like Pullman. It was a feel-good thing for the customer — they felt like they were making a difference," said Laurine Jue, a senior communications manager at Avista.

A dedicated point of contact was critical to answer tough questions about the pilot.

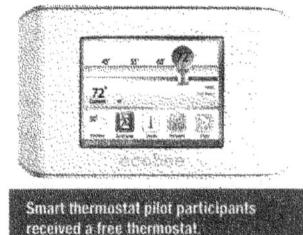

Smart thermostat pilot participants received a free thermostat.

Smart thermostat pilot

The smart thermostat pilot was one of the customer-experience components of the project. Customers who volunteered to participate in the two-year pilot received a free smart thermostat, plus $100 per year in exchange for allowing the utility to remotely adjust the thermostat by 2 degrees Fahrenheit for a period of 10 minutes to 24 hours. The customer could always override the setting at any time.

"You can set up a program to override at any point," said Joshah Jennings, a smart thermostat pilot participant.

Settings, including alerts, can be adjusted directly, over the Internet or with a smart phone. Using the application on a regular basis keeps energy usage "top of mind" for customers. Participants could view energy usage down to the hour, make adjustments, and start saving energy. A price curve was set for hourly consumption. For the Jennings, a fiscally conservative family of five, saving money is important.

"But being a technology buff too, it was kind of fun to play with the new technology," Jennings said.

All customers with the smart thermostats also had advanced meters that provided usage data. At the end of the pilot, data indicated that smart thermostat participants reduced consumption between 4.5 and 9 percent.

AVISTA TIP: The thermostat was designed so that the vendor and its product were not dependent on Avista — the connection was through public Internet. The thermostat read the meter and sent data back to the thermostat vendor through its own mechanism. That means no maintenance for Avista.

Energy Analyzer needs actionable items

Another aspect of the customer experience was giving customers access to information about their energy use. Using a web portal, called the Energy Analyzer, customers could log in to their account to see their energy use patterns and make informed decisions about choices that drive energy costs.

The launch of the web portal was promoted with direct marketing and an online video to help educate customers about how to use it. While some customers looked at the portal frequently, most customers did not find it compelling. Although the average site visit was six minutes and 36 seconds, access to the web portal did not result in a measureable change in consumption.

Avista's marketing campaign used inspirational language to recruit volunteers.

One of the main factors that could have contributed to this is the absence of time-of-use rates, which could directly impact customers' usage patterns. Plus, every customer has a preferred method for accessing information, whether it's direct mail, email, a website or a mobile device.

Avista suspected that if an actionable item doesn't result from the data, it's not something a customer will get excited about. If a customer wants a lower bill, suggestions based on the data provided would be more useful.

"That's why we launched a texting pilot," said Kirkeby. "It allowed customers to opt-in to receive daily or weekly usage updates via text or email, which included usage predictions based on all kinds of factors. Weather factors, household factors and HVAC factors are useful to a customer trying to manage their bill through their own efforts."

"All of the work that we did in Pullman really has helped our customers understand on a personal level what Avista is doing to modernize our grid," said Rosentrater. "What it means for customers is improved reliability; they're going to experience fewer and shorter outages. What used to take hours to restore, can now be done in minutes."

In the long run, fewer and shorter outages, plus options for saving energy, matter most. It's a big win for customers.

A big win for Washington State University

As a key partner in the project, Washington State University brought its best minds and dozens of facilities to the table. The campus can now be operated as a microgrid with the ability to control both loads and generation resources on campus, as well as respond to a transactive control request from Avista based on regional grid needs.

Working with Avista and with WSU professors Anurag Srivastava and Anjan Bose, students helped simplify the process for computing real-time savings from improved power factors and voltage reduction. They also developed new tools for reliability benefit calculations, for data transfer between software tools and for real-time load characteristic estimations.

The project resulted in award-winning work and provided valuable, hands-on experience to prepare students to be leaders in the 21st century power industry.

With its cost-share investment of $2.1 million, WSU also installed 88 smart electric meters, providing direct feedback to Avista for voltage optimization of the campus power circuits, and built sophisticated building control programs to automate its chillers, air handlers and three generators for smart grid operations.

For example, the air handlers in 29 campus buildings used to run without consideration of building occupancy levels. Air handlers ensure the air quality in buildings is consistent. With the new programs, the air handlers automatically ramp down when occupancy levels are low and ramp up before higher occupancy periods. While the technology doesn't change what happens within the building, it optimizes the efficiency of the system by scheduling appropriate actions based upon campus needs.

"Their systems are smart enough to be able to manage all air handlers individually to get to some level of cumulative benefit," said Kirkeby.

The voltage optimization and new air handler programs save the university a lot of money; it expects to save about $150,000 a year. Each one of the controllable assets can be managed in a much more efficient way and fine-tuned as conditions change.

Washington State University was a centerpiece of Avista's smart city in Pullman.

Students helped to simplify the process of validating real-time savings from power factor and voltage reduction. A new tool was built to validate, or at least estimate with a high degree of accuracy, every circuit that's involved in the system for every five-minute interval.

Avista generates a request

These generation assets are also connected to Avista's distribution management system through an Avista Generated Signal. Avista can generate a request to reduce those loads or generate power for various needs.

"If you think back to the energy crisis in 2000 when, in the Northwest, we were all looking for generation, now we have a push button for three of them," said Kirkeby. "All Avista does now is make a request and they can be online."

For the smart grid demo, a transactive control signal request came from Battelle. Avista then translated that into an Avista Generated Request Signal that went to WSU, asking for one of five tiers: one tier for air handlers, one tier for chillers, and three different tiers for generators.

Avista's request asked for deployment of certain assets based on the request from Battelle. A WSU facilities operator decided whether or not to honor that request. If not, software sent Avista a text explaining why. Automatic texts were also generated. Codes were built into the system so that it still functions even without the transactive signal.

"Being a part of this pilot project has really opened doors for improved communications, metering, power system operations, and building controls, which provide the tools for WSU to assist with regional needs while also reducing our operating costs," said Terry Ryan, WSU director of Energy System Operations.

> **"** It's been really exciting! A year and a half after the distribution management system went live, we hit our one-millionth avoided outage minute. That's a tangible benefit for our customers. Avista is also realizing benefits – we've learned so much and we're applying these lessons every day."
>
> – HEATHER ROSENTRATER, AVISTA DIRECTOR OF ENGINEERING

Win! Win! Win!

Leveraging university assets is a model that Avista believes can be used elsewhere. After all, utilities and universities alike are always looking for ways to reduce costs.

"That's exactly what we did," said Kirkeby. "It's a win-win-win for WSU, Avista and for the region."

The project also greatly improved cyber security across the utility footprint. National Institute of Standards and Technology guidelines were used to assess risk. Then a mitigation measure was assigned to each risk to determine whether to proceed or how to proceed.

By participating in the project, Avista learned in a real-world environment the benefit stream from each component in a particular use case, and which pieces will extend well beyond the project and into the future to create an ultimate smart city configuration.

Some pieces of that puzzle are important lessons learned.

Learn! Learn! Learn!

It's not always easy to change an established paradigm, especially when redesigning business processes around technologies such as automating grid control. Even some vendors pushed back.

"Some said: you don't really want automation," said Kirkeby. "To which we responded, 'Yes! We really do!' That's why we're doing this. We don't want an army of people operating it; we want it to be automated."

When working with vendors, "It always comes down to relationships and partnerships," said Kirkeby.

Ironically, the very people expected to push back on automation were the most supportive: the linemen. For them, doing away with mundane tasks was positive. For others, it took trust that the system wouldn't make mistakes and that, if it did make a mistake, consequences were managed.

Creating a smart city meant embracing change. It's everywhere.

"We've changed as part of this project," said Kirkeby. "We've revised the design, engineering and equipment standards going forward. We'll add to those standards as new learnings come about and as we add more technologies or benefit streams we find."

WHAT'S NEXT for Avista?

The future is an evolving vision. The grid will be a system that's flexible, scalable and understandable by the people building it. Avista's roadmap includes a grid modernization program budgeted for the next 25 to 30 years.

"We're trying to be really proactive and create the utility of the future now," said Rosentrater. "So we're trying to figure out as a utility where we need to be, what we need to do to provide the most value to customers, and where customers will value us — value what we do and keep our business viable."

With an increased capital budget for grid modernization, Avista is leveraging what's already in place to advance the rest of the system. Armed with a smart grid roadmap from the demonstration, each feeder in Spokane will be modernized. In Pullman, a new battery unit is already in the works as part of a Washington State Department of Commerce grant. Avista will explore how battery storage capabilities can be integrated onto the grid to address the intermittent energy from renewable power.

Participating in the Pacific Northwest Smart Grid Demonstration project gave Avista the opportunity to realize the "endless" possibilities. And the many lessons learned have provided a solid foundation for Avista as it continues to modernize the grid to meet the energy needs of its customers well into the future.

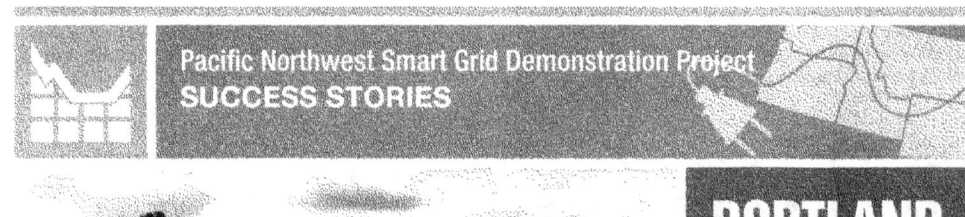

Pacific Northwest Smart Grid Demonstration Project
SUCCESS STORIES

PORTLAND GENERAL ELECTRIC

Smart power in store for the future

Getting smart about electricity has never been more exciting. That's especially true at Portland General Electric — Oregon's largest investor-owned utility. PGE's Salem Smart Power Center is a new five-megawatt battery storage facility that is part of the larger Pacific Northwest Smart Grid Demonstration Project. This first-of-its-kind facility is one of the most advanced electrical systems in the nation and, as such, has inspired the imagination of a region. Energy storage is just one of the new technologies being tested by the project to integrate renewable energy, improve grid reliability and lower costs to customers.

With the foundation of a smart grid already in place — 800,000 smart meters — PGE was inspired to integrate several smart grid programs into one effort. It was an endeavor much larger than PGE could tackle on its own. The heart of it centered on the Salem Smart Power Center.

"A five-megawatt lithium-ion battery system that is grid-tied is very rare in the electricity business," said Wayne Lei, director of R&D for PGE. "It's one of just two owned and operated by an investor-owned utility."

Big – really big – batteries

Lithium batteries are widely used because of their high-energy density — the ability to store a lot of energy in a lightweight, compact form. It's the same battery technology used in laptops and cell phones but on a much larger scale. The key feature of the 8,000-square-foot center is a five-megawatt lithium-ion battery-inverter system. The bank of batteries stores 1.25 megawatt-hours of

Portland General
Electric

PORTLAND GENERAL ELECTRIC
Portland, Oregon
- Founded in 1889
- Oregon's largest invester-owned utility
- Serving 52 communities across Oregon
- 842,000 customers
- 13 power plants with capacity of 2,781 megawatts

INVESTMENT
$6.5 million

HIGHLIGHTS
- Distribution microgrid
- Five-megawatt lithium-ion battery
- Intelligent distribution management
- Commercial demand response
- Demonstrates renewable integration

FOR MORE INFORMATION
Kevin Whitener (503) 464-8219
www.portlandgeneral.com

energy, which allows PGE engineers and planners to demonstrate high-reliability strategies involving intentional islanding of the feeder, distribution automation using smart switches, demand response, renewable energy integration and automatic economic dispatch.

Building the battery facility was an undertaking. But integrating the technologies was the real feat. It required a dedicated engineering team to address the complex challenges that arose in bringing this innovative facility to life. After all, many of the technologies implemented were new to the market.

"We underestimated what it takes to attach a five-megawatt battery to our own system," said Kevin Whitener, the lead engineer for the project. "The complexity and the engineering challenge of doing that is something we hadn't fully anticipated."

Thousands of battery cells are stored in racks and wired together into a single system. Batteries use direct current while the distribution system uses alternating current, so inverters sit between the grid and the batteries. This allows power to flow in either direction, converting from AC to DC and back on demand. Coordinating the communication between the systems and components was substantial and complex.

For example, between the inverters, the battery management system and the other controllers in the facility, there are five different communication protocols. There are sixty-seven separately addressed internet devices communicating on two different networks within the facility. That created a lot of data handling challenges.

"The protocols had to be sorted out and interfaced together," said Whitener. "There's no way to do that short of spending weeks and months struggling to get it to work. But we did it."

The safety of employees and the public is important to PGE, which is why the Salem Smart Power Center was constructed with a focus on safety. Due to the high energy density of the large battery, a unique fire control system was specially designed for the lithium-ion application that includes giant fans that keep the batteries cool at all times.

Creating a microgrid with macro resiliency

A microgrid improves a system's resiliency by allowing the utility to segment a certain part of the feeder and to provide back-up electricity during

an outage. When a substation loses its power supply from the transmission lines, the battery system starts immediately, serving as an uninterruptable power supply.

If an outage were to occur in Salem, all residential, commercial and industrial customers on the circuit can be supplied electricity from the battery's 1.25 megawatt-hours of energy for 15–20 minutes. This is more than enough time to start the six customer-owned distributed diesel generators and synchronize them on the line. Once the feeder is isolated from the utility grid, the generators start up, and the circuit becomes a microgrid.

PGE has been working for more than 10 years to establish cooperative microgrids with customers that own standby generation. Together, they have built the nation's largest distributed generation program which shares customer generation with the utility in times of need. Many of PGE's large customers have local diesel generation on site to prevent a power outage in case of an emergency. By partnering together, PGE is able to tap into this standby generation during an outage situation. The result is a highly resilient system.

WHAT IS A MICROGRID?

A microgrid is a small-scale version of an electrical grid. It can be "islanded," or disconnected from external transmission services. Local distribution provides power for customers' electrical needs with only local generators and battery storage.

> Our electrical grid in the United States is one of the greatest accomplishments of the 20th century. Portland General Electric and its partners are demonstrating new technologies that hold promise for building a more efficient, sustainable and reliable grid. As these technologies became cost effective, they can provide the opportunity to reshape not only the infrastructure that makes up the grid, but the approach utilities take to meeting the needs of our customers, the economy and the environment in the 21st century.
>
> — JIM PIRO, PGE CEO AND PRESIDENT

A High Reliability Zone

PGE named its microgrid a "High Reliability Zone." The HRZ includes the large-scale energy storage system, customer standby generators and distribution automation components. These components, called smart switches, quickly sectionalize the microgrid in case of a fault, like a downed power line. The switches bring an even higher level of reliability to customers.

Unlike a standard feeder switch, which must be manually operated to change or stop the flow of power on a feeder, a smart switch "senses" changes in the feeder, like a fault, and activates the switch automatically. This changes the physical configuration of the feeder within seconds.

It's a microgrid that heals itself.

For solar, it's all in the algorithm

One of the most exciting parts of the project for PGE was exploring solutions to integrate renewable energy into the grid using battery storage. A key challenge to using solar as a power source is that sunshine is intermittent, especially in Oregon. Using an algorithm, PGE demonstrated how solar energy can be combined with a battery to fill in the gaps when the sun isn't shining and offer a seamless power flow.

With more than 6,000 megawatts of intermittent wind and solar power sweeping the Pacific Northwest electrical grid, the project provided an opportunity to learn how to best partner with customers to deliver high reliability. To test the integration, PGE used the solar output from the local potato chip maker, Kettle Brand, and then aimed to levelize, or fill in the blanks of this irregular output, using the battery.

Here is how the process works. First, an instantaneous measurement is taken of the customer demand on the circuit. Then a measurement is taken of the instantaneous power output from Kettle Brands' solar plant. This information is compared to the theoretical ideal load for the utility's circuit. The battery makes up the difference in the output in real-time, either filling in the gaps where the clouds caused output to fall short of the best possible power or charging the batteries when the output from the panels is higher than normal.

"This is one of the few opportunities that the industry has had to prove these concepts and demonstrate that energy storage is indeed a solution to integrating solar energy," said Whitener. "Impacts from what we're doing here are far-reaching."

An interesting exhibit

The Salem Smart Power Center has had many curious visitors. Tours feature a video reviewing the safety of the system, smart grid exhibits and an educational gallery with views into the operations center. Schools, other utilities, industry suppliers, consultants and government representatives all wanted to see this state-of-the-art facility.

"We've had more than 1,200 people visit the facility and learn about the project," said Whitener. "That's pretty astonishing."

Partnering with customers

PGE's smart meters enable a two-way conversation between PGE and its customers, helping the utility to optimize its services, add convenience and lower energy costs. As part of this demonstration project, residential and business customers were enlisted to respond to grid conditions by reducing energy during peak times or during a test.

"The utility can decrease the load at peak times of use, or shift loads from one period to another," says Carol Mills, PGE's senior project manager. "The objective was to offer demand response assets that could respond to the project's integrated systems."

Although PGE installed a demand response management system in Salem, a 'human in the loop' was used to ensure the programs would be initiated and observed carefully when called into action.

An impressive transactive system

As part of this project, PGE is testing ways in which we can automate renewable integration and demand response opportunities to ensure customers receive the most benefit from energy resources for the least cost. The project includes testing a transactive system, an information system that automatically shares real-time data between computers at utilities and the transmission coordinator. Similar to how utilities get information from wholesale power markets today, this system sends out a price signal every five minutes, which reaches a multiple utility footprint at the same time. The signal shows how the price of power is expected to change over the next three days.

Utilities then respond with a load forecast based on that string of future prices. This allows a system coordinator, in this case the Pacific Northwest National Laboratory, to calculate where the entire grid may have congestion issues in advance. The process is then repeated every five minutes, allowing for planning around congestion and prices to occur for everyone in the system.

Using artificial intelligence

Although, automating the electricity market is still in testing stages, strides were made learning about which tools are needed for its development.

"We've proven that we can dispatch resources at the command of the transactive node," said Whitener.

The transactive node, which PGE calls the Smart Power Platform, is the main computer program that optimizes the economic decisions about the smart grid assets: when to dispatch, when to charge or discharge the battery, and when to use the demand response capability. The node responds to a signal from PNNL. To interact with the signal, PGE wrote its own software program using artificial intelligence. Neural networks analyze the thousands of data points in the system and respond to the transactive signal. The computer absorbs all that information, synthesizes it, and makes a decision.

"We were able to demonstrate the ability of the computers on both sides to learn and get better at optimizing power for the least cost to customers," said Lei. "It's literally a monetary estimation in terms of the value to deliver and the value to acquire that power."

Learning from unique systems

Virtually all systems tested by PGE were new and unique. The Salem Smart Power Center demonstrated the ability to island a microgrid with utility-scale storage and customer standby generation, operate demand response, respond to a transactive signal, and how to integrate these complex resources into a single control system. As a result, PGE offered several key takeaways:

- Take full advantage of consulting talent both within and outside the company to assess risk and make plans to mitigate that risk.

- Reduce financial risks by using government funds when possible.

- When it comes to a first-of-its-kind project, testing is your best friend. PGE sought to protect customers by ensuring the systems were reliable and robust. Perform and document lots of testing, especially when there is a potential to impact commercial and residential customers.

- Thoroughly vet vendors' capabilities and financial strength. Smart grid technology is a growing industry, full of emerging companies. Ensure those companies are well-capitalized.

- To ensure the safety of employees and the public, it's critical to have a robust set of safety requirements in place that serve as a system of checks and balances. For example, every test was preceded by a test plan. Test plans were circulated through the project team and various departments within the company for approval.

Finally, assembling a strong, adaptable engineering and project management team makes all the difference.

"The team was able to lead the project over many different hurdles," says Lei. "As the recognized experts in the topic, the team not only had to work with the management and technical aspects, but also to be able to communicate well with everyone in and outside the company."

WHAT'S NEXT for PGE?

Smart grid technologies represent an opportunity to enhance the value customers receive from the electric system. This transition will be a significant challenge — one that involves not only leveraging new technology, but also making major changes in the way electricity is provided and used. PGE is eager to engage in the research and development needed to bring our local and regional grid into the 21st century.

Pacific Northwest Smart Grid Demonstration Project
SUCCESS STORIES

UNIVERSITY OF WASHINGTON

UW's electric grid gets smart with living laboratory

Higher education happens at the University of Washington in more ways than one. For students, it's academics. For facilities, it's smart grid. More than 250 buildings on the university's Seattle campus are temperature conditioned. That's more than 13 million square feet of comfortable space to conduct research. And it goes to good use, because UW has one of the biggest research budgets among schools nationwide.

The scholastic setting provided a unique perspective for the Department of Energy demonstration. More than 20 million people are enrolled in higher-education programs across the country. That's a lot of students, not to mention researchers and faculty. For that reason, college campuses make an ideal environment for advancing smart grid technologies.

"Finding solutions to real work problems is a really important part of today's educational environment," said Norm Menter, UW's energy conservation manager. "Our student population demands that the university be involved in projects like this."

UW is one of two universities, five technical firms and 11 utilities across an unprecedented five-state region selected to participate. The university was awarded $5.1 million in federal funds to complete its $10.2 million project.

By the numbers

The average daily population on campus is 60,000, and the average daily electrical demand is 38 megawatts. This costs the university about $1 million a month — making it Seattle City Light's second largest customer. Even so, the campus also has its own five-megawatt steam turbine generator. The power

W
FACILITIES SERVICES
UNIVERSITY *of* WASHINGTON

UNIVERSITY OF WASHINGTON
Seattle, Washington

- Seattle City Light's second largest commercial customer
- Population of 60,000
- 38 megawatts of demand
- Electricity bill of $1 million a month

INVESTMENT
$5.1 million

HIGHLIGHTS
- Energy dashboard
- Living laboratory
- Five-megawatt steam generator

FOR MORE INFORMATION
Norm Menter (206) 221-4269
www.washington.edu

is distributed through a network of underground utility tunnels.

Energy dashboard drives decisions

Before the project, UW had just seven meters on its Seattle campus to monitor energy use. Now, there are more than 200 smart meters acquiring near real-time data about energy consumption every five to 15 minutes. These meters transmit the data to a central repository. To see the data and more accurately predict future energy use, the project team built a console to analyze and display the information collected.

Data is analyzed and displayed on "dashboards" that provide an at-a-glance, graphical presentation of the energy use, within just minutes of its consumption. Anyone can view the data online and compare a year-and-a-half worth of information on each building's energy consumption during any hour, and see how much that energy costs and patterns of use over time. Engineers at the university also use software tools to analyze the data collected, helping UW eliminate waste and save money.

One feature allows a comparison of energy use by building at different times of the day or year. That information is vital to determining how to reduce energy use and eliminate energy waste.

"The surprising thing we learned was that energy waste is not the same thing as energy consumption," said Menter. "Waste occurs just about everywhere, but just because you have high consumption in a particular building doesn't necessarily mean that it's being wasted. The information is valuable for decision-making, but it's not the whole story."

The dashboard raises awareness about a building's overall efficiency and provides the opportunity to have a conversation about why changes need to be made to a building. This dialogue builds a common understanding so that decisions are made together across campus.

Sharing of data equals a need for improved cyber security. As a public university, UW has an open network. So coming into this project, the university had to overlay a private network onto its existing system. That cost quite a bit of extra money. But it was worth it. After all, having data, and being able to store that data securely and

make it available to the public, is immensely valuable to the research mission of UW.

A student energy intervention

Of those interested in the data were — no surprise — the students. The energy use of freshmen in two residence halls was looked at by fellow students to see if a "technology intervention" could reduce the freshmen's energy consumption, compared to a manual intervention.

In one hall, students received weekly energy-saving tips to encourage conservation and to gain qualitative data. In the other, volunteers received EnergyHubs, which measure, communicate and control individual energy use in real time. When small appliances, electronic devices, TVs or laptops are plugged into EnergyHub power strips or outlets, the cost of using each is displayed on a desktop monitor. The monitor also displays daily energy use, current energy in kilowatt-hours, and monthly cost projections per appliance. And they allow the user to set up schedules for the appliances or remotely turn them off through a smart phone app.

Using its smart meter infrastructure, the research team was able to collect weekly energy consumption data, analyze the students' energy use over 10 weeks, and compare the energy use of students in the two residence halls. But the study yielded inconclusive differences between the two groups, perhaps due to one aspect of studying students' energy use in a university residence compared to adults in their own homes: a lack of monetary motivation.

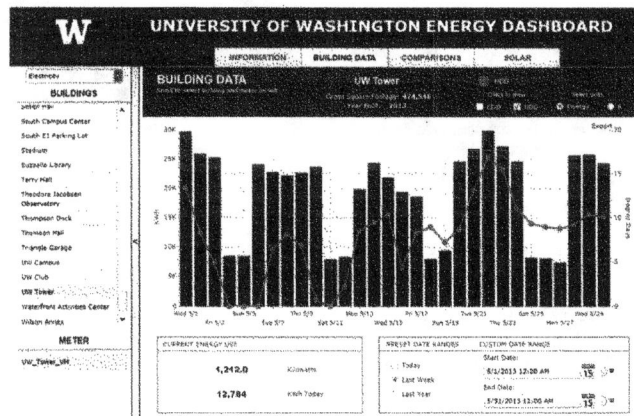

288

"One important long-term outcome of the research was to raise student awareness of the personal energy choices we all make every day," said Kelly Hall, a UW graduate student on the research team.

The team believes the university should not rely on student actions to reduce energy consumption, but move towards using more automated and integrated systems.

A regional transactive role

One part of the project, which spans an unprecedented five-state region and 60,000 metered customers, is automating the system for a regional benefit.

UW connected its steam turbine generator to the demonstration project's transactive control system. Transactive control uses an interactive, market-based signal to increase or decrease energy consumption to achieve greater efficiency in grid operations. The signal is sent over a multiple-utility footprint. Participants in the project test the feasibility of increasing energy use when wind energy is abundant, typically at night, and reducing use during peak hours when energy is most expensive.

"We integrated the steam turbines with the system, so that the turbines would respond and go into a nighttime setback mode when they received a transactive control signal," said John Chapman, UW executive director of Campus Engineering and Operations. "The steam turbine generator concept has potential. I can see how we could use it to vary the operation of our generators to help the region integrate renewables into the grid."

But UW's current generating system is constrained in terms of how it could contribute to a transactive control system. It's a cogeneration system: after the steam goes through the turbine to generate electricity, it then goes on to heat the campus. In the summertime, there's not much steam demand. Only during the winter is there some flexibility on the output.

> **These assets represent an investment that's going to be here for many years. We can use the data to build a greater understanding about how buildings use energy, and to make the entire campus community more intelligent about how we use energy. It's a change in our relationship with electricity."**
>
> — NORM MENTER, UW ENERGY CONSERVATION MANAGER

Conceptually, UW could replace its generator with something that has the ability to interact with new technologies.

"With that type of system, we could bring on maybe an additional 10 or 12 megawatts anytime of the year to feed into the grid whatever time of day it was required," said Chapman. "And we could do that for a reasonable price, I think."

UW also connected two standby diesel generators to the transactive control system as part of this project. But they are more expensive to operate than the steam turbines.

"I don't think the economics would ever pan out to run those and generate electricity," Chapman added.

There are also environmental requirements that limit the use of diesel generators, so UW will focus on its steam turbine.

UW created a demand reduction operation strategy for five buildings on campus. As part of the transactive control system, the university could opt in to reduce electrical demand at three different levels. These strategies included changing discharge air temperature set points, reducing fan static pressure and limiting control valve positions.

5-megawatt steam turbine

17

WHAT'S NEXT
for the University of Washington?

As far as the impacts on operations, the university is just getting started, but a solid foundation is now place.

Tools such as the dashboard allow UW's energy engineers to look for anomalies in energy consumption and determine which buildings they should take a look at first. The dataset builds a greater understanding about how the campus buildings use energy. And, the improved infrastructure makes the entire operation more efficient, which means there is potential to reduce the costs of education.

That makes the entire campus and the entire community more intelligent about how we use energy," said Menter. "It's a change in the relationship with electricity."

Pacific Northwest Smart Grid Demonstration Project
SUCCESS STORIES

IDAHO FALLS POWER

Small city in Idaho gets smarter with automation

Idaho Falls is simply a smart city. Since 1900, the western town has generated electricity by making use of the Snake River, which runs right through it. Today, Idaho Falls Power — which is known for its early adoption of tools to improve system reliability — is testing some hefty smart grid technologies as part of the Pacific Northwest Smart Grid Demonstration Project.

A city-wide wireless network

Idaho Falls Power services 23 square miles within the city, which is framed by wide-street neighborhoods and well-educated residents — many of whom work at the Idaho National Laboratory nearby. So when the utility decided to test smart meters through a city-wide wireless network, most of their technologically-savvy customers didn't blink an eye.

The wireless mesh communications system allowed the utility to test a number of new technologies, such as automation of switches in substations for outage restoration. Another benefit of using the wireless network is improving computer and electronic protections.

"We drastically improved cyber security awareness and increased focus in that area as a result of this project," said Mark Reed, Idaho Falls Power superintendent. "That's something near and dear to my heart."

Having a secured wireless system means the utility can do more with its smart meter system, or AMI — advanced metering infrastructure — and integrate those meters with the centralized computer system of a substation's control center.

iP
Idaho Falls Power

IDAHO FALLS POWER
Idaho Falls, Idaho

- Locally-owned and operated since 1900
- Largest municipal utility in Idaho
- 27,000 metered customers, including 3,500 commercial customers
- Installed 1,500 AMI meters
- Generates 50 megawatts and sells to BPA

INVESTMENT
$3.5 million

HIGHLIGHTS
- Wireless mesh network
- Electric vehicles
- Capacitors

FOR MORE INFORMATION
Mark Reed (208) 612-8234
www.idahofallsidaho.gov

19

Meters talk to in-home displays wirelessly

The progressive utility has 13,250 smart meters, including 1,500 meters that were installed as part of this project. All 27,000 customers are expected to have a smart meter by May 2015.

Eight hundred in-home display monitors communicate wirelessly with the meters of volunteers, allowing a real-time view of their electric use either by the hour, day or month. Usage trends and costs, as well as important utility messages and alerts, are also displayed. All the information will be available on a web portal and mobile app that customers can access by fall 2014, making it even easier for households to calculate daily energy costs and estimate their bills.

For one long time resident, the technology is about much more than crunching numbers.

"I've looked at the climate data," said Jim Seydel, Idaho Falls Power customer. "I believe we're going to have more extreme weather. I believe under those conditions, we're likely to have more outages than we've had in the past because of climate change. So, I'm looking for ways that I can counter that."

In short, these devices empower customers to take ownership of their energy use and planning. Still, about 135 customers have vetoed the meter.

"I think there is just some confusion about what smart meters are," said Matt Evans,

customer relations supervisor at Idaho Falls Power. "When they Google 'smart meter' they find some alarming things."

But the majority of the community, which attracts world-class outdoor recreationists, regional business owners and culturally savvy patrons, has embraced these leading-edge technologies. They can be found just about everywhere ... even on appliances.

Hitting the snooze button on appliances

Back in 1934, Idaho Falls was one of the first cities in the United States to adopt devices that could prevent brownouts. These devices were able to cycle water heaters off, using frequencies, when river flows were low or as households began installing electric appliances and using more energy.

Today, the methodology remains the same — only the communication tools are much more advanced. Loads can now be shifted to later in the day when the cost of electricity has fallen or demand has fallen. For this project, 217 volunteers tried a power control device. These devices cycle off electricity to appliances when the need for energy is highest. This avoids unplanned power purchases, which are more expensive.

Heating and cooling costs account for most of the utility's peaks. Forty-one customers signed up for programmable thermostats. With these "smart" in-home displays, the utility can make

very subtle temperature adjustments within households for brief periods when energy use is highest — the utility equivalent of rush hour traffic.

The newer systems are also more advanced. Idaho Falls Power alerts the customer before it makes any changes. The thermostat indicates when the utility plans to make a scheduled thermostat adjustment, and the customer always has the option to override or opt out of the utility request.

Improving energy use automatically

Another way to reduce energy consumption is by automating the distribution system. One little electronic device, a capacitor, is helping to make this happen. While capacitors have stored energy and stabilized voltage and power flows on distribution lines for decades, the process is getting smarter with two-way communication. The stable and steady delivery of electrons saves time and money. For big industries, that savings can add up.

Making malt for brewing beer, for example, takes a lot of energy. IFP's largest industrial customers are malt houses, which benefit from something called automated power factor control. A power factor is a measure of the efficiency of the power being used. A power factor of 100 percent means the voltage and current are cycling between positive and negative in unison, but when one lags behind the other, the power factor declines. The lower the power factor, the more power the generator has to supply for each watt being consumed.

One way to improve the power factor is by installing banks of capacitors, which can automatically bring the current and voltage closer to unity. Idaho Falls Power purchased two capacitor banks as part of the project — one for each of its large commercial customers. The projected wholesale energy savings is $37,750 per year. But saving money isn't the only benefit.

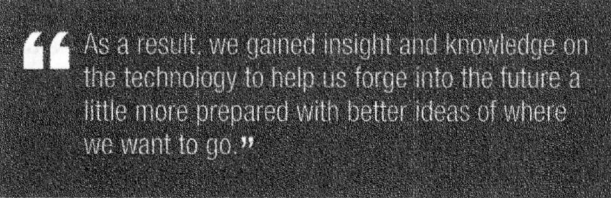

As a result, we gained insight and knowledge on the technology to help us forge into the future a little more prepared with better ideas of where we want to go."

MARK REED, IDAHO FALLS POWER SUPERINTENDENT

"It reduces the electric current on the line," said Reed, "which could defer the large capital cost of an upgrade."

Another way to use distributed automation is through conservation voltage reduction. In this case, CVR reduces the overall voltage on the residential feeder to 1,375 smart-metered customers.

A "self-healing" grid

Let's face it. Lightning causes power outages. That used to mean a utility worker would be dispatched to locate the fault and manually reset switches on the transmission lines to reroute power. Now this can be done automatically through fault detection, isolation and restoration, or FDIR — sometimes referred to as "self-healing" technology. This smart technology can instantly detect a fault and automatically reroute electricity to keep customers from losing power in the first place. The tool uses automated switching between two distribution system feeders and control algorithms to isolate the problem and restore the system.

Saving some for later

Electric vehicles can be great energy storage units. Of course, these mobile batteries also make great transportation tools. Incorporate a solar panel and you have quite a setup.

Here's how it works:
The solar panel charges a stationary battery during the day. That battery, with 10,000 watts of capacity, is hooked up to the car-charging stations. When a car is plugged in, it draws 3.35 kilowatts for four hours from the stationary battery. That means two or three electric vehicles can be plugged in to receive a full charge.

"The integration work was outstanding," said Reed. "The AMI interface was really fantastic."

An electric vehicle charges using stored energy from a nearby photovoltaic panel.

The drive behind the test is to integrate distributed solar with distributed storage to optimize grid efficiency. It works — but with one problem.

The test was nearly complete when the vendor went bankrupt. This has left Idaho Falls Power with no access to the software or any of the data that was collected. If the vendor never returns with the data, the utility is considering working with nearby Idaho National Laboratory to explore possible solutions to retrieving the data.

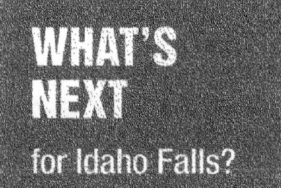

WHAT'S NEXT for Idaho Falls?

For Idaho Falls Power, the next step is evaluation. Assessing the feasibility and cost-benefit of the smart grid technologies tested in the demonstration project is essential.

"It's how we'll determine which technologies have value for expansion across the entire system," said Reed.

And, of course, they'll be sharing what they've learned.

293

AND BY THE WAY ...

Project Landscape

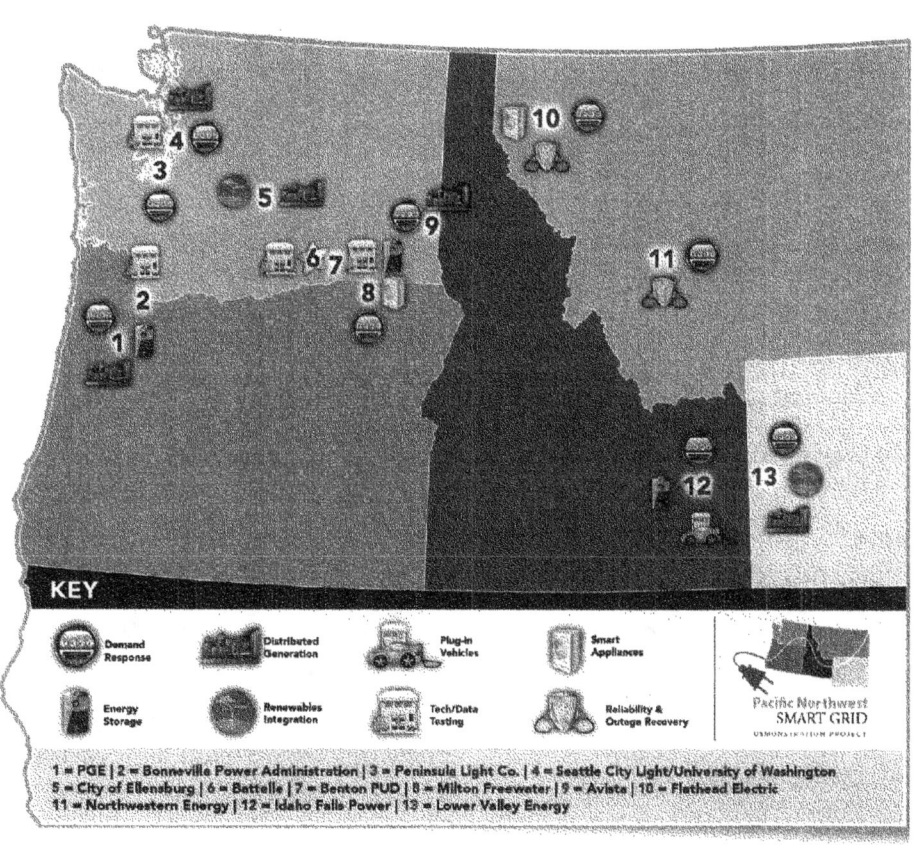

KEY

Demand Response

Distributed Generation

Plug-In Vehicles

Smart Appliances

Energy Storage

Renewables Integration

Tech/Data Testing

Reliability & Outage Recovery

Pacific Northwest SMART GRID DEMONSTRATION PROJECT

1 = PGE | 2 = Bonneville Power Administration | 3 = Peninsula Light Co. | 4 = Seattle City Light/University of Washington
5 = City of Ellensburg | 6 = Battelle | 7 = Benton PUD | 8 = Milton Freewater | 9 = Avista | 10 = Flathead Electric
11 = Northwestern Energy | 12 = Idaho Falls Power | 13 = Lower Valley Energy

Pacific Northwest Smart Grid Demonstration Project
SUCCESS STORIES

FLATHEAD ELECTRIC CO-OP

Building a smart grid the cooperative way

Tucked in the mountains of Glacier Country in Northwest Montana, Flathead Valley's grand landscapes and unspoiled freshwater lake attract recreationalists year-round. Legendary, small-town hospitality appears even in unexpected ways — like the local electric cooperative's participation in the Pacific Northwest Smart Grid Demonstration Project.

With 48,000 members, Flathead Electric Co-op is the second largest electric utility in Montana. Yet it maintains the cooperative spirit of neighbor helping neighbor. When granted the opportunity to help consumers reduce their energy use during periods of peak demand and save money on their monthly power bills, Flathead put its members' needs and interests first.

With the project in its fifth and final year, Flathead is planning for further investment in some of the technologies it has tested. The utility is also teaching others what it has learned.

A solid foundation

Flathead was a prime candidate for the project because it had already invested in developing an advanced metering infrastructure, a crucial component of the two-way communication between the utility and end-users. Households fitted with these advanced meters allow the utility to monitor electricity use in real time and identify peak-use times, when electricity is most expensive.

"This is different from conventional energy conservation because, while participants may not actually use less energy in total, they may choose to use it at times of lower cost to the co-op," says Flathead Regulatory Analyst Russ Schneider. "This has the potential to ultimately reduce power supply expenditures for members and the co-op as a whole."

Flathead's objectives included completing installation of the advanced metering

Flathead Electric *Your Co-op*

FLATHEAD ELECTRIC CO-OP
Kalispell, Montana
- Locally-owned & operated since 1937
- Second largest utility in state
- 3,900 miles of line
- 48,000 customers

INVESTMENT
$2.3 million

HIGHLIGHTS
- Single point of contact
- Completed a systemwide AMI rollout
- Home Energy Network

FOR MORE INFORMATION
Russ Schneider (406) 751-1828
www.flatheadelectric.com

295

infrastructure in northwest Montana, determining member preferences and comparing the cost effectiveness of three program options offered to members who volunteered.

But first the co-op needed community buy-in.

Peak Time

Flathead emphasized customer education and outreach and put tremendous thought into designing a program that its members would support, down to the project's name. Instead of the term "smart grid," Flathead's leaders chose a name they felt would better describe the pilot's purpose and resonate with members: Peak Time.

"I think that worked well for us. We wanted to be very clear about what we were trying to do as a cooperative," says Teri Rayome-Kelly, Flathead's demand response coordinator. "And we also stressed what was in it for them — what they would gain for participating. We basically used any kind of communication tool available and talked to every community group that would listen. We did a lot of boots on the ground stuff."

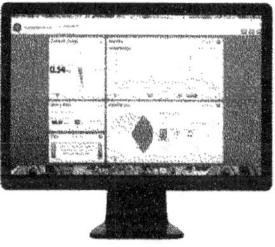

Peak Time aims to help energy consumers reduce energy use when the demand for and cost of power is highest. This type of adjustment in energy consumption is called demand response. Smart grid-enabled demand response requires two-way communication between the utility and the end-users.

The method for carrying out this communication was also an important consideration for Flathead. Based on some initial reactions from members about the use of wireless networks to transmit information, the co-op chose to use an "over the power line" approach, and emphasized that in its communications.

To gather the most information about member preferences, Flathead offered three options:

OPTION 1: In-home display
A free in-home display unit notifies households of peak demand times, signaling them to reduce consumption until demand on the system declines.

Participants receive a $5 monthly credit and an annual rebate determined by their energy consumption. If the participant's highest hour of use during the billing cycle is during a non-peak time, the participant receives $4/kilowatt for the

difference in consumption between the highest non-peak hour and the highest peak hour.

The in-home display was the least-costly option to implement, at about $125 per member. Its purpose was to show consumers how much electricity they were using and when they were using it.

Due to some limitations of the emerging technology, Flathead has been unable to use the tool as it had planned, such as to send volunteers data about their current use or billing information. The only information households receive is an indication that it is a peak time, signaling them to reduce their energy use until the demand on the system decreases.

Keeping the households tuned-in to the tool during the five-year demonstration has been a challenge.

"There is a little bit of attrition on a long project," says Schneider. "Utilities need to have an actionable activity for the members on a regular basis in order to keep them engaged."

> " I think the biggest thing that's misunderstood with smart meters or two-way meters or remote meter reading is there's going to be more privacy intrusions by that than having a person actually walk around your house once a month to check the meter. There's a little bit of disconnect on the privacy/security aspect of it with the public compared to what was done or what they're willing to accept from other technology. "
>
> — RUSS SCHNEIDER, REGULATORY ANALYST

OPTION 2: Water heater demand-response unit

A free demand-response unit automatically cycles off participants' water heaters for up to two hours during times of peak demand to reduce energy consumption.

Members who volunteered for this option receive an $8 monthly credit. The co-op uses over-the-power-line technology to operate each household's water heater in response to peak demands.

The water heater cycling has produced the most reliable savings across most peak demand events. On average, this option reduced energy consumption by 0.58 kilowatts per unit. With an average installation cost of $413, the utility expects it could recover the investment in three to five years.

"The demand response units attached to hot water heaters are very reliable," says Rayome-Kelly. "I can't think of any significant challenges we've had with those. When we started this, we thought this test group would be the hardest one to sign people up for, because you're hooking equipment onto their water heaters. But people really accepted that quite well. It wasn't a problem to get people to sign up."

Flathead received zero complaints from members regarding a lack of hot water.

OPTION 3: Home Energy Network

Volunteers paid $800 for new appliances — a dishwasher, clothes washer and dryer — plus an electric water heater demand response unit and equipment that enables the appliances to communicate with the utility over the members' home wireless internet connection.

Using a signal sent over the power line from the integrated advanced metering infrastructure system to each participant's wireless internet connection, energy-efficient appliances can be cycled off or put into an energy-saving mode as needed to reduce demand on the system. If the participant's highest

Teri Rayome-Kelly chats with residents to rally participation in Peak Time.

hour of use during the billing cycle is during a non-peak time, the participant receives $4/kilowatt for the difference in consumption between the highest non-peak hour and the highest peak hour.

When Flathead offered a Home Energy Network option, it didn't take long for the utility to realize it had taken on more than it had anticipated.

"We hadn't planned on being in the appliance business," said Rayome-Kelly. "But as a small community, we don't have any big box stores, so we had to look for a contractor to install those for us."

Appliance handling and installation scheduling became a newly acquired skill for some.

"I can level a washer with the best of 'em," Rayome-Kelly said.

While the smart appliances proved that they could reduce household peak energy use by up to 2.34 kilowatts, it was the most costly option to implement. It cost about $2,500 more to install the smart appliances and related equipment, compared to the cost of new traditional appliances.

Flathead also learned that the integration of different technologies can be messy. Technical issues arose from the use of interoperable appliances, which struggled to communicate with the Home Energy Network. Vendors had to learn about new technologies and new products and figure out how to make them work together. Flathead also faced challenges integrating its own internal systems with the Home Energy Network.

At the end of the day, the co-op's pilot project gathered important data and learned key lessons to improve future implementation.

WHAT'S NEXT for Flathead?

With a toolkit of expertise and lessons learned, the co-op is ready to get started with a demand response program that makes sense to the bottom line.

"We're already planning to do an extended water heater program," says Schneider. "We're planning to connect 1,000 water heaters each year for five years."

298

Pacific Northwest Smart Grid Demonstration Project
SUCCESS STORIES

NORTHWESTERN ENERGY

Small steps to a smarter grid

Listen to a planning meeting at NorthWestern Energy and you'll likely hear: *deploy at the speed of value, and stay on right side of the repair-versus-replace curve.* Decisions here are made very carefully. After all, this award-winning, investor-owned utility serves one of the largest, most geographically diverse territories in the region. With an infrastructure that spans over 28,000 miles of transmission and distribution lines across three states, planning ahead is important. Especially as the 500,000 poles, components and wires get older.

A plan to upgrade its basic distribution system was already in the works when the opportunity arose to take part in the $178 million Pacific Northwest Smart Grid Demonstration Project. Improving upon existing infrastructure using smart grid technologies just made business sense.

"We weren't quite ready for it," said George Horvath, manager of automation and technology for NorthWestern. "We expected that the technologies would advance, change and be improved over two to three times during the course of the project."

So going small-scale was NorthWestern's solution.

With its $2.1 million investment, North-Western also planned to learn from the other participants.

Customer side of the meter

A perfect urban area to test new technologies turned out to be Helena, Mont. With 30,000 customers and an electric load of 90 megawatts, Helena had

NorthWestern Energy
Delivering a Bright Future

NORTHWESTERN ENERGY
Butte, Montana
- Serving 349 communities across Montana, South Dakota and Nebraska
- One of the largest service territories in the Northwest
- 28,000 miles of lines
- 400,000 metered customers

INVESTMENT
$2.1 million

HIGHLIGHTS
- Demand response program
- Distribution Voltage Reduction
- Advanced AMI communications network
- In-home energy displays

FOR MORE INFORMATION
Claudia Rapkoch (406) 497-2641
www.northwesternenergy.com

27

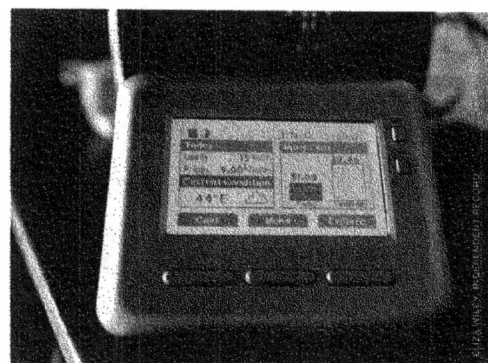

the right mix of customers and basic systems.

FIRST STEP: recruit participants. Around 200 residential customers and two commercial buildings from the State of Montana were enlisted to take part in the nation's largest test of new smart grid technologies. It took two marketing campaigns and extending the area beyond Helena to reach recruitment goals.

NEXT STEP: install equipment. Residents' homes were fitted with switches to control appliances, outlet-type switches that turn regular electrical outlets on and off, as well as programmable thermostats and energy system display devices.

Installing the equipment was easy.

Educating customers and learning from the experience took more time.

Working closely with customers

"We hired a company to work directly with our customers and teach them how to benefit from the equipment in their homes and to learn to use it effectively," said June Pusich-Lester, NorthWestern's demand side management engineer.

Training included how to program the equipment and use a web portal. A web portal is another name for a dedicated website that has special functionality.

The portal showed past energy use, as well as the energy consumption of every device connected to the network. A monthly electronic newsletter was also sent to educate customers.

The benefits were twofold. Customers could see their energy use to better understand ways to save, and North-Western gained insight into what customers want and what they are willing to do to conserve energy.

Would customers respond to a reward?

Time of use and demand response

Residential customers were set up for time-of-use pricing. These programs help the utility to control some of the consumers' electrical load in response to grid conditions and the price of electricity. Here's how it worked:

Montana has a flat residential rate, but the demonstration project offered a regional price. So for testing purposes, the rate fluctuated. Each customer received a signal that displayed the price of electricity

as low, medium or high depending on the time of day, the day of the week and the month or season.

Customers responded by adjusting and programming the equipment, attached to a home area network, based on the pricing schemes. As a result, load decreased during peak times of use.

"We rewarded our customers for energy savings," said Pusich-Lester. "With the smart meters and the communication network working together, we could read energy usage in 15-minute increments."

If a customer used less energy when prices were high, NorthWestern credited the customer's monthly bill. As of September 2014, total savings from the time-of-use pricing totaled $13,787 for all customers.

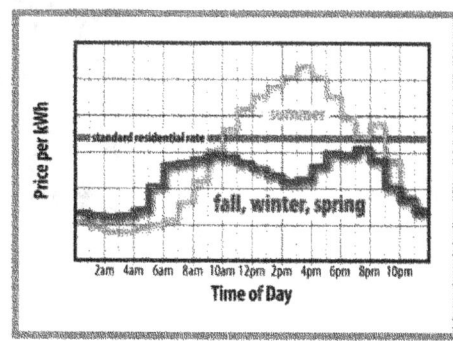

300

The system also gave the utility direct control over some residential customer loads.

"Now we are able to send demand response events directly to the homes," said Pusich-Lester.

With the ability to control home appliances using two-way communications, North-Western reduced home temperatures or turned off appliances when the price was high in the middle of the day. Demand response and time-of-use methods provide flexibility while saving money.

Other technologies focus on overall efficiency.

Voltage reduction = efficiency

We're all familiar with the concept of energy efficiency. Consumers reduce power consumption through choices in lighting, insulation, appliances and many other methods. Utilities have been working with customers to improve energy efficiency for more than 30 years.

But there's a new player in town.

Distribution voltage optimization, or DVO, lowers the energy consumption on a whole feeder — the line that delivers electricity from a substation to a home. By dropping the voltage on a circuit — while staying above the minimum level necessary to operate electric devices — the customer's energy costs decrease. "We flatten the voltage profile, make voltage adjustments, and save energy for the whole feeder," said Horvath.

According to industry data, a potential exists to shave one to three percent of circuit load using this technique.

Utility side of the meter

Keeping the lights on is a mission of every utility. Until smart grid technologies came along, distribution systems were in the dark. Utilities were unaware of an outage until a customer called to report it.

Now, new technologies on the utility side of the meter use automation to improve reliability by detecting a problem, isolating it, and then restoring as many customers to service as possible.

This type of system is called distribution automation or self-healing technology. Computer systems quickly react to electrical issues in the system, like a fault in a feeder, without intervention from an operator or line worker. NorthWestern tested Fault Detection, Isolation and Restoration, or FDIR, software.

"We configured circuits with remote capabilities, to monitor circuits with central software, and to be able to reconfigure circuits in case of issues," said Horvath.

Since October 2012, the system has already automatically reconfigured and mitigated customer service on the feeder for two outages in Helena. That means shorter outages for customers and resource savings for NorthWestern.

Testing technologies to solve real-time, real-world problems is what the demonstration is all about. Still, many lessons were also learned in the research and development initiative.

Lessons to share

NorthWestern's goal was to make slow improvements to prepare for larger business objectives and to learn how to invest in products and services that have longevity.

"We definitely realized our objectives," says Pusich-Lester.

One unexpected lesson was in selecting vendors. Some vendors evolved or went out of business, leaving the utility stranded with products that didn't work. The complexity of integrating components from different vendors while building the systems was also unexpected.

Other notable lessons from NorthWestern:

- Start with a small project — it makes the business case analysis easier
- Emphasize the importance of customer education with project stakeholders
- Integrate a customer information system to smart grid at the start
- Work closely with customers to understand new system enhancements and in-home display features

> "We're going to have the benefit of all the other, much larger projects from the demonstration, reading through their evaluations, and learning from them. It's an important part of our project, what we've learned and accomplished, so now we can better communicate about the smart grid with our regulators and customers in the future."
>
> — GEORGE HORVATH, MANAGER OF AUTOMATION AND TECHNOLOGY

29

- Billing system integration for new programs requires significant planning effort

- Build a distribution management system first, then add smart grid

- Allow sufficient time and money cushion for communication backbone

- First-time equipment rollouts have engineering, IT system, communication and business program learning curves

- In your risk analysis, consider the possibility of a vendor going out of business during your deployment

"The project has helped to mold our thinking on how we plan on a larger scale," said Horvath. "We might do things differently now from the big picture perspective."

WHAT'S NEXT for NorthWestern?

During the project, NorthWestern kept a clear focus on its basic infrastructure and worked to remain on that right side of that repair-versus-replace curve. Outcomes from the project will inform future smart grid improvement processes and projects.

"We will continue to invest in the basic infrastructure and incorporate new technologies where they make sense," said Horvath. "This project provided a foundation for us to evaluate something much larger going forward."

That includes keeping customers engaged.

For both the utility and the customer, the chance to become better informed, educated and experienced with the technologies will prepare everyone for the utility of the future, whatever that may bring.

Pacific Northwest Smart Grid Demonstration Project
SUCCESS STORIES

MILTON-FREEWATER

Milton-Freewater: A frontier for new technology

Oregon's oldest city-owned utility is far from old-fashioned. In fact, it's a power pioneer in the Pacific Northwest. Since 1985, Milton-Freewater City Light and Power has reduced peak energy use with a technique called demand response. Demand response may not be the talk of the town, but it's still a big deal to this homegrown utility. And it's one of the many technologies the Pacific Northwest Smart Grid Demonstration Project intends to advance.

Milton-Freewater is the only public utility in Oregon chosen to take part in the nation's largest smart grid test.

With Milton-Freewater's $1.8 million investment and DOE's matching funds, the rural utility upgraded its historic demand response program and tested some newer technologies, such as voltage reduction and voltage-sensing water heaters.

Of the 60,000 metered customers involved in the regionwide project, Milton-Freewater City Light and Power is the smallest participant with only 4,550 customers. The utility did not hire any additional staff except for a contractor to perform installations, because the utility's only electrician had retired.

Blazing the trail for demand response

When not at his cherry orchard, retired city electrician Bill Saager enjoys the cowboy shooting range just outside of town. The 75-year-old bandana-wearing quick-draw installed the city's original demand response units — all 500 of them.

But getting and using a contractor's license wasn't easy. Saager convinced the State of Oregon Construction Contractors Board to forgo the insurance and bonding requirement because he was only doing work for the utility, not the customer. Clearly, Saager is a fan of smart technology.

"I think that utilities are really missing the target if they don't pursue some type of DR

MILTON-FREEWATER CITY LIGHT & POWER
Milton-Freewater, Oregon

- Locally-owned and operated since 1889

- Average annual sales: 118,000,000 kilowatt-hours

- 4,550 customers

INVESTMENT
$1.8 million

HIGHLIGHTS
- Completed AMI rollout
- Direct load control/DR
- Five-megawatt reduction
- Conservation voltage reduction

FOR MORE INFORMATION
Tina Kain (541) 938-8238
www.mfcity.com/electric

in the future," said Saager. "You can really extend the capabilities of your system."

Milton-Freewater's original demand response program used a radio energy management system, or REMS, to control electric water heaters and electric heating and air conditioning systems. A centralized computer system monitored the power demand and sent a radio signal to the REMS units to cycle off connected loads to reduce energy when the peak demand set-point was reached.

The city's goal of the Pacific Northwest Smart Grid Demonstration Project has been to enhance its already highly successful demand response program.

Listen. Respond. Repeat.

To do that, the newer technology must go one step further, by listening and responding from both sides of the communication. With a smart meter system that uses two-way communications, the utility confirms that the demand response units receive the signal and controls the amount of time the electricity is shut off. The goal is for the entire demand response process to go unnoticed.

Water heaters, electric heaters and air conditioners are connected to the demand response system to trim energy use when a certain set-point is reached. Up to three megawatts can be reduced by shaving the energy used in 754 customers' homes, businesses and even churches.

Replacing the old units with the newer models spurred a lot of questions. Many customers didn't know the units existed. A few were suspicious of the utility installing new, more intelligent units. But only a few opted out.

"We found that many homes had changed tenants," said Tina Kain, engineering technician for the city. "New residents were unaware of the DR units, which is a great sign. The undisruptive units helped residents save money without even knowing it."

To entice participation, a rate discount was offered to customers:

* 2.5 percent for water heaters
* 2.5 percent on electric heating
* 1 percent for air conditioning

DR is used as a last resort in the peak shaving process. When the system begins to approach its peak energy use, the first step is to reduce voltage by 4.5 percent.

Next, the city shuts off the wells that are connected to the centralized computer system. Finally, DR units are employed. Reverse order restores the system to its original state.

The incredible value of the smart meter

Every customer of Milton-Freewater City Power and Light is now set up with a smart meter that uses two-way communication over the power line. With these more advanced meters, energy use is monitored every fifteen minutes. That means better customer service, such as helping homeowners troubleshoot high bill complaints and remote meter reading. On selected meters, a device called a disconnect collar, which allows for a remote disconnect or reconnect, was installed.

One of the biggest benefits is a smart meter's ability to quickly detect an outage. Previously, the utility wasn't aware of an outage until a customer called it in. A crew was then dispatched to investigate and do the repairs. Often, the process could take hours. Now, many outages are fixed in minutes, saving the utility both labor and transportation

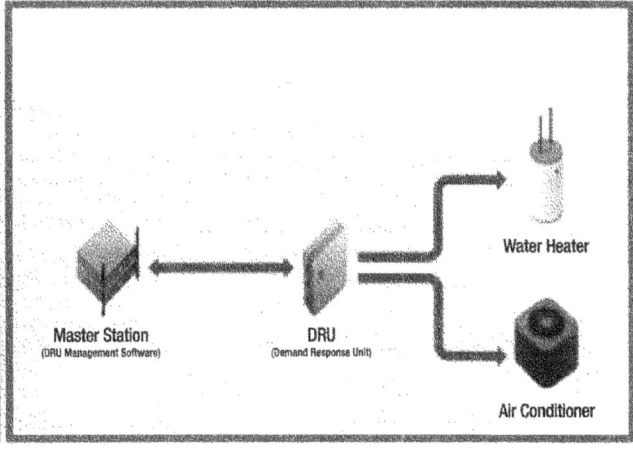

Master Station
(DRU Management Software)

DRU
(Demand Response Unit)

Water Heater

Air Conditioner

> **❝** You can see from our history that Milton-Freewater City Light and Power is innovative and forward thinking. It's about planning ahead. We all know that electric rates probably aren't going to be going down any time soon, so if a new technology pencils out now, it's going to be more beneficial in years to come. **❞**
>
> — RICK RAMBO, THE CITY OF MILTON-FREEWATER ELECTRIC SUPERINTENDENT

expenses as well as providing better service to the customers.

Conservation voltage reduction

Conservation voltage reduction is Milton-Freewater's first step to address a peak on the system. Substation voltage regulators lower the system voltage by 1.5 volts on four feeder lines out of the Milton Substation. This reduces the megawatts used on the entire system while still maintaining adequate distribution voltage.

To test the theory that every 1 percent in voltage reduction leads to a 1 percent reduction in kilowatt-hours used, Milton-Freewater reduces voltage one week, and then returns it to the status quo the next. By alternating weeks, the utility will be able to compare the two datasets and calculate the benefits once the demonstration is complete.

Specialized water heaters

To further test possibilities with conservation voltage reduction, Milton-Freewater installed 100 demand response units on water heaters that operate when the city's voltage reduction occurs. The demand response units are programmed to sense voltage and turn off connected load.

Although the water heaters worked well with the voltage reduction system, they didn't work so well with another part of the project — the transactive control signal. This signal is being tested as part of the demonstration across a multiple utility footprint to assess the feasibility of automating the trade of energy based on many conditions, such as the availability of wind or solar power, for a regional benefit.

When the transactive control signal activated a voltage reduction, the demand response units turned off the connected water heaters, and they stayed off until the voltage reduction ended. When the signal lasts longer than five or six hours, the customer may run out of hot water, resulting in customer complaints and countering the city's goal of ensuring demand response goes unnoticed.

Sizing up the study

Even though the city has been doing demand response for three decades, testing new technologies presented challenges.

First, differences between the original and new demand response systems were disappointing. The old REMS units operated 20 minutes on, 15 minutes off, so Milton-Freewater could reassure customers that their water heaters would be turned off no longer than 15 minutes. But the new units turn on and off in an inconsistent, unpredictable pattern, which is a little more difficult for customers to accept.

And that wasn't the only challenge with the units. They all cycled on or off at the same time, creating their own peaks and valleys. Innovative solutions were needed to stagger the units.

Data storage is also a problem. Unless the data is manually extracted from the unit before its next operation, the data is lost. That means the utility cannot determine how well the unit worked for that cycle. To work around this issue, the research laboratory will analyze meter data to determine load fluctuations from the demand response units.

Overall, lack of resources has been one of the largest challenges for the small rural utility — the employees had to learn a whole new system while doing their regular jobs. But that also demonstrates the city's dedication to keeping rates low.

"Every little bit counts," said Bill Saager, the retired electrician-turned cowboy.

WHAT'S NEXT

for Milton-Freewater?

As a result of the lessons learned from the project, the city plans to discontinue conservation voltage reduction. Even though it has provided some benefits, operating more than one voltage reduction system at the same time on different feeders was confusing for work crews.

Furthermore, once the project concludes, the city will explore transactive control opportunities based on the needs of the city and the potential benefits.

Other ideas for the future include a customer pre-pay system so that energy use can be managed based on an individual family budget.

'We're at the point of trying to learn what we can do with the system," said Rick Rambo, the electric utility's superintendent. "I know we're not using it to its full potential."

Still, the tried and true prevails. Demand response will continue as it has since 1985 across the city's entire electrical system, with fine-tuning of the operation as needed and with very little impact on the customer.

Pacific Northwest Smart Grid Demonstration Project
SUCCESS STORIES

LOWER VALLEY ENERGY

Cold-climate co-op heats up with smart grid

Lower Valley Energy provides electricity to one of the biggest resort towns — Jackson Hole, Wyo. — in one of the coldest climates in the Northwest. At the southern base of the Grand Teton and Yellowstone national parks, out-of-town visitors and residents alike rely on the home-grown electric co-op for heat, hot water and light — especially during cold snaps when the demand for power is highest.

That's one reason Director of Engineering Rick Knori wanted to complete the utility's deployment of its smart grid metering system and help its more than 26,000 electric customers better understand their energy use. That opportunity surfaced with the Pacific Northwest Smart Grid Demonstration Project.

With a focus on exceptional customer service, reliability and low rates, Lower Valley Energy jumped on the chance to improve the way it provides services.

Smart meters and demand response

Before the project, more than 12,000, or nearly half, of Lower Valley's members

had smart meters that allow for two-way communications between the utility and end-user. As of March 2014, 100 percent of its members had smart meters installed in their homes. This technology is a necessary component of many smart grid technologies, including water heater demand response, which allows a utility to cycle off participants' water heaters during times of peak demand to reduce energy consumption.

High-end tourism is big business in the Grand Teton area. One county in the Lower Valley service area is consistently rated one of the top five wealthiest counties in the United States. So, it's no surprise that hot water is a hot commodity. Getting a cold-climate community warmed

LOWER VALLEY ENERGY
Afton, Wyoming
- Began in 1937 with 10 members
- Provides electricity, natural gas and propane
- 5,616 square-mile territory
- 26,406 electric customers
- 195 megawatts

INVESTMENT
$1.2 million

HIGHLIGHTS
- Adaptive Voltage Control
- Complete AMI rollout

FOR MORE INFORMATION
Rick Knori (307) 739-6038
www.lvenergy.com

35

> Lower Valley consumers have been very excited to see the new technologies related to smart grid, and that we have been doing pilot projects to improve their power quality and efficiency."

— RICK KNORI, LVE DIRECTOR OF ENGINEERING

up to a program that even hinted at the risk of a cold shower was a tough sell.

A $15 a month incentive to participate in the demand response program just wasn't enough for some members, but for others it was worth a try. Ultimately, the co-op deployed more than 500 demand response units and used them to temporarily turn off customers' water heaters during periods of high demand, when energy prices are highest, thereby reducing energy consumption.

"They worked excellent," said Knori. "These are going to be long-term assets that we keep and control."

Now there are 100 people on Lower Valley Energy's waiting list for a water heater demand response unit.

Adaptive voltage control a surprise

The most successful assets Lower Valley deployed were tools for adaptive voltage control, which can reduce a customer's overall voltage during brief high-demand periods and result in short-term demand reductions.

Using regular feedback from its customers' meters, Lower Valley reduced voltage — and therefore demand — during peak load periods at the utility's East Jackson Substation. This technology provided the greatest benefit for the least investment of time and money.

Warren Jones, Lower Valley's distribution engineer, programmed the adaptive voltage control signals.

"I think it was a surprise for us how well the adaptive voltage control worked," said Jones. "That's the one thing I would suggest to other utilities that have advanced metering infrastructure in place. It does open up some opportunities for a utility to use that intelligence to actually lower your voltage."

Adaptive voltage control has the potential to greatly lower future monthly demand charges paid by Lower Valley to Bonneville, reducing energy costs to customers. The test case also proved that Lower Valley can easily expand adaptive voltage control to all of its distribution system substations in the coming years.

Solar success

Lower Valley wanted to capture wind and solar power during the day, store it in batteries and discharge it during the utility's two-hour morning peak. The purpose was to test whether new technologies could reduce transmission system losses and improve voltage stability on a 60-mile distribution line to Bondurant, Wyo. At its Hoback Substation, Lower Valley installed a system of renewable energy resources and battery storage, including one 15-kilowatt solar photovoltaic, one 20-kilowatt windmill set and a 200-kilowatt-hour battery.

"The solar and inverters worked flawlessly," said Knori. "We're getting about a 17 to 18 percent capacity factor out of those units. And they're working trouble-free. But the windmills were kind of a bust."

After two years of operation, the total output of the four windmills was about 80 kilowatt-hours. Lower Valley installed an anemometer on the windmills to prove to the vendor that the turbines were not producing to the expected capacity. But by the time the data was available from the anemometer, the vendor was out of business.

The battery storage — the newest of the technologies — also presented some challenges. The batteries arrived damaged and had to be replaced, which caused a delay. After a couple of programming issues were worked out with the vendor, the batteries were up and running, but at half the expected 120 kilowatts of storage capacity.

In-home displays, in the drawer

Lower Valley also installed in-home display units to provide consumers information about their energy consumption. It was a lot of work for crews to install the 400 devices. Yet with minimal customer feedback, the tool's impact on consumer behavior is unknown.

Knori believes that few customers are paying attention to the display units because newer tools, such as smart phone applications that perform the same function, outpaced the in-home display technology.

308

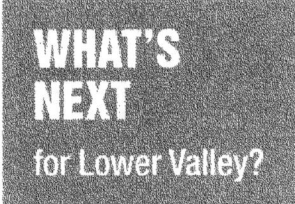

WHAT'S NEXT

for Lower Valley?

Continued use of the hot water heater demand response units will help Lower Valley keep electric bills affordable. As energy costs continue to increase over time, Lower Valley might decide to expand its demand response program.

By installing the adaptive voltage control technology to additional substations, the co-op will take advantage of even more energy savings.

Finally, with the help of the solar array, Lower Valley Energy looks forward to keeping more electrons where they belong by avoiding the losses that typically occur over long distances.

AND BY THE WAY ...

Profile of Battelle

In the Pacific Northwest Smart Grid Demonstration Project, Battelle leads a collaborative effort that includes the Bonneville Power Administration, 11 utilities and six technology companies. Battelle also was a partner in the AEP Ohio gridSMART Demonstration Project.

Battelle has been headquartered in Columbus, Ohio, since its founding in 1929 and is a research and development organization that also designs and manufactures products and delivers critical services for government and commercial customers. Battelle's contract research portfolio spans consumer and industrial, energy and environment, health and pharmaceutical, and national security. As part of its government-related work, Battelle manages national laboratories, including Pacific Northwest National Laboratory in Richland, Wash.

Battelle is the world's largest nonprofit research and development organization, with over 22,000 employees at more than 60 locations globally. A 501(c)(3) charitable trust, Battelle was founded on industrialist Gordon Battelle's vision that business and scientific interests can go hand-in-hand as forces for positive change. Battelle's strong charitable commitment to community development and education includes a focus on staff volunteer efforts; science, technology, engineering and mathematics (STEM) education programs; and philanthropic projects in the communities Battelle serves.

Pacific Northwest Smart Grid Demonstration Project
SUCCESS STORIES

PENINSULA LIGHT CO.

Smart grid provides power bridge to Fox Island

Nearly half of Fox Island's 1,200 residents moved here to retire by the water. They sail. They hike. They kayak in scenic Puget Sound. But the highly educated maritime residents also expect the same reliable electricity they once enjoyed in the city. That was a challenge for the utility that relied on aging cables to deliver the island's electricity.

Yet it was also an opportunity to test new technologies to improve service, and the reason Peninsula Light Co. applied to participate in the Pacific Northwest Smart Grid Demonstration Project.

PenLight invested $1.2 million in the $178 million cost-share project. Just in time, too.

A critical need

PenLight serves all of Fox Island, which, given its watery surroundings, has no gas service. That means many residents and businesses are dependent on electric power for everyday needs. Two cables deliver electricity to the island: one submarine cable and the other attached to a bridge.

"During the summer of 2010, the submarine cable that was installed in 1970 started to fail," said Jonathan White, PenLight director of Marketing and Member Services. "An analysis determined that if the temperature fell below 20 degrees Fahrenheit, the ability to maintain load on the island's remaining circuit would be difficult."

To cope with potential outages, a comprehensive strategic plan was developed. The plan included rolling blackouts, backup diesel distributed generation and a smart grid program.

"Fox Island became a smart grid test bed," said Mike Simpson, PenLight's manager of Engineering. "Knowing there was an aging issue with one cable, we thought: let's look at demand response to reduce the load."

Peninsula Light Co.
The power to be...

PENINSULA LIGHT CO.
Gig Harbor, Washington
- Member owned since 1925
- Second largest membership of any co-op in Northwest
- 997 circuit miles of line
- 112 square-miles of service territory
- 31,000 metered customers

INVESTMENT
$1.2 million

HIGHLIGHTS
- Power Sharing
- Power line-carrier system
- Self-healing network
- Integrated volt-VAR control

FOR MORE INFORMATION
Mike Simpson (253) 857-1580
www.PenLight.org

Failing cable leads to fast launch

By September 2010, with winter quickly approaching, a demand response marketing program for electric water heaters called Power Sharing was launched. The program would reduce electric demand when needed by controlling residential water heater operations during peak load time periods.

Community meetings were held as construction started on a new cable 40 feet below Puget Sound. The participation

goal was to get 400 of the 1,700 metered customers on the island to participate in the program. Volunteers were offered a $5 credit on their monthly power bills.

"You don't have to tell the light company about this, but I would do it without a discount," said Dick Olszewski, a PenLight customer. "Electricity is something you must have to survive. Being on an island makes it more difficult, but it's worth it."

A few months later, the old cable partially failed. That's when the program fast-tracked its way to load reduction during the coming winter months. By the end of December 2010, approximately 25 percent of the targeted customers were on board with the project. A month later, an additional 10 percent had joined. By February 2011, the goal was met.

During several low-temperature days, a phone message was sent alerting island residents of the low-temp risk and that the utility would begin cycling off water heaters.

"We were sending communications out to the members that this was a crisis situation and that we needed their assistance," said White. "Rolling blackouts were imminent if the demand

response program and voluntary curtailment failed to meet the need."

The emergency plan worked to get the island through the winter.

One bit every 20 minutes

Communication sent to the water heater controllers, turning them on or off, worked well. But getting data from the controllers back to the utility over its power line-carrier data collection system was another story.

The technology sent the data along with electricity through the power line. While these systems are generally known for being low-cost, they have significant bandwidth limitations. A smart phone or streaming TV connection delivers 10 megabits per second — PenLight's system could only deliver one bit every 20 minutes. Hourly meter reads created challenges for the required project reporting.

With 500 homes on the island participating in the program and with the low speed of data transmittal, it was hard to tell if the program was working.

"We realized we had to get higher speed communications or it would be difficult to determine how effective it would be to use the system throughout the rest of our service area," Simpson said.

A cellular solution

Cellular-based transformer monitors provided a perfect work-around.

The devices — part of PenLight's systemwide monitoring program — were installed to measure, in 15-minute intervals, transformer loads and conditions. This monitoring was critical to the ongoing reporting requirements of the demonstration, allowing the utility to

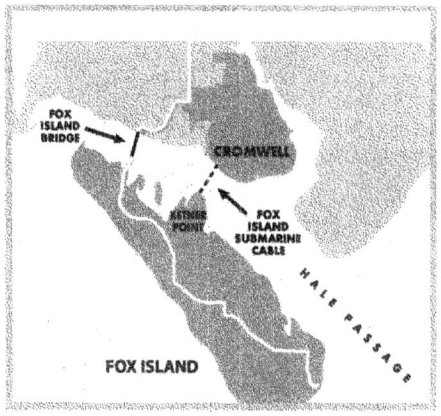

know if a hot water tank was turned off and if so, precisely when.

In addition, gathering data at the transformer meant that a demand response event could be validated at one point for multiple participating homes. That means the utility now knows, on a cost-per-unit basis, whether it's practical to deploy the system elsewhere.

Shorter and shorter and shorter outages

Imagine a grid that fixes itself when a car hits a power pole or a storm trips a wire. That's exactly what a self-healing grid does. The system detects the fault and then isolates the problem quickly — within minutes — by automatically deciding which switches on the transmission system to open and close. The result: fewer members affected. In addition, work crews know the precise location to investigate. Eliminating the need to patrol the entire circuit saves time and also reduces the length of the outage.

For some, adopting such technology without hands-on experience is tough to embrace. That's why PenLight is taking time to learn about the operation of the system and how to get the most value from the demonstration project for its customers.

"It's difficult to embrace what the system can do," said Simpson. "You're allowing it to exert control automatically. It's one thing to accept the technology; it's another thing to trust it."

The system — installed in strategic locations on the island — will ultimately be allowed to minimize the impact of an outage on its own. But until that trust is established, a person still needs to hit the button to execute the fault isolation. Still, it didn't take long for the system to flex its muscle.

When a large tree fell on a major feeder and disrupted power to 1,300 customers, the system quickly identified the location of the problem. Then a foreman opened up a switch to isolate the damaged section. The crew then restored the rest of the circuit, which brought about 1,000 customers back on line within 30 minutes. It took four hours to bring the remaining 100 customers on line. That reduced the number of customers who were affected by the longer outage by 80 percent.

Eventually, the utility may give the system actual control, but that will take time and a clear understanding of how well the system works. "It's about having complete confidence in the system," said Simpson.

Volts-VAR

Every home typically gets its electricity from what's called a feeder — the wire outside a home connected to a substation. The voltage is higher at a substation than it is at the other end of the line. Utilities work to flatten these voltages so they are more consistent using devices called regulators and capacitors. When the voltages are about the same, the utility has more flexibility to raise or lower voltage, a tool that can increase or decrease energy consumption during a demand response event, without the voltage getting too low for homes further down the line. This is called Integrated Volt-VAR (volt-ampere reactive) Control, or IVVC.

PenLight initiated a Volt-VAR control project but couldn't automate it as planned, due to the impact to capacitor switches on the power-line-carrier system, as well as software and monitoring concerns. In addition, the technology requires very accurate voltage measurements, which the utility's current technology isn't able to provide.

The smart grid IVVC solution was built to address a global marketplace which requires data measurement at five-minute intervals. PenLight has a small distribution network with very short feeders, so there is not a lot of variation across the feeder, particularly over five-minute intervals.

"You see a little voltage fluctuation at the beginning of the day and a little at the end," said Simpson. "Five-minute timeframes were a little off-putting because you need

measurement devices in the field and data transport mechanisms, which is costly."

As a compromise, PenLight used the transformer monitors to get very accurate voltage measurements, capture them over 15-minute intervals along with the load data, and then used a man-in-the-middle approach to adjust the voltage.

More lessons to share

Some lessons learned in demonstration projects are qualitative, like in managing change.

"It's important to ensure that everyone participates in the decisions about operational change so the entire company supports it,"

said Simpson. "If new processes and procedure are thrust upon employees who have been doing business the same way for a very long time and trust that process, it's difficult to make determinations about the new solution. Because you're really dealing more with perceptions and how comfortable people are with something rather than the value you've built into it."

Other lessons to note from the team include:

- Have operational personnel involved in equipment selection. The overhead switches are awkward and difficult to manually operate for some line crews.

- Ensure that the geospatial database is very accurate if purchasing a system model out of the box.

- Use more than one vendor when integrating software. With just one vendor, it's really difficult to identify the root cause when there's a problem. When integrating tools from different vendors there is an integration point and a boundary that assists in problem-solving.

- Be as clear as possible about team responsibilities — be strategic in assignments according to the functionality of a device and the role of a team member.

'Often, sharing the lessons learned is the most valuable part of a project like this," Simpson said.

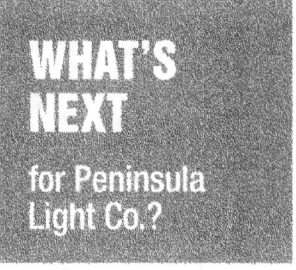

WHAT'S NEXT
for Peninsula Light Co.?

The member-owned utility intends to ensure that resilience is built into the system. One way to do that is by being very proactive. They will take a step forward by taking a look back. Technology platforms installed over 10 years ago may need to be upgraded. And if so, PenLight will determine how those technologies will be integrated into other platforms.

After all, many customers are already on board.

"This type of technology is not a bother," said Olszewski. "It only operates when there's a peak that's taking place."

In today's world, there's really not a single right path from one utility to another. Every utility has its own unique challenges. PenLight's included being surrounded by water.

Pacific Northwest Smart Grid Demonstration Project
SUCCESS STORIES

CITY OF ELLENSBURG

Renewable expansion for a historic utility

There's a lot of sunshine in the heart of Washington State. So much so that the City of Ellensburg uses the area's most abundant natural resource — the sun — to help meet the needs of its energy consumers. In 2006, the nation's first city-owned community solar array was installed north of Interstate 90. The successful system brought with it an audience of onlookers and inspired lawmakers. So, when the chance to expand its renewable resources came along, the utility sought funding from the Pacific Northwest Smart Grid Demonstration Project.

Washington State's oldest municipal utility invested $850,000 to test the effectiveness of a variety of wind and solar systems and to gather information to share with the public.

What the customer wants

In this tight-knit town of just under 10,000 people, when the customers speak, the utility listens. One message was loud and clear: Offer more distributed energy options.

"There was a lot of interest in residential wind and solar generation," said Larry Dunbar, the city's director of Energy Services. "So we wanted to test which ones would work."

After all, the city had already achieved great success with the 300-watt solar array at the Ellensburg Community Renewables Park, one of the first of its kind in the nation.

"Residents were invited to purchase a (solar) panel, which ranged from $250 and up," said Beth Leader of Ellensburg's

CITY OF ELLENSBURG LIGHT DIVISION

Ellensburg, Washington
☀ Founded in 1891
☀ 9,300 electronic customers

INVESTMENT
$850,000

HIGHLIGHTS
☀ Renewable energy park expansion
☀ Residential wind turbine demonstration
☀ Advanced SCADA communications network
☀ Real-time data system for education and research

FOR MORE INFORMATION
Beth Leader (509) 962-7124
www.ci.ellensburg.wa.us

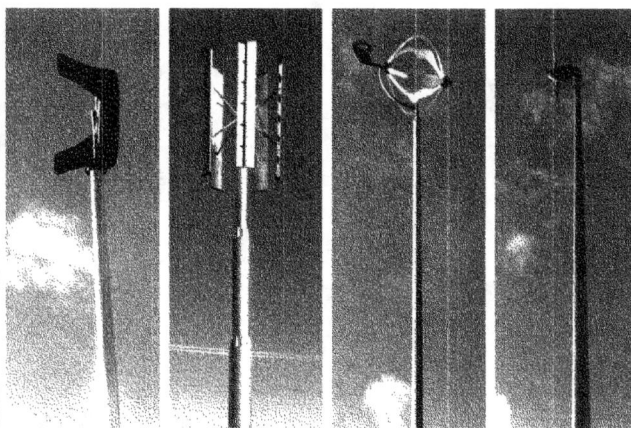

Energy Services group. "Then they receive a percentage of that generation in the form of a rebate on their utility bills."

With that kind of community investment, the city knew it needed to make very careful decisions.

Research and development

The city conducted a significant amount of research before selecting the technology for the project. Nine residential-class wind turbines were selected, ranging from

1.0 to 10 kilowatts, with a total output of about 45 kilowatts. The turbines were purchased from separate vendors to allow for a performance comparison. A meteorological tower was also installed to capture wind and temperature data in real-time. In addition, the existing solar array was expanded with an additional 40 kilowatts of thin-film panels. Finally, the city's fiber communications system was connected to tie all of the resources together.

The resulting array could have won awards for its artistic appeal.

Small wind a blow

The different turbine designs, staggered at differing heights, were visually appealing. But their performance was problematic. The smaller turbines required frequent maintenance.

The first tower that we installed was a great producer with no issues," said Dunbar. "Then one of the turbines failed."

That was just the beginning.

Months after installation, several other turbines stopped producing energy. Of the nine turbines, only five produced data significant to the project.

Then, a tower structure failed under high winds.

It turned out that four of the tallest tower structures, at nearly 80 feet, actually posed a safety risk, due to sustained winds of 30 to 40 mph. This is one of the windiest locations in the state — a good thing for wind generation, but not if your turbines aren't designed for it.

"That prompted a safety review of every turbine mounting," said Shan Rowbotham, the city's power and gas manager.

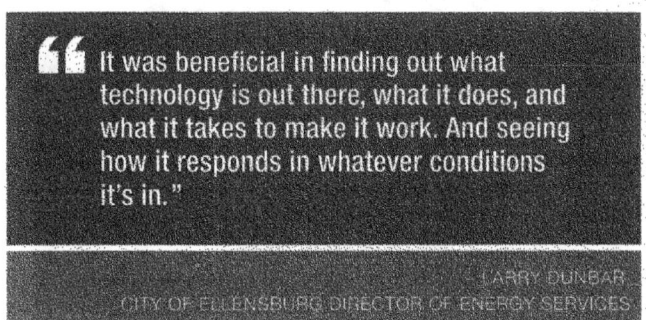

" It was beneficial in finding out what technology is out there, what it does, and what it takes to make it work. And seeing how it responds in whatever conditions it's in. "

LARRY DUNBAR
CITY OF ELLENSBURG DIRECTOR OF ENERGY SERVICES

Ultimately, two blade casings came apart and fell to the ground. Windy conditions sent the blades airborne — threating public safety at the park, where a trail wanders through the base of each turbine. Although no one was hurt, all towers were quickly deconstructed and removed.

"Having to take down the wind turbines was a very sensitive matter for us," said Dunbar. "We don't want to be portrayed as being anti-wind, but public safety is the most important thing."

Although the wind element was a blow to the project, the utility still saw solar success and fiber-optic fame of sorts.

Fiber fame and a school tool

The city's fiber system has drawn the attention of much larger cities, like Seattle. In fact, the City of Ellensburg fiber-optic network is well known throughout the state. Millions of points of data are delivered in real-time each second through five or six streams of information. The fiber network is connected to the utility's centralized computer system to capture data from the turbines and the solar array. That data will be used to help Central Washington University develop a K-12 renewables curriculum.

Learning from the project is important to the city.

Lessons to share

For the small utility, with less than 10,000 meters, buying into the technology and installing it was one thing. Keeping it operating and generating was another. Special products and special tools were required.

'It's imperative to do that research," said Dunbar. "Expect to take chances with new technology, expect things aren't always going to work out as planned. There's going to be some trial and error. That's the whole point of doing a project like this."

Other lessons learned from the project include:

- A fleet of small wind generation is very costly to maintain
- Carefully vet small vendors before making payments
- Use extra due diligence with ever-evolving experimental software and control systems
- Get expected costs right up-front

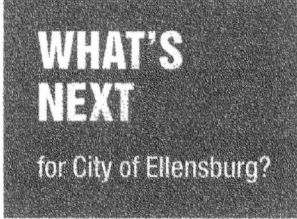

WHAT'S NEXT
for City of Ellensburg?

After uneven results from the demonstration project, the utility plans to march forward with planning for the future. With energy costs rising, looking at alternative energy options is top-of-mind. And the solar array isn't going anywhere.

"We're excited as we move forward to have the solar photo-voltaic array operating until the end of its life," said Dunbar.

That's in 11 years.

AND BY THE WAY ...

Profile of BPA

The Bonneville Power Administration is a federal nonprofit power marketing administration based in the Pacific Northwest. Although BPA is part of the U.S. Department of Energy, it is self-funding and covers its costs by selling its products and services. BPA markets wholesale electrical power from 31 federal hydro projects in the Columbia River Basin, one nonfederal nuclear plant and several small nonfederal power plants. The dams are operated by the U.S. Army Corps of Engineers and the Bureau of Reclamation. About 30 percent of the electric power used in the Northwest comes from BPA. BPA's resources — primarily hydroelectric — make its power nearly carbon free.

BPA also operates and maintains about three-fourths of the high-voltage transmission in its service territory. BPA's service territory includes Idaho, Oregon, Washington, western Montana and small parts of eastern Montana, California, Nevada, Utah and Wyoming.

BPA promotes energy efficiency, renewable resources and new technologies that improve its ability to deliver on its mission. BPA also funds regional efforts to protect and enhance fish and wildlife populations affected by hydropower development in the Columbia River Basin.

BPA is committed to public service and seeks to make its decisions in a manner that provides opportunities for input from stakeholders. In its vision statement, BPA dedicates itself to providing high system reliability, low rates consistent with sound business principles, environmental stewardship and accountability.

Pacific Northwest Smart Grid Demonstration Project
SUCCESS STORIES

BENTON PUD

Stepping into smart grid

Flanking the Columbia River in Washington are three sunny cities — Richland, Pasco and Kennewick — with roots in many things from vineyards to energy. Challenged by being one of the fastest growing areas in the state, utility providers here work together to make sure the lights stay on. They also team up to find new, innovative energy solutions to meet the area's growing energy needs. That's why Benton PUD opted to take part in the Pacific Northwest Smart Grid Demonstration Project. The goal: investigate new technologies and prepare for a more efficient future.

The project — in its fifth and final year — is led by Battelle Northwest, about 15 miles upriver from Benton PUD's headquarters at the Pacific Northwest National Laboratory.

The public-power utility district focused on two test cases: energy storage and a web platform to work with smart meter data. A smart meter uses two-way communications between a customer and the utility to improve services. This framework is called an advanced metering infrastructure, or AMI.

An AMI opens up a world of useful data to make the operation of the grid more efficient.

For Benton PUD, knowing how to apply the data was the first step.

A data training tool

A web platform called DataCatcher™ used real-time communication to acquire alarm data from the wireless AMI system. These alarms notified the utility of potential problems on the

BENTON PUD
Kennewick, Washington
- Founded in 1934
- 939 square miles
- 1,626 circuit-miles of line
- 49,816 metered customers

INVESTMENT
$512,500

HIGHLIGHTS
- AMI event data
- Energy storage

FOR MORE INFORMATION
Blake Scherer (509) 585-5361
www.bentonpud.org

47

system, such as high voltages, low voltages, hot sockets and outages. This off-the-shelf but customizable software enabled operations groups to know what's going on with the system. Any potential problem could then be actively addressed.

Learning the dos and don'ts of handling the alarm data demonstrated how to best maximize the AMI meters in two ways. First, identify which department would best benefit from that information. Second, learn how to most effectively implement the information into a future data management system that will eventually replace the DataCatcher™ demonstration.

"We now know the basics about collecting the AMI meter alarm data and how to present it. That will help us define our requirements for future integration with our other systems," said Blake Scherer, project manager with Benton PUD.

Ultimately, learning about those future requirements was the biggest benefit for Benton PUD from the demonstration.

Energy storage partners

Forging an energy storage partnership with Franklin PUD and the City of Richland was conceptually unique. Each utility installed an energy storage device that would be controlled by Benton PUD. The battery-based 10-kilowatt system would store electricity during off-peak periods when the price is cheap, and then distribute the energy later when the demand is high.

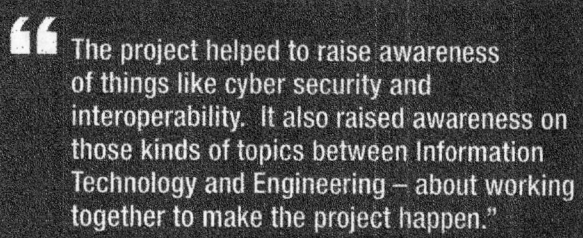

> " The project helped to raise awareness of things like cyber security and interoperability. It also raised awareness on those kinds of topics between Information Technology and Engineering — about working together to make the project happen."
>
> — BLAKE SCHERER,
> PROJECT MANAGER WITH BENTON PUD

"One goal was to test software that uses a transactive incentive signal provided by Battelle," said Scherer. "Then, wirelessly direct operations of storage devices from neighboring utilities and dispatch them as load or generation, as circumstances dictated."

Transactive control uses an interactive, market-based signal to increase or decrease the energy consumption of households and industries to achieve greater efficiency in grid operations. In the Pacific Northwest, the signal is sent over a multiple utility footprint. Participants in the project test the feasibility of increasing energy use when wind energy is abundant, typically at night, and reducing use during peak hours when energy is most expensive. The integration of the technologies was a challenge for some

participants in the project because of the complexity of the project and the cost.

Unfortunately for Benton PUD, the energy storage vendor went out of business while a contractor was trying to implement the transactive control software.

"We were unable to have it working before the vendor went out of business," said Scherer.

Lessons learned

The biggest lesson learned: structure vendor contracts using milestones for certain accomplishments.

"We initially needed flexibility in the scope of work," said Scherer. "But as the project became clearer, we

should have established new payment milestones for our vendor. As it was, our contractor struggled with the complexity of the project, which used up our project budget and resulted in several deliverables not being completed by the contractor."

Other lessons included properly understanding the level of effort required for a federal demonstration. The project reporting and project management requirements were more than expected.

On the plus side, the project improved awareness of cyber security best practices.

WHAT'S NEXT
for Benton PUD?

With an AMI foundation in place, Benton PUD wants to maximize the value of this investment by applying its lessons learned. It was also continue to organize and present AMI meter event data to utility personnel to improve system operations, reliability and power quality. With regards to energy storage, Benton PUD will continue to monitor developments in the industry, looking for the technology to mature and the costs to decrease.

GLOSSARY

ADVANCED METERING INFRASTRUCTURE (AMI): an integrated system of smart meters, communications networks, and data management systems that enables two-way communication between utilities and customers.

AMERICAN RECOVERY AND REINVESTMENT ACT OF 2009: an economic stimulus package signed into law on Feb. 17, 2009, in response to the Great Recession. The primary objective for ARRA was to save and create jobs almost immediately. Secondary objectives were to provide temporary relief programs for those most impacted by the recession and invest in infrastructure, education, health, and renewable energy.

AUTOMATED POWER FACTOR CONTROL: consists of a number of capacitors that are switched by means of contactors. These contactors are controlled by a regulator that measures power factor in an electrical network. Depending on the load and power factor of the network, the power factor controller will switch the necessary blocks of capacitors in steps to make sure the power factor stays above a selected value.

CAPACITOR: a passive two-terminal electrical component used to store energy electrostatically in an electric field.

CELLULAR-BASED TRANSFORMER MONITORS: a device that monitors the transformer activity using cellular technology.

CONSERVATION VOLTAGE REDUCTION (CVR): a technique for improving the efficiency of the electrical grid by optimizing voltage on the feeder lines that run from substations to homes and businesses. Adaptive voltage control or dynamic voltage control is controlling voltage at specific times.

DEMAND RESPONSE: a resource that allows end-use electric customers to reduce their electricity usage in a given time period, or shift that usage to another time period.

DISTRIBUTION AUTOMATION (OR SELF-HEALING TECHNOLOGY): when computer systems quickly react to electrical issues in the system, like a fault in a feeder, without intervention from an operator or line worker.

DISTRIBUTION VOLTAGE OPTIMIZATION (DVO): lowers the energy consumption on a whole feeder — the line that delivers electricity from a substation to a home.

FIBER COMMUNICATIONS SYSTEM: a method of transmitting information from one place to another by sending pulses of light through an optical fiber.

HIGH RELIABILITY ZONE (HRZ): what PGE calls its microgrid.

IN-HOME DISPLAY UNIT: provides consumers with real-time hourly, daily, weekly and seasonal energy consumption information.

INTEGRATED VOLT-VAR (VOLT-AMPERE REACTIVE) CONTROL: a tool that can increase or decrease energy consumption during a demand response event, without the voltage getting too low for homes further down the line.

INVERTER: an electronic device or circuitry that changes direct current (DC) to alternating current (AC).

MICROGRID: a small-scale version of an electrical grid. It can be "islanded," or disconnected from external transmission services. Local distribution provides power for customers' electrical needs with only local generators and battery storage.

PEAK TIME OR PEAK DEMAND: when energy consumption is highest.

PREDICTIVE APPLICATIONS: software applications that learn by doing and become better at solving problems as they collect more and more data.

PROGRAMMABLE THERMOSTAT: a thermostat designed to adjust the temperature according to a series of programmed settings that take effect at different times of the day.

RADIO ENERGY MANAGEMENT SYSTEM (REMS): A centralized computer system monitors the demand, sending out a radio signal to the REMS unit and cycling off connected loads in order to reduce energy when the peak demand set-point is reached.

REGULATOR: is designed to automatically maintain a constant voltage level.

SMART EQUIPMENT: technology that can be controlled or managed regardless of location.

SMART METER: an electronic device that records consumption of electric energy in intervals of an hour or less and communicates that information back to the utility for monitoring and billing purposes. Smart meters enable two-way communication between the meter and the central system.

SMART SWITCH: detects changes in the feeder, like a fault, and activates the switch automatically.

SMART THERMOSTAT: a Wi-Fi enabled thermostat that can be controlled regardless of location.

SMART TRANSFORMERS: work independently to constantly regulate voltage and maintain contact with the smart grid in order to allow remote administration if needed and to provide information and feedback about the power supply and the transformers themselves.

TRANSACTIVE CONTROL SIGNAL: incentive (price) and feedback (load) signals exchanged between active components in the electric power system. These signals forward forecasts and enable a process of negotiating future consumption against future price.

TRANSACTIVE NODE: which PGE calls the Smart Power Platform, is the main computer program which optimizes the economic decisions about the smart grid assets.

TRANSFORMER: an electrical device that transfers energy between two or more circuits through electromagnetic induction.

VOLTAGE OPTIMIZATION: uses a smart transformer to provide the exact amount of power that is needed and responds instantly to fluctuations within the power grid, acting as a voltage regulator to ensure that the optimized voltage is undisturbed.

WATER-HEATER DEMAND RESPONSE: controls when the water heater cycles on and off so usage can be shifted to off-peak times.

WIRELESS MESH SYSTEM: a network connection that is spread out among dozens or even hundreds of wireless mesh nodes that communicate with each other to share the network connection across a large area, such as an entire city.

Acknowledgement & Disclaimer

This material is based upon work supported by the Department of Energy under Award Number DE-OE0000190.

This report was prepared as an account of work sponsored by an agency of the United States Government. Neither the United States Government nor any agency thereof, nor any of their employees, makes any warranty, express or implied, or assumes any legal liability or responsibility for the accuracy, completeness, or usefulness of any information, apparatus, product, or process disclosed, or represents that its use would not infringe privately owned rights. Reference herein to any specific commercial product, process, or service by trade name, trademark, manufacturer, or otherwise does not necessarily constitute or imply its endorsement, recommendation, or favoring by the United States Government or any agency thereof. The views and opinions of authors expressed herein do not necessarily state or reflect those of the United States Government or any agency thereof.

324

www.bpa.gov

DOE/BP-4654 • December 2014

Dr. TAFT. But what has been demonstrated there is that you can coordinate very large numbers of consumers and their appliances and equipment in such a way that they reduce the peak demands on the utility significantly. They operate in a more efficient mode overall, and they do a better job of coordinating with variable energy resources.

So the impacts have been significant. The final details are coming out shortly, and we'd be happy to supply them for this hearing.

Senator CANTWELL. My understanding is that you were able to demonstrate you might be able to get as much as double digit savings out of current supply.

Dr. TAFT. The indications from those experiments——

Senator CANTWELL. I mean I'm not talking like 20 or 30, but, you know, 10, 11, 12?

Dr. TAFT. Yeah. The indications are, in fact, that such results are practically achievable. The results from that program showed that on scales that were larger than any that have been tried anywhere else. So that's a solid justification for saying that some level of double digit improvements are, in fact, achievable using these methods.

Some of the methods involve new forms of control and new forms of coordination necessary in order to make all those pieces work together, but when you can do that there's an enormous synergy from all those pieces and that's why you can get the double digit savings.

Senator CANTWELL. I know the people that will discuss what does that mean to the consumer and as you said in this case, they buy in. I'm waiting for the smart appliance that says turn my dish washer on at the lowest megawatt rate today. And obviously some intelligence for these, you know, big energy users in the household.

Dr. TAFT. Well and that's a crucial issue is how in fact can you make that work and make it work in an unobtrusive manner? People are not interested, probably, in spending a lot of time trying to micromanage the thing, so it needs to be automatic. And a lot of the work done on these programs is to show how to automate that.

Senator CANTWELL. Thank you.

Senator King.

Senator KING. Thank you. Following up on that discussion.

If you have ever toured a sawmill most modern sawmills now have a device called an optimizer which takes a computerized picture of the log as it comes in, calculates the most effective way to cut it and checks with the market at that moment to see whether the market will value two by fours rather than two by sixes. It seems to me that is exactly what we are talking about here. We are talking about dishwashers and clothes dryers and heaters that will check the grid, understand what the needs are, what the price is and will automatically optimize, if you will, I mean, that's sort of what we were talking about.

That leads me into a, sort of, general observation here. We are talking about a disruptive technology. We are talking about a fundamental change in an early 20th Century model for how electricity is generated and delivered. If you look back, I can remember the phone company. I am old enough to remember the phone company saying it's impossible. It will be disruptive. It will ruin the system if people can choose their own phones.

Then, of course, it was cellular phones. You've got to have a phone in your home. That's gone by.

Cable TV, you know, you can only get your news through cable TV. All of a sudden people are getting—streaming.

I believe that distributive generation a, is going to happen, b, it's a disruptive technology. The only question is how do we manage it?

I think the real question is who will supply the battery? Who will supply the battery for distributed generation on your roof? Will it be the grid as the backup or will it be batteries in your basement?

The grid, speaking largely as a large institution, has a choice to make. They can price themselves out of that market and thereby make batteries more economic which is what people will then choose or they can choose to adapt to this disruptive technology and figure out ways to make it work.

I think one of the great questions is how do we price store it? How do we price backup? How do we price the backup charge?

I think it is reasonable that utilities should, if they have to be there and maintain the wires and backup capacity, how do we price that both in terms of benefits and costs? It seems to me that is the really fundamental question here, and the price should be reasonable and not so high as to be punitive in order to ward off this change which is going to come anyway.

Mr. Taft, what are your thoughts about how you price? How do you price backup?

Dr. TAFT. So, pricing is not a specialty of mine, but I do have a couple of comments that are related.

You're probably familiar with the effort in the State of California where they determined that they'll have 1.3 gigawatts of storage on their system. And the California ISO has produced a road map for achieving that.

One of the things that's going on though now is exactly that question of how to determine the pricing for the services that can be supplied by storage, in particular. There's not a settled answer to that yet, so they're working hard with vendors and others in the State of California to see if they can develop a model for that.

Separately, but in a sense related, is the activities going on in the State of New York with the New York rev process where they're looking at restructuring the rules and responsibilities of distribution companies, and a lot of it has to do with the penetration of distributed energy resources and storage. And so they're looking at changing the very structure of the industry in their state so that they can facilitate that but are faced with the same question ultimately, and they're asking themselves how are these going to be valued? The answer is not known.

Several states are working on trying to determine value propositions for storage, in particular, that includes the State of Washington, for example, Oregon, Hawaii and so on. So we don't have a settled answer for that yet.

And, you know, most people think that a way to get at that is to make it a market function and let a market determine those values.

Senator KING. Well, you are talking about a classic non-market situation. You are talking about monopolies who deliver the power whether they're monopolies and generations. So saying to let the market do it and then allow the utility to impose a $100 a month backup charge—that's not the market. I am all for markets, but only if they are real markets.

Dr. TAFT. Well and I think that's what some of the states are trying to sort out now is how they can actually accomplish that. And I reference California ISO trying to understand how they can allow third parties to offer services based on storage and figure out what those values would be. So there is a complex question there. I don't think it's settled.

Senator KING. One market related point, and I will end with this. This is not a central part of the discussion, but it seems to me that time of day pricing is one thing that would make some sense in order for the market to drive more efficient utilization of the grid. There's a lot of excess capacity both in terms of generation and transmission at night, and if you had time of day pricing and people just routinely saying oh, I'm not going to turn the dryer on until after nine.

I can remember making those decisions about long distance phone calls, you know, looking at my watch and saying I am going to make my call at 9:05. If people started making those decisions about utilization of the grid that would be much more efficient, but it won't happen unless people get the price signals. It seems to me time of day pricing is something that we ought to be thinking about because of the fact that there is so much excess capacity.

Madam Chair, I am very pleased that you are having this hearing. I think this is a fascinating topic, but the bottom line for me is we are talking about a disruptive technology. We have got to figure out how to adapt to it, not fight it and strangle it in its crib because that isn't going to happen anyway. I think this is a very important subject. Thank you.

The CHAIRMAN [presiding]. Well it is, Senator King. I certainly agree with you.

When you mentioned the point about making telephone calls based on the time of day, I think we're both dating ourselves there. But that was absolutely——

Senator KING. I do that all the time, but I have given up on worrying about it.

The CHAIRMAN. Yeah, yeah, I suppose. [Laughter].

Let's go to Senator Heinrich.

Senator HEINRICH. Thank you, Madam Chair.

I am not going to try and date myself that hard, but I will say my father was a utility linemen, and I appreciate some of the comments about the work that they do. One of the things that I had when I was a kid was a t-shirt that said, I turn on after seven, and I don't think I got the joke.

The underlying message from the utility he worked for was about peak load management and about trying to shift loads to later in the evening just in the way that the long distance pricing regime shifted people to using their long distance later in the day.

I think that Senator King is right. Anything we can do to, sort of, actively make the market work in our favor in terms of both loads and then hopefully storage and utilization of storage over time.

So I want to start out with a quick question for Dr. Littlewood. I am a big fan of the work that is being done at DOE's joint center for energy storage research. And while I think most of us here agree that the pricing regimes and the market to the extent that we can have an efficient and transparent market will settle out the winners and losers with regard to energy storage over time.

I am wondering if you can talk a little bit about the state different technologies for utility scale storage and the challenges inherent in scaling that. What are you excited about? What do you see rising to the top of that very exciting, emerging technology sector?

Dr. LITTLEWOOD. Thank you very much, Senator. I'm glad you like the program because we do too.

What else? I mean, I think our view of storage is that it is not going to be a single solution.

There are solutions which may come through improved storage which is actually mobile. We were discussing earlier that if we could actually make electric vehicles more effective they are affected. You also have storage medium on the grid. That will be one class of storage. There will be classes of grid storage which will involve very different technologies that can be very large and cumbersome, not mobile, but have to be made extremely cheap.

One should also remember that there are storage on the grid which is distributed across the grid in a way that you can't see it.

To be a little bit philosophical and look very far ahead, I will say that our goal, actually, is to turn electrons into a medium of exchange.

The thing that will happen in the very long term, eventually, is that indeed this will be distributed where you do not know how. But that you will be able to go there in the same way that there's a bank. That means that I would also expect that we would have a system which looks like the banking system. There will be big banks. There will be small banks. There will be local communities with local community banks. And all of those will be based around different technologies and different ways of doing it.

I think that we are doing very well in developing new technologies which push the price down in the area of large scale grid storage, and that's partly because this is an area which has not been explored very much.

I think that there are vast opportunities, more broadly, for moving forward even with conventional technologies based around lithium ion, and the reason is that the community which has been pushing that technology forward, which was your cell phone, is now beginning to understand that there's a business opportunity which involves something different.

So it's a very exciting time to be in there. I think it's really important to be engaging in all of this. You know, one shouldn't make predictions about the future, but I'm very positive about it.

But just a quick philosophical comment on phones. I used to work for Bell Labs, and I worked for Bell Labs at the time when I was part of the long distance network and we were inventing the cell phone. While my colleagues were doing that we were kind of aware that we were putting our company out of business, but it turned out that we were all so confident that that was going to happen anyway we might as well be the ones that did that invention.

Senator HEINRICH. It's interesting that you bring that up because it's always interesting to look back at what was said about the adoption of cell phones early on and the penetration and numbers of that adoption and how dramatically we all underestimated that technology early on.

It's not unlike when, you know, 20 years ago in terms of people looking at a technology that hasn't changed much in terms of photovoltaics and what the eventual generation from that would be. Looking last year at the gigawatts that were installed in this country alone, I think, we are realizing now that we are on the edge of a very changed landscape. We have a lot of work to do to manage the grid and to manage the transition in all of this into some very new territory.

Thank you, Madam Chair.

The CHAIRMAN. Thank you, Senator Heinrich.

You used the terminology, Ms. Barton, that the grid can be a natural enabler of new technology. I think that this is what we see play out all the time.

Let me ask just one last question for you and then I'll ask if others have a desire for more. I know that we have some panelists that need to leave at noon.

How can we do a better job in so far as the utilities identifying or quantifying the costs of the ancillary services that are provided by the grid and ensuring then that these costs are perhaps born proportionately by all the customers that are utilizing the grid? How do we do a better job here? I'd ask Ms. Barton and then Ms. Edgar, if you'd like to comment as well.

Ms. BARTON. I think that's an excellent question. We've talked a lot about distributed generation, and we've talked about it as being a disruptor. I really think it is more of a complementary technology to the grid.

And if you really think about the origins of the grid where it started off as, quite frankly, a series of small microgrids, is what you really could have called them back in the day. They got stronger by being networked together.

There is an incredible value associated with having that grid as a backup, and I'll just use the example of super storm Sandy. While there are a number of customers who were out of power for a few weeks, that's nothing compared to if their only option was a roof top solar facility to provide them with generating capacity.

So the grid has an incredibly inherent value, and a lot of it has to do with the line workers and the tree crews who, quite frankly, can travel across the country and get to any place where they need to be through our mutual aid agreements. There is value to that as well.

I think that that becomes part of the philosophical debate as to what is the value of the grid, and I would say the value of the grid is not defined as backup power. The value of the grid is that it is an enabler and therefore you have to support the costs and the infrastructure of that grid and so that those costs are not disproportionally spread to other consumers. Unfortunately that's some of the things that we're seeing happening. If you advocate the view that I can just have my solar roof top, solar generating facility and unplug from the grid, you know, that's different than wanting it there as a resource. And the vast majority of Americans have become very relying upon a very reliable grid and an affordable grid.

It's that balancing act that we need to pay attention to, but there's a lot of fantastic technologies out there. Right now a lot of them come at a fairly significant cost and that affordability feature is key.

The CHAIRMAN. Let me ask you for your comments, Ms. Edgar.

Ms. EDGAR. Thank you, Madam Chair.

So many issues here and this kind of brings it all home, the value of the grid.

Something that a number of states are looking at is trying to put, for example, a value on solar as different states work their way through their different processes, and the other states can learn as they're moving through is part of what makes all of this so exciting.

Many cost tools that are out there, capacity charge, service charge, fee based, cost causer, time of use, all of this, I think, needs to come into play as we recognize, again, yes, the value of the grid, but also the very essential nature for public health, for public safety, for community interaction.

We haven't talked much today about economic development, but the key for the grid and new technologies for job creation, for work force. And then, I think, to bring it all back is, as we have these discussions, what our members would ask of Congress and governors and state legislators and other policy makers, is that we have transparency as we are talking about those cost allocation issues and cost burdens recognizing that how that falls, for instance, on a very large industrial customer is going to be different than how it falls on, for instance, a retiree on a fixed income and how we try to make sure that everybody is aware and is served well.

The CHAIRMAN. Very important considerations.

Senator King.

Senator KING. Ms. Barton, I wanted to follow up. You made one statement that I thought was very important. You said diversity and redundancy are at the heart of reliability, and it seems to me that that's part of what we're talking about here, the ultimate in diversity and redundancy is distributed generation. It's people having their electricity on their roof as part of the system.

So it just seems to me that we are really on the edge of a very important discussion to get back to what I mentioned before. How do you value the plusses of distributed generation in your own words, diversity and redundancy, versus the costs which is the maintenance of the grid and the backup generation in order to have customers ultimately make rational, economic decisions, based upon the true costs of whatever it is that they're using?

I just think that this is a development that's going to happen anyway, and the real question is how is it adapted to and how does the grid facilitate rather than block what I think, in the long run, will be very salutary developments on behalf of all of our citizens?

Ms. BARTON. I think the key is that there's value on both sides of the transmission system is maybe one way of looking at it.

Senator KING. Absolutely.

Ms. BARTON. In terms of distributed generation can be certainly valuable in terms if you aggregate. The grid is really an aggregator. It's an aggregator in days of old, of large generating plants and taking that power through substations to consumers, and today it's basically an aggregator in both directions. So it's aggregating those distribution resources and distributive technologies as well as the larger scale generating facilities.

When you look at the utility infrastructure, the math is maybe a little bit easier in the sense that it's a cost-based model. And for, you know, a reasonable level of return the investors in the utility industry that we go to to borrow the capital really require that level of transparency to say I know I'm going to give you this dollar and you are going to give me a return on this dollar, plus this dollar back in a reasonable period of time. And so, I think, making sure that that is in place helps to take care of the utility side of the equation.

With respect to the distributed energy side of the equation, I think that there's a lot of different ways of valuing that. Part of it can be from what does it do for the benefit of that consumer? For example, the pharmaceutical industry often needs to have a heightened level of reliability that, quite frankly, they're willing to pay for that heightened level of reliability. And so it's very valuable to them.

There's also benefits with respect to the ancillary services market in various market driven ways of estimating the value of those resources. And, you know, that can be done through the RTOs and how they have established their energy markets and how they're valuing those resources. So I think it's really a balance of that.

Ultimately the states play an incredibly important role in determining what level of energy efficiency, what level of distributed generation that they want to advocate and how it's paid for.

Senator KING. Of course, there are two pieces to this which is generation as backup and transmission and distribution to deliver the generation. So there are two pieces of analysis that have to be done. And part of the analysis of how much backup capacity, generating capacity, do you need, is a question of a complicated analysis of what are the resources on the grid.

Insurance companies don't have all the money in the bank for people's accidents or health insurance or particularly, life insurance. They invest it. They have actuarial studies, engineering studies, if you will, as to when the draw will be. And so you don't have to have every bit of capacity there to backup if every solar insulation goes out at the same moment.

And of course, as you distribute these kinds of facilities, solar and wind, across the landscape you mitigate somewhat, it's not going to be cloudy everywhere, for example. So I just hope that we can develop policies that will be fair to all rate payers, fair to investors, but also not inadvertently obstruct what could be and can be and will be, I believe, very positive developments both for rate payers, for individuals and we haven't mentioned the environment much, for the environment in terms of moving away from fossil fuel generation.

Thank you very much, Madam Chair.

The CHAIRMAN. Thank you, Senator King.

Senator Cantwell, any final questions, comments?

Senator CANTWELL. Well I would just, again, thank all the witnesses and Ms. Edgar said we really didn't talk about the whole plethora of economic issues, but I do think it's something we should work with our colleagues on. I went recently to the 110th anniversary of the IEEE in Seattle. And the fact that that organization has been around that long, of electrical engineers and all the things that they have done and then the next generation of things that they're creating I think is important for us to remember that the workforce here within the United States on these issues is a great potential for us.

So, thank you and thanks for holding this hearing.

The CHAIRMAN. Well, thank you, and thank you to all of our panelists. This has been very helpful and informative as we are building out some of the discussions that have gone on with some listening sessions and figuring how we move forward with a broader energy vision, an updated vision. Clearly in the electricity space what we have been talking about here this morning is so key.

Again, a reminder that while there is so much happening out there in terms of the technologies and the advancements and the modernization, we can remain excited about it, but we also need to make sure that the boring work of turning on the lights around the country, every day, stays in place. So it's pushing out on the technology side, but it's performance as well and performance at a time that we are focusing on security, absolutely the cyber piece, on reliability, on customer affordability, on environmental sustainability and financial viability as we have been reminded by Ms. Edgar.

It's a big challenge, but thank you for what you do within your respective spheres to make it happen. We appreciate it, and we appreciate the time that you have given us.

I know that members will likely have questions so we will submit them to you and we look forward to those replies as well. So thank you for being here before the Energy Committee.

And with that, we stand adjourned.

[Whereupon, at 12:01 p.m. the hearing was adjourned.]

333

APPENDIX MATERIAL SUBMITTED

U.S. Senate Committee on Energy and Natural Resources
March 17, 2015 Hearing: Electric Grid Innovation

Questions for the Record Submitted to Ms. Lisa Barton
from Senator Lisa Murkowski

Question 1: Please describe what ancillary services are and why they are needed to maintain grid reliability. How can distributed energy sources contribute to these needed grid reliability services, and under what time frame?

Ancillary services are necessary to support reliable transmission of electric power and help deal with unforeseen or unplanned yet credible events on the grid. Even the best planned electricity grid can experience short-term changes in capacity or demand. As an industry, we must be prepared for when a power plant or major transmission line goes out of service unexpectedly or for unanticipated changes in demand. Ancillary services help to mitigate short-term imbalances in electricity markets after an unacceptable disparity. There are many types of ancillary services, but the most notable are Reserves, Regulation and Black Start Capability. In an organized market, a Reserves Market ensures that ample resources are available to provide fast ramping in the event of a facility outage. Similarly, a Regulation Market helps keep generation and load in constant balance by relying on resources that can adjust very quickly in response to changing demand. Black Start Capability provides the energy and power needed to restart the grid after an extensive outage.

Distributed generation generally does not contribute to ancillary services. In fact distributed generation relies on these resources being provided by the market to ensure acceptable levels of reliability. Although batteries installed in conjunction with distributed generation could provide certain ancillary services, a decision to install batteries in this situation would be made by the RTO and would depend on the quality and value of the resource.

Question 2: Politicians often talk about not picking winners and losers. This is particularly important with rapidly developing technology. If, for example, the government had mandated Napster and Myspace as the only options in their realms, the world would look very different today. What happens if the government picks the "wrong" technologies to pour money and time and regulations in to? In the same way that we like the App Store because it gives us lots of options, can you describe the importance of letting new technologies develop?

The primary concern of a government selecting or picking a particular technology is that this inhibits the natural process of various technologies competing to find the best solution for a defined problem and substitutes a political decision for a scientific, engineering or customer decision.

Legislators should also avoid picking winners and losers because giving subsidies to one business puts other businesses that do not receive subsidies at a disadvantage, distorting investment and other economic activity; and government subsidies increase the incentive to focus resources on government relations, instead of focusing resources on innovation. Lastly, it is important to recognize the long-lived nature of utility investments when setting policies.

U.S. Senate Committee on Energy and Natural Resources
March 17, 2015 Hearing: Electric Grid Innovation

Frequent changes in policy direction in this industry can lead to significant cost increases for consumers. It is for this reason that I support allowing the market to truly identify the winners and losers of a particular technology rather than technologies being artificially advanced.

Question 3: Can you explain why renewables such as solar and wind need the grid to operate? What services specifically does the grid provide to consumers who are producing their own electricity?

Utility scale wind and solar need to interconnect into a robust grid because 1) the most efficient and productive facilities are typically located at a considerable distance from load and need a robust transmission system to get that power to market, and 2) because they are intermittent, consumers rely on the availability of other dispatchable resources on the grid in order to provide reliable service.

The US economy and security is dependent on the availability of reliable and cost effective electric power. The grid is designed to meet demand during extreme conditions such as a hot humid summer day or a cold winter night. For this reason, to ensure continuity of service even small scale wind and solar facilities that are connected to the distribution system must rely on the grid. In essence, the grid enables these technologies to be used together in a complimentary fashion with the rest of the system. Without the grid, the level of reliability that we have become accustomed to today would be significantly compromised, unless these variable resources are supported by a redundant supply of resources. Overall, since the grid serves as a natural enabler of these technologies, it is essential that there be adequate compensation for the value that the grid provides.

Question 4: What do you believe the biggest game changer will be for the future of the grid?

Cost effective, efficient large capacity electric storage would be the biggest game changer. The second biggest game changer could be cost effective modular nuclear generation.

Question 5: What technologies are you most excited about and why? Also, what is the timeframe for development and at what cost?

While "smart grid" technologies receive much of the attention, we believe there are significant efficiency gains that can be accomplished through engineering of next-generation transmission grid infrastructure. AEP's BOLD transmission design is one such example, where three times the capability of an existing line could be constructed within the same right-of-way, with lower energy losses and more aesthetic appeal. As existing grid infrastructure ages, there is a tremendous opportunity to replace it with much improved designs. AEP has already developed the BOLD technology at 345 kV, with the first project under construction. We have plans to further refine the technology and develop designs for other voltages over the next 2-3 years, but this will be dependent upon the level of receptivity within the industry and available opportunities for its use. This technology would enable utilities to dramatically increase the capability of lines, increase efficiency of the system through significant reduction in line losses, all without the need to increase the size of an existing right of way,

U.S. Senate Committee on Energy and Natural Resources
March 17, 2015 Hearing: Electric Grid Innovation

Question 6: Can you explain in layman's terms the technical difficulties that utilities face in integrating renewables? Can you also explain some of the technical benefits of bulk baseload generation, both for the grid and for consumers?

Often large scale renewable resources are abundant in areas where the transmission network is scant or weak. As a consequence, transmission needs to be planned and constructed to enable these resources to get to the market. FERC Order 1000 has changed the way RTO's plan for the future needs of the system and now RTO's need to consider state renewable portfolio standard (RPS) requirements in their planning methodology. This over time should enable the development of a more robust transmission grid. In the meantime, the challenge with these resources is their variable nature. In essence, we cannot control when the wind will blow and how much sun will shine. Therefore, renewables have to be linked via robust transmission to hedge the intermittency risk and ensure adequate supply to meet demand.

Maintaining a balance between dispatchable resources (gas, coal and nuclear) and variable resources (wind/solar) is necessary for stable grid operations. In Texas where wind resources are abundant and to date over 14,000 MW have been successfully interconnected, it was transmission expansion that enabled the integration of these resources coupled with dispatchable generation to moderate the system.

Question 7: You spoke about the need to avoid picking "winners" and "losers" regarding technological innovation. What do you see as the appropriate role for the federal and state governments in this regard?

Federal and state governments should develop and determine even-handed regulatory policies and the choice among competing technologies should be driven by performance and cost, with winners determined by the market itself rather than through long standing subsidies. As federal policy makers develop policies, the cost and benefit associated with various technological innovations should be considered.

The Quadrennial Energy Review (QER) was created to provide a comprehensive look at the nations' energy industry, including transmission, with the expectation to create technology and policy recommendations to improve transmission across the country. This is the ideal platform to shape policy that facilitates and encourages technological innovations that support grid reliability and resiliency.

Question 8: You testified regarding the need to expand transmission investment. Please be more specific on the types of transmission investments that are needed. Could you elaborate on the role for state and federal governments respectively in this area?

Investment in transmission, for the most part, is being driven by the age of existing transmission infrastructure, integration of renewables, and mitigation of generation retirements. As our nation shifts from reliance on baseload generation to a diversified mix of resources, we will need a more robust transmission network to ensure continued delivery of reliable and cost effective

U.S. Senate Committee on Energy and Natural Resources
March 17, 2015 Hearing: Electric Grid Innovation

electricity. Also, we will need to address the existing infrastructure - a significant portion of which will surpass its life expectancy in the coming years.

At the Federal level, strong direction and incentives are needed to plan, site, permit and construct the transmission infrastructure required due to generation retirements and renewable development. Transmission development is a capital intensive business and relies heavily on capital from the market to enable the construction of these resources. Participants in these markets like predictable and fair returns for their investments and stability in federal policies supporting the need and importance of transmission investments. The greater the level of stability in returns and recoverability of the investment, the lower the cost of debt, and the more security and predictability there is for the debt and equity participants will encourage investments in this type of infrastructure.

Lastly, protecting investors from abandonment of projects for reasons outside the utility's control is essential to enable the continued support of capital to these long lived assets. Because it takes up to five years to complete a project, if there is a cancellation of the project by an RTO during those years, it is essential that developers recover their full investment. Federal policies that protect these investments is essential for long-term support of these investments.

At the state level, expeditious siting and permitting processes and the support for multi-state robust transmission solutions over localized Band-Aid projects is needed. State siting laws that enable expedited reviews of facilities needed to support the interconnection of large scale renewables, customer interconnections and projects needed to support generation retirements can be very helpful in enabling new transmission resources to be in place sooner and when needed. Adoption of trackers at the state level which ensure the recoverability of needed transmission investments can also be helpful.

Of course, the RTOs play a big role in transmission expansion and with the help of the Federal Energy Regulatory Commission, we will need to craft regional planning processes and RTO tariffs that help shape our energy future by focusing on long-term solutions rather than simply a series of short-term stopgaps.

Question 9: In your testimony, you spoke of a number of new technological innovations for the grid. What will the magnitude of their impact be on grid?

The magnitude will primarily be a function of scalability. At the distribution level, the impacts may be in smaller increments (measured in kilowatts) but the opportunity is quite large. For example storage such as batteries deployed individually at the distribution level will have limited impact on the overall grid. However, if deployed widely and/or with larger sizes, that technology could provide a significant tool for managing the grid and facilitating integration of more variable energy resources. Another example is the aforementioned BOLD technology. This technology will provide significant capacity throughput improvements over existing lines (measured in many megawatts), and could provide a real opportunity to more efficiently and effectively reach new resources in a significant way.

**U.S. Senate Committee on Energy and Natural Resources
March 17, 2015 Hearing: Electric Grid Innovation**

Question 10: What is the biggest challenge facing utility companies today? What are the respective roles for utilities, the market, technology, and consumers in addressing this challenge?

The biggest challenge facing utility companies today is the potential vulnerability of the transmission grid. This could be from extreme weather events, cyber-attacks or physical security threats.

We all have a role in strengthening the nation's electric transmission and distribution grid. Utilities need to understand where the areas of vulnerability exist and enable the integration of new technologies; the market needs to be agile and capable of responding when necessary; the technology sector needs to continue to provide and improve upon innovative cost effective solutions that result in making the grid more resilient and reliable.

Question 11: Germany is often cited as an example of a country with high penetration by renewables. They may experience reliability issues in the face of a partial solar eclipse later this week. In your testimony, you noted that Germany still relies on conventional power sources for reliability. Can you elaborate on the risk of retirements and how that potentially impacts grid reliability?

The intermittent nature of renewable wind and solar generation means that conventional fossil generation must be kept online to serve as backup and/or to balance the impacts of renewables entering and leaving the grid. Renewables have very low marginal costs because they have no fuel costs, thus they are bid into the market at low prices, earning them an early place in the dispatch priority, depressing wholesale market prices, and making it increasingly difficult for conventional fossil generating plants to economically operate. Germany has installed enough wind and solar capacity to fully meet the country's peak electricity demand. However, to maintain reliability they must maintain an equivalent amount of conventional generating capacity on hand to serve as backup for the intermittent renewables. In recent history there have been days and even weeks in Germany where very little renewable energy was available and greater than 85% of the demand for electricity was served by conventional fossil and nuclear generation. The graph below represents one week in December of 2013 where renewables provided less than 12% of the electricity generated during the week.

U.S. Senate Committee on Energy and Natural Resources
March 17, 2015 Hearing: Electric Grid Innovation

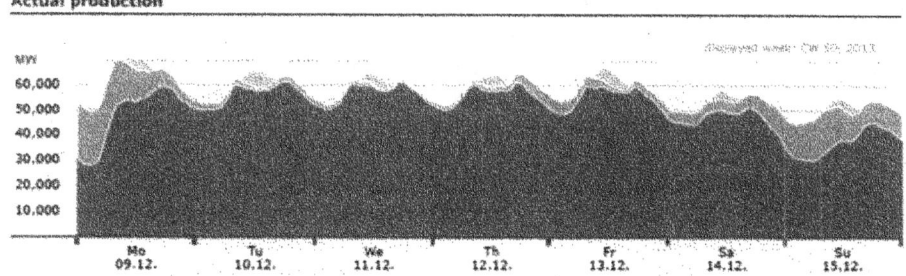

Actual production

	max. power	date max. power	weekly energy
Solar	5.9 GW	13.12., 12:15 (+1:00)	0.14 TWh
Wind	21.9 GW	09.12., 01:15 (+1:00)	0.95 TWh
Conventional > 100 MW	61.7 GW	12.12., 17:00 (+1:00)	8.6 TWh

Graph: Bruno Burger, Fraunhofer ISE; Data: EEX Transparency Platform

Source: Fraunhofer ISE "Electricity Production from Solar and Wind in Germany in 2013." www.fraunhofer.de

It is not uncommon for wind and solar to diminish in this manner for significant durations and Germany is effectively maintaining double the generation capacity to meet its demand and maintain reliability in these instances. This is proving to be unsustainable as electric rates have skyrocketed to roughly $0.40/kWh and generators are struggling to recover costs on conventional generation as it is displaced by highly subsidized renewables. Without the ability to economically operate conventional fossil fueled generation, German utilities are suffering substantial financial losses. The German utility RWE recently recorded a $3.8 Billion loss, its first full-year loss since 1949 due in large part to the impact of rapid subsidized renewable penetration.

Question 12: In your filed testimony you said that constructing the infrastructure necessary to integrate new technology "require[s] significant lead times to obtain approvals, permits, and rights-of way." Can you give me a concrete example? How long does this process take? Can you detail some of the specific steps required?

One of the changes that has occurred in transmission development is that RTO's are now responsible for determining the need for transmission upgrades and determining the best solution to address those needs. RTO's make this determination by modeling the system. Through this process, they project future loads, consider available generation and evaluate the system and its ability to response to transmission and generation outages. All such studies are designed to ensure that the system is reliable at all times. These studies take up to a year to perform for an entire RTO and once completed, a list of transmission upgrades are assigned to transmission owners to construct. Once assigned, the projects need to be engineered, designed, planned and constructed. Typically it takes a minimum of 5 years to construct a project.

U.S. Senate Committee on Energy and Natural Resources
March 17, 2015 Hearing: Electric Grid Innovation

Significant drivers of a construction schedule include long lead times to order and secure on site equipment, which can take over 2 years, as well as the need to secure the necessary land rights and certificates of necessity from state or local siting authorities. These state and local authorities evaluate the need for the project, evaluate land owner rights, evaluate the environmental impact of the project, determine the best route for locating these facilities and enable, if necessary, the condemnation of land. This process can often take two years or longer to complete. Finally the project needs to be constructed which typically takes 18 months to 24 months.

In short, a series of question must be answered before construction may begin –

- What are the future needs of the system?
- What transmission is needed to support this system?
- What is the route and design needed?
- Where will the project be located?
- What is necessary for siting approval and land acquisitions

Question 13: In your testimony you provided eight examples of new technologies that AEP is deploying. Can you further describe which one you think is most promising, and what is needed from the government to make it a reality?

AEP is extremely proud of our work on BOLD™ technology. The development of BOLD™ technology was driven by a heightened public interest in more efficient transmission solutions. Our engineers recognized the numerous challenges associated with transmission construction to support increased renewables and the limited availability of right-of-way for new construction and went to work. BOLD™ technology will, among other things, use less right-of-way than higher voltage facilities that would transport the same amount of power; and increase the capacity available when replacing existing older lines, while also using the same right-of-way. Of equal significance, while conventional transmission lines require complex and costly terminal equipment for long-distance bulk power delivery, the configuration of the BOLD™ technology also eliminates the need for this equipment by improving the intrinsic loading capacity of the line. Although still a relatively new technology, BOLD™ has already been awarded seven patents and seven patents are pending. As I noted in my testimony, debut of the technology is planned for 2016 in Indiana and this was all accomplished without Federal subsidies.

Battery technology and its potential is also an exciting prospect. At present, however, the commercial viability of this technology appears to be limited.

Question 14: What can the government do to help educate the public and policymakers about essential but obscure issues like voltage support, frequency regulation, and ramping?

All of the items listed above are complex topics and represent the physics behind the grid. Maintaining each element is essential to making our nations' grid more secure, more reliable and more resilient. Some efforts are underway with the WIRES organization which holds an annual

U.S. Senate Committee on Energy and Natural Resources
March 17, 2015 Hearing: Electric Grid Innovation

WIRES university session to educate policy makers in transmission related issues. The next university session is scheduled for April 21, 2015 at location just east of the Capitol, https://www.eventbrite.com/e/wires-university-electric-transmission-the-survey-course-tickets-15787458714 .

Another opportunity is for utilities representatives to speak with their respective State commissions on these issues. Government agencies could also encourage town hall meetings or public service announcement to get the conversation started with the general public.

Question 15: In your opinion, how should infrastructure investment decisions be prioritized?

The grid itself serves as an enabler of technologies and generation resources both for large scale renewables and distributed generation. As such the investment in the grid itself is essential and construction is prioritized by need. Specifically, changes in the generation resource mix with respect to the Mercury and Air Toxics Standards (MATS) plant retirements is changing flows on the grid significantly. As additional changes occur due to environmental rules or for other reasons having a robust grid is essential and will determine where infrastructure is needed.

Question 16: Can you explain how and whether net metering accounts for the value of ancillary services?

Net Metering tariffs generally 'net' the production from a customer's distributed generation (DG) against their consumption of energy. During production the meter spins more slowly as the production offsets some of the usage, and in some cases the meter spins backward as the energy produced exceeds the local usage causing energy to flow back into the delivery infrastructure. The customer pays/receives payment for the balance.

While specifics vary, a very significant amount of the costs to serve most residential and smaller business customers are fixed (i.e., don't vary with the volume of consumption). These charges include the transmission and delivery infrastructure (substations, towers and poles, meters, etc.), as well as the generation 'capacity' (the fixed generating plant that needs to be available whenever the customers wants to use energy and his system is not producing). However, retail service tariffs recover most costs volumetrically (through per-KWH charges), rather than through fixed charges. Therefore, when a customer reduces their consumption significantly through the use of distributed generation, they avoid paying their fair-share for these fixed facility costs, even though they continue to utilize them.

Ancillary services (such as voltage support, frequency regulation, etc.) are services that are provided to the market, typically at the ISO level, to maintain proper conditions for the reliable and efficient delivery of energy to customers, and are part of the retail rate charges. Thus, to the degree that net metering customers avoid paying the all-in retail rate, they also avoid paying for ancillary services, the costs of which are included in that rate.

In addition, as the amount of DG increases on the grid, it is likely that additional investments in the infrastructure (and a corresponding increase in fixed costs) will be required to monitor and control the operation of the electric system. The distribution system was designed to distribute

U.S. Senate Committee on Energy and Natural Resources
March 17, 2015 Hearing: Electric Grid Innovation

power from the substation to customers along a circuit with a one-way flow of power. Larger amounts of DG challenge that one-way power flow. The technical effort, time, and costs required in redesigning this system while simultaneously maintaining constant reliable service to customers should not be underestimated. One of the first impacts is voltage rise caused by the DG, particularly at times when the DG production is high and the load is low. Each circuit has limits as to the amount and location of DG that can be accommodated without impacting the reliability and power quality of the customers.

Question 17: How do you capture the appropriate range of values in regulatory treatments for energy storage projects?

This is an important question that regulatory agencies are just now beginning to examine. As described in the Brattle Group report The Value of Distributed Electricity Storage in Texas: Proposed Policy for Enabling Grid-Integrated Storage Investments **http://www.brattle.com/system/news/pdfs/000/000/749/original/The_Value_of_Distributed_Electricity_Storage_in_Texas.pdf** electricity storage is driven by a range of potential applications that include avoiding power outages for customers, reinforcing the grid, reducing other transmission and distribution (T&D) costs, shifting power consumption away from costly peak-load periods, balancing intermittent renewable energy resources, and providing ancillary services and emergency response service in the wholesale power markets. A regulatory framework that enables investors in the storage devices that capture both the wholesale market and the T&D system values would make a significant step in providing the necessary transparency and ability to capture the benefits it provides.

U.S. Senate Committee on Energy and Natural Resources
March 17, 2015 Hearing: The Electric Grid

Questions for the Record Submitted to Ms. Lisa Barton
from Senator Ron Wyden

Question 1: As many of you have testified, energy storage is the Swiss Army knife of the electric grid. Energy storage has the potential to solve many of the problems discussed in this hearing. Storage helps integrate renewables, it can help keep the lights on during extreme weather events, and it makes the whole grid operate more smoothly. It's no secret that I think energy storage is a good deal for the American public; the tool just needs sharpening to live up to its potential to reshape the grid. In particular, the recent progress to bring down costs and increase efficiency must continue. The barriers to storage aren't just cost, though, utilities also aren't used to working with storage yet, and regulations for storage vary from state to state.

I continue to favor providing tax incentives for energy storage to encourage its deployment, but what else should be done to give energy storage the boost it needs? Please don't restrict your answer to the jurisdiction of the Senate Energy Committee, please also comment on useful policy solutions within the jurisdiction of the Senate Finance Committee.

As indicated in my testimony, AEP has extensive experience with utility-scale energy storage using battery technology.

In 2006, AEP's Appalachian Power Company (APCo) subsidiary commissioned the first megawatt-class NaS battery to be used in North America at its Chemical Station in Charleston, W.Va. This advanced energy storage technology can supply 7.2 megawatt-hours of energy. This technology allowed APCo to defer building a new substation for several years. In 2013, this battery was no longer needed and was removed from service.

In 2008, a 2-MW NaS battery was installed at a new station, called Balls Gap, in West Virginia, and is being used to supply energy on the distribution system to help relieve the load burden at another substation. In addition, this battery has the capability to provide service to up to 700 customers for up to 7 hours when power is interrupted due to an outage. AEP has also installed two other 2-MW NaS units with similar capabilities in Ohio and Indiana.

In 2010, AEP's Electric Transmission Texas installed a 4-MW NaS sodium-sulfur battery system in Presidio, Texas, to provide transmission backup in the event of a transmission line outage. It was designed to improve power quality and reduce voltage fluctuations in the Electric Reliability Council of Texas region. Presidio is a small, remote community bordering Mexico along the Rio Grande River – the only load at the end of a single radial transmission line. Previously, when Presidio's line encountered an outage, the town had an immediate blackout and its only alternative electricity could come from Mexico.

U.S. Senate Committee on Energy and Natural Resources
March 17, 2015 Hearing: The Electric Grid

As part of our gridSMART® initiative, AEP Ohio initiated a research project to test Community Energy Storage (CES), which are lithium-ion battery-based energy storage systems designed to provide 2-5 houses with 2-3 hours of backup power during an outage. A CES system has sensing technology to sense when utility power is lost, automatically provide battery-based power to the home, and then reconnect customers automatically to grid-supplied power when available. AEP tested the performance of CES for various other functions, such as islanding load, peak shaving, and VAR support. AEP continues its evaluation of this technology in our testing lab.

As I indicated in my testimony, in most circumstances, the cost of energy storage currently exceeds what the market will support for most broad-based applications. Storage costs need to come down, and that can be aided through increased research and development. The Energy Storage Program, led by the Department of Energy, is studying a wide range of energy storage technologies and high voltage power electronics to demonstrate their benefits. Additional funding would help further that program's research.

Further, as you indicate, tax incentives could further support development and deployment. These tax incentives should be available to utilities to encourage further deployment of batteries on the grid.

In addition, there are regulatory factors that currently present challenges to one entity aggregating the dispersed benefit streams from the variety of benefits that batteries can provide. For instance, as described above, there are benefits to utilities and their customers of utilizing storage technologies for ensuring reliability and avoided incremental costs on transmission and distribution circuits. Additionally, there are opportunities to use storage to arbitrage between peak and off-peak energy markets. Batteries also work well in providing certain ancillary services (e.g., frequency regulation and contingency reserves), with value that can be realized in some wholesale markets. Further, there are other applications that provide value, such as firming-up variable output of renewable resources and providing back-up power during outages. Finally, end-use customers could utilize batteries to reduce their peak-demand, and participate in demand response programs. However, where a particular battery project provides transmission or distribution grid benefits in some circumstances that would support inclusion of the battery investment in cost-of-service utility rates, and is available to provide ancillary services or energy market arbitrage in other circumstances that could support market based rate sales, the owner may not be able to realize both types of revenue streams.

Optimizing storage's potential is a challenging situation which could require addressing significant market and regulatory rules. This will likely involve an incremental, evolutionary approach via multiple rulemakings and stakeholder discussions to address the numerous issues identified. Some market participants are initially advancing the potential of energy storage through providing frequency regulation services in PJM, as battery technology is particularly well-suited to provide this service, and the valuation of that service is relatively high.

U.S. Senate Committee on Energy and Natural Resources
March 17, 2015 Hearing: The Electric Grid

Many have speculated that the enormity of the potential opportunity to electrify the transportation sector is the 'holy-grail' for batteries. As production volumes would drive efficiencies and cost-reductions, many other applications for battery technology would likely proliferate. Therefore, further support of electric vehicle technology would almost certainly further battery deployment in the broader energy sector.

Question 2: What opportunities are there for the federal government to play a role in further unlocking the potential of demand response in America using the smart grid? Is it simply setting communications standards for appliances? Is it providing funding for pilot projects? I'm particularly interested to hear your thoughts on how to help grow demand response opportunities for homes and small businesses.

Additional research and development efforts at the Federal level would provide opportunities for new and effective technologies, such as energy storage and other technologies that address deficiencies in reliability, improve system efficiencies and reduce consumer demand. Once a particular technology is demonstrated, the states are in a better position to determine when and where these technologies are best employed based on relative performance and cost. Similarly, additional Federal funding such as was provided by the American Reinvestment and Recovery Act of 2009 (ARRA) grant would allow the states to approve implementation pilot and other projects.

One of AEP's largest technology initiatives, called gridSMART®, integrates a host of advanced grid technologies into the existing electric network that can improve service quality and reliability, lower energy consumption, and provide additional customer benefits. The new technologies can help us improve our efficiency, identify and respond to outages more quickly, and better monitor and control operation of the distribution grid. As part of this initiative, AEP has deployed over 1.2 million smart meters, and has proposals pending with state public service commissions to install over 1.4 million more.

gridSMART® also provides customers with new and innovative programs and pricing options that allow them to monitor and control their own energy use, saving resources and money.

Through this initiative, AEP Operating Companies have initiated numerous creative customer programs to offer customers new pricing mechanisms in order to save energy and money. These pricing programs typically employ some form of time-of-use rates, and often couple that with enabling technologies (such as a communicating programmable thermostat) that allows them to respond to pricing signals provided either when the grid is physically stressed or wholesale prices are peaking. Enabling technologies which allow easy or automated participation and response are important for success. Nevertheless, even in areas where newer smart grid technologies are not deployed, AEP companies utilize other less-sophisticated methods of pursuing the same effect. These programs effectively allow residential and small businesses to participate in the demand response realm, structured in a manner that is accessible and simple for them

U.S. Senate Committee on Energy and Natural Resources
March 17, 2015 Hearing: The Electric Grid

to use. It is our experience that these residential and small business customers are much more receptive to this approach rather than passing the real-time volatility of wholesale markets through to retail prices. It should be noted that in restructured states, the roles of market participants (utilities, retailers, etc.) can vary regarding responsibilities for providing pricing options and enabling technologies.

In addition, AEP has maintained demand response programs for larger, more sophisticated industrial, commercial, and institutional customers linked more directly with wholesale markets.

AEP Ohio was a recipient of an American Reinvestment and Recovery Act of 2009 (ARRA) grant that allowed testing of many iterations of such pricing programs and enabling technologies among residential and small business customers. AEP Ohio equipped residences with auxiliary devices (in-home display devices, communicating thermostats, smart appliances, direct load control devices, etc.) coupled with several time-of-use pricing tariffs, all designed to provide usage, pricing, and event information, as well as the technical capabilities to respond to that information. The initiative also included an innovative program where we worked with Battelle Memorial Institute and Pacific Northwest National Laboratory (PNNL) to test linking real-time wholesale energy costs with consumer preferences for comfort levels. A comprehensive report analyzing the findings from that initiative, including customer acceptance, capacity and energy impacts, and so forth is available: (https://www.smartgrid.gov/sites/default/files/doc/files/AEP%20Ohio_DE-OE-0000193_Final%20Technical%20Report_06-23-2014.pdf).

Continued support for such programs through grant funding and/or tax incentives would help further the objective of these initiatives.

In addition, AEP continues to partner with NIST and DOE within the Smart Grid Interoperability Panel (SGIP), a public/private funded non-profit organization that supports the work behind power grid modernization through the harmonization of technical interoperability standards to advance grid modernization. Continued and increased support is needed for this effort, specifically in the form of funding, in order to accelerate interoperability benefits for grid modernization and, in the process, bring down costs through economies of scale. Federal assistance in the form of ARRA or tax incentives can be helpful in advancing the deployment.

U.S. Senate Committee on Energy and Natural Resources
March 17, 2015 Hearing: The Electric Grid

Questions for the Record Submitted to Ms. Lisa Barton
from Senator James Risch

Question 1: The nation has multiple agencies doing an excellent job monitoring and advising on cyber threat to the nation's power grid; the DHS ICS-Cyber Emergency Response Team at INL is one of them. How are you using this information to strengthen the grid?

AEP entered into a CRADA (Cooperative Research and Development Agreement) with the Department of Homeland Security's NCCIC (National Cybersecurity and Communications Integration Center) in 2012 to enhance threat and information sharing with the DHS ICS-CERT (ICS-Cyber Emergency Response Team). This enables AEP's Cyber 24x7 Operations Desk to monitor and respond to industry and governmental alerts for new cyber vulnerabilities and threat vectors and to actively interface with threat sharing organizations such as US-CERT (United States – Computer Emergency Readiness Team), and the ICS-CERT (Industrial Control Systems - Computer Emergency Readiness Team).

In addition, AEP routinely sends their cybersecurity analysts to industrial control systems cybersecurity training and Red Team / Blue Team exercises conducted by the DHS ICS-Cyber Emergency Response Team at INL (Idaho National Laboratories). Lastly, as appropriate, AEP will call on the expertise of the DHS ICS-CERT to assist in any significant cyber event.

As a nation, we are seeing an escalation in threats to our national security associated with cyberattacks. We ask ourselves -- What improvements can be made from the standpoint of enhanced coordination between the federal government, the private sector and the utility industry to ensure that as a nation, we are best positioned to guard against these attacks? Are there any shortcomings in the current process?

As cyber and physical threats increase across the US, enhanced coordination between the federal government, the private sector and the utility industry is increasing to ensure the secure operation of the electrical grid. AEP works with a consortium of utilities across the country and the Electricity Sub-sector Coordinating Council (ESCC), a CEO-led industry group that meets three times a year with senior officials from the Department of Energy, Department of Homeland Security, Department of Defense, White House, FERC, NERC, and the Federal Bureau of Investigation. This collaboration has produced advanced tools and technologies to improve situational awareness, threat and information sharing, and development of coordinated plans to respond to an attack on the grid. As an industry, we also continue to improve threat sharing by interfacing and reporting new and emerging security threats to Electricity Sector Information Sharing and Analysis Center (ES-ISAC).

U.S. Senate Committee on Energy and Natural Resources
March 17, 2015 Hearing: The Electric Grid

The ES-ISAC establishes situational awareness, incident management, and coordination and communication capabilities within the electricity sector through timely, reliable and secure information exchange. The ES-ISAC, in collaboration with the Department of Energy and ESCC, serves as the primary security communications channel for the electricity sector and enhances the ability of the sector to prepare for and respond to cyber and physical threats, vulnerabilities and incidents. As new industry processes or tools are identified to further enhance the security of the grid, those industry enhancements are addressed by the ESCC.

Question 2: What resources could enable you to be more effective and timely in adapting your networks to evolving threats?

At AEP, we protect our system in three key ways:

- We partner with government agencies, utility industry and non-utility industry partners,
- We share threat information and best practices, and
- We stay current with emerging threats and risks identified by several threat-sharing collaborations and our own team of analysts.

As events emerge, we continually assess our own cybersecurity tools, processes, and defenses to determine where we can improve our defenses, response readiness and recovery processes.

Threat information sharing is inherently a continuous improvement process to ensure utilities receive timely and actionable threat intelligence data. Those sharing processes and mechanisms can be further enhanced by providing liability protections to utilities for sharing threat information. For example, utilities should not be liable for sharing information unless willful misconduct causes injury.

U.S. Senate Committee on Energy and Natural Resources
March 17, 2015 Hearing: The Electric Grid

Question for the Record Submitted to Ms. Lisa Barton
from Senator Al Franken

Question 1: Is there any effort underway in the utility sector to assess which regions in the country would benefit most from dynamic line ratings?

AEP is not participating in any formal effort at this time.

In general, dynamic line ratings are used to relieve constraints during real-time operations, thus alleviating the need to curtail generation in certain instances. In areas where curtailments occur with frequency, or where generation output is generally less predictable, dynamic line ratings may provide temporary relief of transmission constraints while permanent solutions are pursued.

Let me note that line ratings are also typically adjusted in real-time by transmission operators based on ambient temperatures, providing additional capacity beyond static ratings. The difference is that dynamic line rating devices allow for adjustment based on additional variables (e.g., actual conductor temperature, wind speed, line sag) and can adjust more frequently and precisely than the calculated method.

Note that forward-looking planning analyses must consider static ratings that are established based on North American Electric Reliability Corporation (NERC) standard practices. This is necessary to ensure that the transmission grid can operate reliably in a variety of conditions, including worst-case conditions. Thus dynamic line ratings cannot be considered as an alternative to physical infrastructure upgrades for the purposes of long-term planning.

U.S. Senate Committee on Energy and Natural Resources
March 17, 2015 Hearing: The Electric Grid

Questions for the Record Submitted to Ms. Lisa Barton
from Senator Joe Manchin III

Question 1: Ms. Barton, has AEP done any studies or other analysis to determine the effects of the EPA's proposed Clean Power Plan on the reliability of the grid? If so, what did you find?

AEP ran a load flow analysis of the PJM region using EPA's modeled unit retirements. The retired generation was replaced with proposed new generation currently in the PJM generation interconnection queue. The results were sobering. Our analysis identified severe, widespread transmission reliability concerns consisting of thermal overloads, low voltages, and voltage collapse leading to cascading outages. Significant new transmission enhancements would be required to address these reliability concerns and allow the generation to be retired. Given the amount of construction required, I question the industry's ability to meet the initial deadline of 2020.

Question 2: In your testimony, you discuss the need to maintain an "adequate level of generation resources." Are you concerned about base load generation retirements that are not being replaced by new plants?

I am concerned about the retirement of base load generators even when those units are expected to be replaced with new plants of equal capacity. First, the location of new generation matters. A generator that is close to the load center and connected to multiple transmission facilities cannot be equated to a unit with the same capacity that is far from the load centers and connected to a weaker transmission network. Large base load generation is tightly connected to the grid. Changing the location, the size or the type of generation resource requires transmission upgrades to redirect the power on the grid. Second, energy supplied by base load generation is not the same as energy produced by an intermittent resource, such as wind or solar. Base load generators not only provide electricity, but also provide the reactive power and dynamic support required for the transmission of electricity to load centers. Similarly, base load generators provide the support needed to react quickly to emergency conditions such as an outage of a major plant or major transmission line.

Question 3: Technological innovation in the electric grid will be critical to ensure that electricity remains abundant and affordable. As this Committee considers new energy legislation, what statutory or agency rules or regulations that are currently in force now, if any, are stymieing innovation and infrastructure development for the electric grid?

As described in the MIT Future of the Grid report that was published in 2011, http://mitei.mit.edu/publications/reports-studies/future-electric-grid, decision makers in government and industry took important actions in recent years to guide the evolution of the U.S. electric power system to address the challenges and opportunities the new technologies will provide. Yet, the diversity of ownership and regulatory structures

U.S. Senate Committee on Energy and Natural Resources
March 17, 2015 Hearing: The Electric Grid

within the nation's grid complicates policy-making, and a number of institutional, regulatory, and technical impediments remain that require action.

AEP would like to make the following recommendations that would enable the investment in the right kinds of technology for use on our electric grid.

- Ensure that the regional planning and interregional transmission planning policies contained in Federal Energy Regulatory Commission (FERC) Order No. 1000 are effectively implemented to ensure that wise transmission grid investment is made in a timely manner.

- Encourage FERC to implement the direction of the Energy Policy Act of 2005 with regard to incentive transmission rate treatments in a manner that effectively encourages needed transmission investment.

- Encourage FERC to guard against regional and state policies that impede the use of competitive forces to pick the most advantageous regional transmission solutions.

- Ensure that FERC and RTO policies and practices create a level playing field in which the incumbent transmission owner and new competitors can fairly compete on the merits to develop transmission infrastructure.

- Encourage FERC to resolve uncertainty with respect to return on equity policy in a manner that supports and encourages new transmission infrastructure investment.

- To facilitate integration of renewables, the FERC should be granted enhanced authority to site and grant certificates of public convenience and necessity for major transmission facilities that cross state lines.

- To cope more effectively with cybersecurity threats, a single federal agency should be given responsibility for cybersecurity preparedness, response, and recovery across the entire electric power sector, including both bulk power and distribution systems.

- To make effective use of new technologies the electric power industry should fund increased research and development in several key areas, including computational tools for bulk power system operation, methods for wide-area transmission planning, procedures for response to and recovery from cyberattacks, and models of consumer response to real-time pricing.

- To improve decision making in an increasingly complex and dynamic environment, more detailed data should be compiled and shared, including information on the bulk power system, comprehensive results from "smart grid" demonstration projects, and standardized metrics of utility cost and performance.

**U.S. Senate Committee on Energy and Natural Resources
March 17, 2015 Hearing -The Electric Grid**

**Responses to Questions from Senators Murkowski, Risch, Flake, Wyden and
Manchin for the Record by Commissioner Lisa Edgar
President, National Association of Regulatory Utility Commissioners
Commissioner, Florida Public Service Commission**

**I.   Questions for the Record Submitted to President Edgar from Senator
Murkowski:**

Question 1: Please describe what ancillary services are and why they are
needed to maintain grid reliability. How can distributed energy sources contribute
to these needed grid reliability services, and under what time frame?

Commissioner Edgar response: NARUC as an organization has taken no
position nor have we developed a strict definition on what ancillary services are
and how they relate to grid reliability.[1] However, from a very general perspective,
a number of components may be considered. In brief, ancillary services are needed
to support the transmission of high-voltage electric power from the generating
resource to the consuming load while maintaining reliable operation of an
interconnected grid in accordance with prevalent engineering standards.[2] FERC
includes the following six services:

---

[1]   The association did pass a November 20, 2013 *Resolution Encouraging State Commissions and Policy
Makers to continue to Engage in Collaborative Dialogue Regarding Distributed Generation Polices & Regulations,*
available online at: http://www.naruc.org/Resolutions/Resolution-Encouraging-State-Commissions-Policymakers-
to-Continue-to-Engage-in-Collaborative-Dialogue-Regarding-Distributed-Generation-Policies-Regulations1.pdf,
that encourages State commissions and policymakers to continue…collaborative discussions regarding DG so that
State(s)… have the benefit of key stakeholder input and are better prepared to - Evaluate the system-wide benefits
and costs of DG (including costs and benefits relating to the investment in and operation of generation and the
transmission and distribution grid) so that those costs and benefits relating to DG can be appropriately allocated and
made transparent to regulators and consumers; Ensure that all necessary consumer protections are maintained and
assist consumers as they consider or invest in DG technologies and services; (and) Facilitate the continued provision
of safe, reliable, resilient, secure, cost- effective, and environmentally sound energy services at fair and affordable
electric rates as new and innovative technologies are added to the energy mix."

[2]   PJM Interconnection, LLC. 2014. *PJM Manual 35: Definitions and Acronyms.* Revision: 23;
MacDonald, Jason, et al. 2012. *Demand Response Providing Ancillary Services: A Comparison of Opportunities
and Challenges in the US Wholesale Markets.* Report No. LBNL-5958E. Lawrence Berkeley National Laboratory,
Berkeley, CA.E. Hirst and B. Kirby. 1996. *Electric Power Ancillary Services.* Oak Ridge National Laboratory,
Technical Report ORNL/CON 426; Federal Energy Regulatory Commission. 1995. *Promoting Wholesale
Competition Through Open Access Non-discriminatory Transmission Services by Public Utilities,* Docket RM95-8-
000, Washington, DC.

(1) Scheduling and dispatch – the "control room operation of all generation and transmission resources and transmission facilities on a real-time basis to meet load within the transmission provider's designated service area;"

(2) Load following – the "continuous balancing of resources with load . . . to match moment-to-moment load changes;"

(3) System protection – the "operating reserves . . . available in order to . . . maintain the integrity of its transmission facilities;"

(4) Loss compensation – the "capacity and energy losses [that] occur when a transmission provider delivers electricity across its transmission facilities for a transmission customer."

(5) Energy imbalance – the "difference [that] occurs between the hourly scheduled amount and the hourly metered (actual delivered) amount associated with a transaction;" and,

(6) Reactive power and voltage control – the "reactive power support necessary to maintain transmission voltages within limits that are generally accepted in the region and consistently adhered to by the transmission."

Distributed energy resources (DER) are flexible resources that can provide some of these services to enhance grid reliability; however, such support from DER depends on the technology available at specific locations.

Question 2: As new technologies are being incorporated, the grid is being used in ways for which it was not designed. How do pricing mechanisms need to be adjusted so that all the value generated from both the grid and distributed generation is capture by the market? How do we ensure regulations capture all of the different values provided by the grid? Can you explain how and whether net metering accounts for the value of these ancillary services?

Commissioner Edgar Response: As with any industry when technology advances, business models in the electric utility industry need to adjust or adapt to the effects of the technology changes. Policy makers and regulators should be mindful of these advances in order to consider whether revisions to pricing mechanisms are warranted. Net metering policies, which can include consideration of the value of distributed generation and ancillary grid services, differ from State to State. In Florida, investor-owned utility customers participating in net metering receive the full retail value for excess energy. This policy does not explicitly address the value of any specific ancillary grid service beyond those necessary to complete project interconnections.

Question 3: How might grid integration vary by State and by region? What are some of the advantages of different States having different policies? Can you give some specific examples of policies that work well in one region, but would be problematic in another? What do you see as the appropriate role for the federal government?

Commissioner Edgar Response: In the U.S., "the grid" consists of vertically integrated States and restructured States. In the restructured category there are seven RTOs. This includes three single State RTOs—Texas, California and New York—along with regional markets in the Midwest, Mid-Atlantic, and New England. Florida and the rest of the Southeast are not participants in an RTO, and neither are utilities and States in the desert Southwest and Pacific Northwest.

Each State and region is responsible for determining what kind of wholesale market works best for their consumers. For the Southeast, a traditional, vertically integrated market continues to serve the public interest. Adopting market mechanisms that are not suited to a traditional utility model is an example of policies that would not work well in similar regions.

Defining grid integration within the confines of the distribution level, States and regions have different approaches. The advantage in the State level implementation is two-fold. First, each State has unique demographic, potential generation source, economic/financial and political constraints that should lead to different policies. Second, the varying approaches provide policy makers in other States useful information on the impacts of those approaches on prices, grid reliability and resiliency, different consumer classes, etc.

Currently, States vary in their judgment of the relative benefits of some new technologies like DG and the smart grid. An aggressive DG policy, *e.g.*, New York, Hawaii and California approaches, may not be attractive, economically viable, or efficient in States facing different circumstances. Over the next few years a few electric utilities may undergo a significant change, while others will see only incremental change.

Active integration can provide distribution voltage support, optimize distribution operations, improve power quality and reduce system losses, and improve grid resiliency. A policy question that will challenge many States in the years ahead is how to accommodate an increased number of network users while maximizing the network's value to utility customers as a whole. The dynamics of coordinating new market players, technologies and business models makes this a complex task. For example, the California Public Utilities Commission recently observed:

> California needs to consider a more advanced and highly integrated electric system than originally conceived in many smart grid plans. This integrated grid will evolve in complexity and scale over time as the richness of systems functionality will increase and the distributed reach will extend to millions of intelligent utility, customer and merchant devices.[3]

Physics requires that the distribution network keep the system in balance and confine factors such as voltage and frequency levels within tolerable ranges. Operators must also respect contingency limits, meaning no violation of a line's physical limit if some other line or generator goes out of service unexpectedly. The operation of an interconnected electric network has to be monitored in real time to assure that: (1) production always matches consumption, and (2) power can flow across the network within established reliability and security constraints. Initially, integrating DG makes these tasks more complex.

The federal government should continue its oversight of grid reliability in its current form, with States determining how best to serve and protect end-use customers. Essentially, local distribution is a State matter that conforms to legal precedent, decades of successful experience and good economics. The electric distribution system differs greatly in various ways from transmission systems. The latter appropriately falls under federal jurisdiction while distribution is appropriately a State matter.

Question 4: What kind of consumer protection issues have been encountered with distributed generation?

NARUC is not aware of any studies explicitly analyzing consumer protection issues associated with distributed generation. However, NARUC has been a collaborative partner in the Critical Consumer Issues Forum. This forum issued a report, *DG: A Balanced Path Forward*, in July 2014 that addressed aspects of consumer protections and education.

---

[3] California Public Utilities Commission, "Order Instituting Rulemaking Regarding Policies, Procedures and Rules for Development of Distribution Resources Plans Pursuant to Public Utilities Code Section 769," *Order Instituting Rulemaking*, August 14, 2014, at 24.

As with any other consumer purchase, there can be problems associated with vendors over-promising and under-delivering. Anecdotal reports of problems such as inflated promises, premature product failures, lengthy and costly delays in affecting repairs, and difficulties completing installations because of non-standard and frequently changing rules and regulations have been heard. Additionally, complications with residential sales can occur.

Under the purview of State regulatory authorities, many State and utility funded energy efficiency programs are establishing quality control and quality assurance practices that screen competitive service providers, requiring pre-certified vendors to act in accordance with prescribed standards of practice. Best regulatory practices are still evolving.

The Florida Public Service Commission's (FPSC) net metering and standard interconnection rule for customer-owned renewable generation recognizes safety standards for the interconnection and equipment required to interconnect with the grid. The rule also addresses insurance requirements, limits fees and charges, and establishes an expedited interconnection process. To date, the consumer complaint process has not resulted in any customer-owned renewable generation issues requiring resolution through formal FPSC action.

Question 5: What do you believe the biggest game changer will be for the future of the grid?

Commissioner Edgar Response: Future pressures on the operators of the electric grid and distribution system may include substantial growth of distributed generation, the sale of electricity by third parties, and the effects of carbon regulation. Operators of Florida's grid may be challenged by sudden changes that have not allowed for adequate study, planning, and development of mitigating reliability measures. Consequently, regulators may be faced with new and different solutions to maintain safe, reliable and affordable electric service.

Question 6: What technologies are you most excited about and why? Also, what is the timeframe for development and at what cost?

Commissioner Edgar Response: Microgrids, small nuclear reactors, inverter technology, wave energy and battery storage, to name a few. Federal support of R&D is crucial. The FPSC expects utilities under our jurisdiction to explore opportunities and integrate developing technologies that are cost-effective in order to maintain safe, reliable, and affordable electric service.

Question 7: Can you explain the most pressing specific challenges faced by State public utility commissioners? If you were speaking to the market and innovators, what is the most pressing problem you would ask for help solving?

Commissioner Edgar Response: The most pressing challenge State public utility commissioners address is balancing the need to maintain affordability while encouraging utility management to pursue innovative approaches to the generation and provision of electricity. Long-term planning and cost-effective investment is essential, yet difficult during a time of uncertain federal and State policy directives.

Question 8: In your testimony, you noted that new transmission will be necessary to address retired generation. Can you explain why new transmission will be required to preserve grid stability? What role will transmission play in integrating distributed generation?

Commissioner Edgar Response: Transmission is just one component of this puzzle. If the U.S. wishes to become more dependent on renewable energy, we will need more transmission to deliver power across States from these generation sources to the load. Utilities will need new transmission to get their product to the market. But building transmission comes with uncertainties; how is it priced? Will consumers in one State pay for electricity delivered in another? Will federal agencies be able to site power lines on federal lands timely enough so as not to cause market imbalances?

Additional environmental requirements, in particular the Mercury & Air Toxics Standards (MATS) and the Clean Power Plan, along with low natural gas prices, will continue to place pressure on some coal-based and nuclear units to consider retirement. The existing transmission system was not built to accommodate a generation fleet shifting to natural gas combined cycle, wind and distributed solar generation.

Transmission services such as regulation, frequency response and voltage control will be required to maintain grid balance and system reliability as the generation fleet shifts from fossil and nuclear-based base load plants to variable energy resources (VERs). FERC defines Variable Energy Resource as a device for the production of electricity that is characterized by energy source that: (1) is renewable; (2) cannot be stored by the facility owner or operator; and (3) has variability that is beyond the control of the facility owner or operator. This includes, for example, wind, solar thermal and photovoltaic, and hydrokinetic generating facilities.

With respect to the relationship between transmission and distributed generation, DG is a distribution-level resource that falls appropriately under the jurisdiction of the State public utility commissions. While it is true that constructing transmission lines may create additional opportunities for generation resources, including DG, to be delivered to loads that need them, decisions about transmission needs and DG resources are often made independent of each other. However, large, grid-scale distributed generation, such as large solar farms and micro-grids will benefit from increased transmission infrastructure, because the best areas to place grid-scale distributed resources are often not currently served by the transmission infrastructure needed to deliver those resources to load.

Having said this, many predict the grid to become more localized and less dependent on long-distant power lines. In this case, the focus will be on bolstering the distribution system and making sure it can handle a more reactive consumer base.

Question 9: Regarding the possible issues with regulatory federalism and the need to expand transmission, could you elaborate on the role for State and federal governments respectively in this area?

Commissioner Edgar Response: State agencies should remain the first and primary decision-maker regarding expanding the transmission system. States site and FERC sets tariffs for wholesale transactions. The grid is regional, not national. State commissions have the resources and knowledge of local needs that the federal government does not. And unlike federal agencies, we are directly accountable to our citizens and our legislators.

At the same time, State Commissions know the grid is regional and are able to site transmission, if deemed necessary, as quickly as due process allows. State decision-makers are participants in an interconnection-wide dialogue on transmission planning issues. Through the Eastern Interconnection States Planning Council—funded by the Department of Energy—States in the Eastern Interconnection are looking at planning with a holistic approach. These discussions are proving to be essential. This is a great example of federalism at work.

Question 10: In your opinion, how should infrastructure investment decisions be prioritized?

Commissioner Edgar Response: For any investments, the prime criterion is whether the benefits exceed the costs. Because of capital limitations, however, utilities as well as other private entities often must choose among investment opportunities that meet this criterion. Some investments may require more immediate attention because they address serious problems, such as inadequate generation or transmission capacity and prolonged outages from extreme weather. Regulators and consumers place a high priority on utility service reliability and safety. If the absence of investments jeopardizes these goals, regulators will press utilities to make those investments. In those cases, all other investments assume a lower priority even though they pass a cost-benefit test.

The FPSC review of utility investment decisions includes factors such as impacts on system reliability and overall cost-effectiveness. Each utility faces different challenges and therefore prioritization of the needed infrastructure investment is not identical in all cases.

Question 11: How many States currently utilizing time-of-use pricing?

Commissioner Edgar Response: NARUC was unable to locate an up-to-date listing of States using time-of-use (TOU) pricing. However, we know that many States do have such pricing, at least as an option available to larger commercial and industrial customers.[4] Additional States, partly because of the installation of smart meters, have initiated proceedings to consider the merits of TOU pricing. Almost all residential (as well as small commercial consumers) in the U.S. buy electricity on rate structures with "flat" prices that remain uniform across time periods irrespective of supply and demand conditions, marginal costs or wholesale market prices. Indeed, a survey conducted by FERC indicates that only about one percent of residential consumers pay TOU rates.[5] The percentage for large commercial and industrial customers is much larger. The lack of necessary technology has become less of a factor limiting the adoption of time-of-use rates. According to FERC, by the end of 2015, 50 percent of U.S. households are projected to have a smart meter.

---

[4] Analysts usually consider TOU pricing to include many variations, such as on-peak and off-peak pricing, varying seasonal rates, critical peak pricing, real-time pricing, and peak-time rebates. See also, the Massachusetts Department of Public Utilities' recent endorsement time-varying rates as a default option, which is uncommon. Massachusetts Department of Public Utilities, *Investigation By the Department of Public Utilities on Its Own Motion into Time Varying Rates, D.P.U. 14-04-B*, June 12, 2014.

[5] Federal Energy Regulatory Commission, "2010 Assessment of Demand Response and Advanced Metering," Staff Report, December 2014, at 31.

Some regulators, consumer advocates and others still have concerns about the effects of TOU pricing on certain customers, especially those who, for a variety of reasons, do not change their usage behavior. While studies on real-time pricing generally show that the benefits outweigh the costs, most of the benefits go to a small number of consumers. Although some customers will likely benefit from such pricing, other customers will see higher bills. The prospect of a large number of losers is a political obstacle to widespread adoption. This posture prevails notwithstanding evidence that TOU pricing has potentially large benefits, but *only* if the majority of customers react to its incentives. These benefits include a more efficient allocation of electric power across time periods (lowering peak), and potentially lower utility energy and capacity costs.

Question 12: Can you explain how and whether net metering accounts for the value of ancillary services?

Commissioner Edgar Response: NARUC does not have a position or consensus on this issue. However, generally speaking, net metering does not explicitly account for all of the value that utilities produce and provide to net metering customers, nor for all of the values that net metering customers produce and provide to utilities and non-net-metering customers. Although the net metering rules differ in some important ways in different States and in specific jurisdictions, as a general concept net metering policies seek to simplify the relationship between customers and utilities. Most often, net metering has used the retail price of electricity as a rough proxy exchange value, without trying to accurately quantify specific costs and benefits. Some recent analyses suggest that the retail price might overstate the value that net metering provides to the host utility and non-net-metering customers, but many other studies find the opposite, that net-metering under-compensates participating customers (see, for example, Hansen et al., 2013,[6] for a review and comparison of over a dozen studies).

Smart-inverters and related communications and control technologies are enabling small, distributed generators to produce and deliver ancillary services, and new modeling capabilities are making it more practical to assess the costs and benefits associated with all kinds of distributed energy resources and ancillary services. Therefore, information will gradually become available to help with the

---

[6]   Hansen, Lena, Virginia Lacy, and Devi Glick. (2013). A *Review of Solar PV Benefit & Cost Studies, 2nd Edition*. Rocky Mountain Institute, eLab. http://www.rmi.org/elab_empower

design of net metering or possible replacement approaches, to more accurately reflect all costs and benefits.

As I mentioned in my testimony 43 States use net metering and all their programs operate differently depending upon the specific local situation and need, so these questions are being actively investigated.

Question 13: How do you capture the appropriate range of values in regulatory treatments for energy storage projects?

Commissioner Edgar Response: NARUC has no position or consensus perspective on this issue.

II. **Questions for the Record Submitted to President Edgar from Senator Risch:**

Question 1: The nation has multiple agencies doing an excellent job monitoring and advising on cyber threat to the nation's power grid; the DHS ICS-Cyber Emergency Response Team at INL is one of them. How are you using this information to strengthen the grid?

Commissioner Edgar Response: NARUC has partnered with first DHS and then DOE to provide cybersecurity trainings to 41 States and DC. These trainings have resulted in orders and dockets in 19 States and DC, and ongoing information exchange between States and the private sector in 9 more States (likely more, since these interactions can be informal and undocketed.) Funding for these trainings from DHS ran out in 2014; DOE funding concludes this September. We also engage in unclassified information sharing through biweekly conference calls and a monthly unclassified threat briefing from Energy Sec / the National Electricity Sector Cybersecurity Organization.

Question 2: What resources could enable you to be more effective and timely in adapting your networks to evolving threats?

Commissioner Edgar Response: Three areas would help. First, as standards for new Information Sharing and Analysis Sharing Organizations (ISAOs) develop, an interest by States in forming an ISAO should be incorporated to enable one to be formed among PUCs. Second, more funding to conduct clearance investigations for Commission staff. Currently only 6 Commissioners and as many staffers have clearances at SECRET or better. These clearances are essential to have context to effectively review and encourage utility cyber-security activity, cost-benefit analysis, and cost allocation. Finally, federal money through DHS and DOE to provide strategic technical assistance and training to State regulators on cyber-security through NARUC is important and should be renewed.

III. Questions for the Record Submitted to President Edgar From Senator Flake:

Question 1: In the recent report, *The Integrated Grid: Realizing the Full Value of Central and Distributed Energy Resources*, EPRI shows that the costs to maintain the transmission and distribution system represent roughly half of an average residential bill and change very little with actual energy consumption. Consumers with distributed generation (DG) systems who consume no net energy still demand benefits of reliability, startup power, energy transaction, voltage quality, and efficiency from the grid. What challenges do utilities and regulators face in designing and approving rate structures that recognize these benefits?

Commissioner Edgar Response: At NARUC's February 2015 meetings, the association adopted a resolution that "recognizes the contributions of EPRI's 'Integrated Grid' for evaluating the value of energy resources and grid connectivity, and commends EPRI for its beneficial analytical framework and communications outreach to stakeholders."

This EPRI report is one reflection of rapid changes in electric utility markets. Driven in part by policy-makers' interest in encouraging the use of distributed energy resources (DER), distributed solar photovoltaic electric generating systems are proliferating. This increased emphasis on DER presents rate design challenges for policy makers. There is no "right" answer. The challenge is to design rates that balance the costs of providing electricity to customers while sending price signals that should encourage customers to more efficiently manage their use of those services. Regulators must also ensure that all necessary consumer protections are maintained and assist consumers as they consider or invest in DG technologies and services. Moreover, we must facilitate the continued provision of safe, reliable, resilient, secure, cost-effective, and environmentally sound energy services at fair and affordable electric rates as new and innovative technologies are added to the energy mix.

State utility commissions have decades of experience with rate design issues. Those decisions are necessarily limited by an evidentiary record created through a process that permits all interested parties to make proposals and challenge/rebut the arguments of others. In addition, State legislatures sometimes prescribe an approach limiting or restricting the State Commission's consideration of certain rate design options. For example, in some States, the legislature has specified the applicable standards, which include both rates for net excess generation and charges, if any, that can be assigned for utility standby service.

These State experiences provide useful information on the value, utility and effectiveness of various approaches. This is because there is no purely objective means of establishing the ideal proportions of utility charges based on the three major components of:

(1) numbers of customers,

(2) average and peak system infrastructure usage by each customer; and,

(3) energy usage by each customer.

As a matter of due process, and economic efficiency, utilities must be provided a reasonable opportunity to recover the fixed costs of service. However, if fixed charges for "back-up" service are too high, some argue it could reduce the incentive for customers to conserve energy and could depress growth of some distributed energy technologies.

Understanding the precise nature of ongoing market changes and modeling the likely effects of rate design changes is no simple matter. Research is ongoing to more clearly describe and understand the nature of current concerns and potential problems and to explore various proposed rate design solutions.

There have been many collaborative efforts involving regulators and various stakeholders addressing the multitude of regulatory and other issues relating to the potential and challenges of DG in providing safe, reliable, affordable, cost-effective, and environmentally sound energy supply. We encourage these kinds of discussions so all stakeholders can learn more about the costs and benefits of broad DG development. The result of these collaborations is that some States have already made rate design changes and many others have decisions pending. Certainly, in the near future, all policy makers will have more information from these early State experiences with different rate designs and market conditions.

IV. Questions for the Record Submitted to President Edgar from Senator Wyden:

Question 1: As many of you have testified, energy storage is the Swiss Army knife of the electric grid. Energy storage has the potential to solve many of the problems discussed in this hearing. Storage helps integrate renewables, it can help keep the lights on during extreme weather events, and it makes the whole grid operate more smoothly. It's no secret that I think energy storage is a good deal for the American public; the tool just needs sharpening to live up to its potential to reshape the grid. In particular, the recent progress to bring down costs and increase efficiency must continue. The barriers to storage aren't just cost, though, utilities also aren't used to working with storage yet, and regulations for storage vary from State to State.

I continue to favor providing tax incentives for energy storage to encourage its deployment, but what else should be done to give energy storage the boost it needs? Please don't restrict your answer to the jurisdiction of the Senate Energy Committee, please also comment on useful policy solutions within the jurisdiction of the Senate Finance Committee.

Commissioner Edgar Response: NARUC has not adopted a specific policy position on this topic. On behalf of NARUC, I believe letting States move forward with varying approaches is the most effective way to proceed. That approach gives policymakers the opportunity to learn from jurisdictions trying various strategies for incorporating different storage solutions into the grid.

Moreover, the federal government plays a critically important role through research, development, and demonstrations. For example:

- USDOE's ARPA-E has engaged in research to support further development of cost-effective battery technologies

- ARRA funding has been instrumental in supporting battery manufacturers who are actively engaged in research to lower costs and improve efficiency and long-term reliability

- Many of the national energy laboratories are engaged in important RD&D activities for the support of public utilities and their regulators

It can be important and helpful to provide general incentives for all kinds of businesses to engage in research and development, and incentives for depreciation can similarly play an important role in encouraging investments.

As efficient storage options emerge, utility rate designs are likely to be adjusted to allow storage to participate in markets where it is cost-effective. If that happens, then incentives may not be needed to initiate and sustain rapid growth.

Question 2: What opportunities are there for the federal government to play a role in further unlocking the potential of demand response in America using the smart grid? Is it simply setting communications standards for appliances? Is it providing funding for pilot projects? I'm particularly interested to hear your thoughts on how to help grow demand response opportunities for homes and small businesses.

Commissioner Edgar Response: From the NARUC perspective, we believe the States are and should remain the main driver of demand response and proliferation of, at least, distribution level smart grid technologies. When lights go out or rates go up, consumers hold State regulators directly accountable. Although there are similarities, every State is different and has unique needs and customer demographics.

Generally, the demand for distributed generation and demand-response technologies comes from the end-use consumers. Regulators are accountable for any complications or problems associated with new programs, particularly when it comes to technologies or policies that alter how consumers use and consume electricity. There is great promise in demand response and smart grid technologies, but it is up to State regulators to make sure the grid is reliable, rates are fair and reasonable, and cost allocation is equitable.

NARUC has a long and successful history of collaboration with the Department of Energy and the Environmental Protection Agency. The Association has engaged in joint federal-State initiatives like the National Action Plan for Energy Efficiency, now the State and Local Energy Efficiency Action Network (SEE-Action). Our members have benefited from federal educational grants that allowed NARUC to hold educational workshops and publish reports on these and related issues. Moreover, the seed money provided by the American Recovery and Reinvestment Act is just now bearing fruit in the smart-meter space. The first wave of smart-meter deployments continues, and it'll be a few years before we see the results. Again, letting the States proceed with staggered rollouts will let all policymakers learn from these experiences, limiting implementation problems and increasing cost-effectiveness.

NARUC has no specific position endorsing the following suggestions for communications standards or pilot project funding. However, I offer them as a potential starting point for discussion.

Setting communications standards for appliances and funding pilot projects may promote growth of DR. There are examples of both. A 2009 FERC Policy Statement on Smart Grid identifies DR as one of four key grid functionalities for which the National Institute of Standards and Technology (NIST) should develop and recommend interoperability standards. FERC determined that smart grid technologies have considerable potential to promote DR, which can reduce wholesale prices and wholesale price volatility and reduce potential generator market power, and enhance the application of demand response to accommodate the integration of variable generation.[7]

An advanced grid technology that may benefit from further pilot funding is grid integrated vehicle (GIV) technology – which allows electric vehicle charging stations to support grid reliability. Regulation, managed by the regional grid operator, is the method by which proper balance is maintained on the electric grid. This balancing service is typically delivered by natural gas generators ramping up or down in response to a regional grid operator dispatch signal based on fluctuations in demand, but emerging research on grid integrated vehicle (GIV) technology suggests that electric vehicles may also contribute to this function. Electric vehicle battery systems can provide near-instantaneous response to grid operator signals, while other generation technologies may require several minutes to hours to react to the dispatch signal.

Another area that could benefit from pilot funding is dynamic pricing of retail electric rates. In 2009, the FERC released a National Assessment of Demand Response Potential. In this report, FERC found that under a "full participation" scenario, peak load could be cut by 20% or 188 GW from a no-DR baseline, by 2019 (NADRP, p. x). The "full participation" scenario is an estimate of how much cost-effective DR would take place if advanced metering infrastructure were

---

[7]   NARUC passed a resolution June 15, 214 encouraging NERC "to include with all releases of *electric reliability* standards a statement about the extent to which costs were considered, and if costs were not considered, an explanation for why not." (emphasis added) See, *Resolution Requesting Ongoing Consideration of Costs and Benefits in the Standards Development Process for Electric Reliability Standards*, online at: http://www.naruc.org/Resolutions/Resolution%20Requesting%20Ongoing%20Consideration%20of%20Costs%20and%20Benefits.pdf

universally deployed and if dynamic pricing were made the default tariff and offered with proven enabling technologies. *It assumes that all customers remain on the dynamic pricing tariff and use enabling technology where it is cost-effective* (NADRP, p. xii). As periods of peak electric usage drive up energy costs and stress the grid, funding to determine the viability of dynamic pricing programs, and determine whether peak shaving is a potential outcome, could be important.

Growth in DR by homes and business should be driven in large part by load serving entities that take demand response programs seriously and engage in the customer education necessary to deliver successful programs. Where load serving entities do not offer DR programs, aggregators of retail customers could drive the DR market and take up the role of customer education.

Finally, it is important to note that the ability for the FERC to allow demand response to "bid-in" to restructured wholesale market will be determined by the Supreme Court's decision on a petition seeking review of <u>Electric Power Supply Association v. FERC</u>, 753 F.3d 216 (May 23, 2014). For more than a decade, FERC has permitted demand-side resources to participate in organized wholesale markets, allowing Independent System Operators (ISOs) and Regional Transmission Organizations (RTOs) to use demand-side resources to meet their systems' needs for wholesale energy, capacity, and ancillary service. *NARUC has not taken a position on that Court decision.* In that case, the D.C. Circuit vacated FERC Order 745, which set compensation for DR participating in organized wholesale markets. The D.C. Circuit held that DR is a retail product and the FERC lacks jurisdiction. In January 2015, FERC filed a petition for certiorari with the U.S. Supreme Court. We are all waiting to see what the Supreme Court decides to do.

V. Questions for the Record Submitted to President Edgar from Senator Manchin:

Question 1: Ms. Edgar, in your testimony, you point out the vital role of States in implementing new grid technologies, a position I share. In your opinion, what policies are required at a national level and how can Congress help States implement new technologies that ensure continued reliability and affordability?

Commissioner Edgar Response: NARUC has no official position on this issue. However, many States would welcome additional funding and grants to help pay for expansion and integration of new technologies. The smart-grid grants provided under the American Recovery and Reinvestment Act helped States and utilities expedite the deployment of smart-meters throughout the country. Increased technical assistance from the DOE laboratories could prove useful.

Question 2: As a State regulator, do you have concerns about new EPA regulations threatening reliability? What role do new grid technologies play in reducing our carbon output while ensuring reliability?

Commissioner Edgar Response: NARUC has passed three policy resolutions on the EPA Clean Air Act regulations. All were reflected in NARUC's November 2014 comments to the EPA, available at: http://www.naruc.org/Testimony/14%201019%20%20NARUC%20Letter%20to%20McCarthy%20on%20111d%20guidelines1.pdf. However, the association has not conducted any specific analysis of the reliability impacts of the Clean Power Plan and our members have divergent views. Some believe it can be implemented without impact on reliability and effectively endorse the proposal, while others believe it will have a negative impact locally in certain regions and have joined in lawsuits to block its implementation.

As you know, specifics vary State to State, and region to region. A number of our members in their comments to EPA raised reliability concerns, citing, *inter alia*, the NERC study on Potential Reliability Impacts, which details a number of concerns about the timing and feasibility of the proposal. Additionally, a number of State Commissioners recently testified at FERC hearings on this topic.

Many States have questions about infrastructure and whether the U.S. has the pipeline capacity needed to accommodate an anticipated increase in natural gas use to comply with the proposal. These infrastructure concerns also extend to the electricity transmission system. In some cases, additional renewable energy will require additional transmission lines. New linear infrastructure is expensive and often takes years to navigate the permitting requirements. These costs will likely land on ratepayers, many already overburdened. Utility rates and reliability are the responsibility of State commissions and FERC.

Speaking specifically for Florida, the FPSC has exclusive jurisdiction over the planning, development, and maintenance of a coordinated electric power grid throughout Florida to assure an adequate and reliable source of energy. Intrusion by EPA into these matters could interfere with our jurisdiction over the generation and distribution of electricity, Florida's electricity grid, and the economic regulation of electric retail service. In addition, the proposed rule compromises Florida's ability to maintain a diversified generation fuel source mix and the rapid addition of large scale intermittent generating resources may compromise grid reliability. Finally, without knowing the final requirements of a State implementation plan, individual utilities will not be able to determine their most cost-effective compliance path nor be able to develop aggregate costs resulting from consolidation and coordination of each utility's compliance plan across the State.

UNITED STATES SENATE
COMMITTEE ON ENERGY AND NATURAL RESOURCES

Hearing Questions for the Record
The Honorable Ron Wyden

Advancing an Integrated Electric Grid
Tuesday, March 17, 2015

Questions for Dr. Howard

1. *As many of you have testified, energy storage is the Swiss Army Knife of the electric grid. Energy storage has the potential to solve many of the problems discussed in this hearing. Storage helps integrate renewables, it can help keep the lights on during extreme weather events, and it makes the whole grid operate more smoothly. It's no secret that I think energy storage is a good deal for the American public; the tool just needs sharpening to live up to its potential to reshape the grid. In particular, the recent progress to bring down costs and increase efficiency must continue. The barriers to storage aren't just cost, though, utilities also aren't used to working with storage yet, and regulations for storage vary from state to state.*

   I continue to favor providing tax incentives for energy storage to encourage its deployment, but what else should be done to give energy storage the boost it needs? Please don't restrict your answer to the jurisdiction of the Senate Energy Committee, please also comment on useful policy solutions within the jurisdiction of the Senate Finance Committee.

As you illustrated through the apt analogy of the Swiss Army Knife, energy storage has significant potential to provide a variety of functions that can improve the flexibility, reliability and resiliency of the power delivery system. EPRI, along with DOE, ARPA-e and many other research organizations have been working on advancing the applications of energy storage for over 40 years.

In 2013, EPRI studied the value of energy storage in a variety of different uses in the state of California, with inputs from the California Public Utility Commission, the three California investor-owned utilities, and other industry stakeholders. The study, documented in EPRI Report No. 3002001162 titled *Cost-Effectiveness of Energy Storage in California*, found that in all but 3 of the 31 cases studied, the multiple values and uses of energy storage outweighed the cost of the storage system.

Despite the demonstrated value of energy storage, the technology still faces significant technical, economic, and market barriers which may slow its adoption. Several of these barriers are described below.

Market barriers:
Many of the identified benefits of storage described in the abovementioned EPRI report accrue to different stakeholders. To extend the analogy of the Swiss Army Knife, it is as though several people were sharing the same Swiss Army Knife but using it for different functions – one using

the blade, another the corkscrew, still another the screwdriver – each unable to justify personal ownership of the knife but eager to borrow it for their own purpose. Value in ownership can arise only when the owners of the asset are able to defray the expense of ownership with payments from others who benefit, through an efficient market with low transaction costs.

Integration:
With the variety of different value streams that can theoretically be achieved with energy storage, the question of how to actually achieve these benefits still requires substantial development and industry coordination. This involves standard control system functions that the industry will need to agree on and then methods for integrating these functions with the operation of the grid. As we support early deployments and demonstrations of next generation storage systems, significant focus needs to be placed on the standards and integration approaches to scale these applications for broader deployment.

Safety and Reliability:
The industry would benefit from greater clarity on safety and reliability requirements for energy storage systems. Some equipment providers of energy storage systems have struggled with the stringent safety and reliability requirements for modern grid-connected assets, and negative experiences have led to hesitancy in more aggressive deployment. Many of these issues are delineated in EPRI Report No. 3002003675 titled *Energy Storage Integration Council 2014 Update*. EPRI has attempted to address these issues by working with the U.S. Department of Energy's Office of Electricity Delivery and Energy Reliability and with other research organizations to develop common approaches to safety, reliability and integration of energy storage. In response to utility industry requests, the Office recently formed the Energy Storage Safety Working Group. This working group, in which EPRI is participating, will bring together stakeholders in a common understanding of energy storage safety issues and produce guidelines to help the industry with safer energy storage deployment.

Demonstration:
Even with the development of common approaches and guidelines, the proof of safety, reliability and performance is possible only with practical experience in the field. Developers and users of energy storage will be reluctant to invest in this technology without field experience. Deployment programs, including those conducted by EPRI and EPRI members, as well as projects funded by the Department of Energy's Office of Electricity Delivery and Energy Reliability, have been an important source of information for informed deployment decisions of energy storage technologies. Demonstrations will continue to be important, particularly related to understanding value propositions and dealing with integration issues.

Basic Research:
Continuous work is needed to develop storage technologies that are more efficient, more durable, and less expensive. This work is going on around the world due to the tremendous potential market for technologies that can make significant gains in these areas.

2. *What opportunities are there for the federal government to play a role in further unlocking the potential of demand response in America using the smart grid? Is it simply setting*

*communications standards for appliances? Is it providing funding for pilot projects? I'm
particularly interested to hear your thoughts on how to help grow demand response
opportunities for homes and small businesses.*

There is no doubt that one step in the process is setting the standards so that smart devices
(appliances, thermostats, electric vehicle chargers, etc.) can communicate and be integrated with
grid management systems. These standards can be implemented in home or building energy
management systems (HEMS and BEMS), in devices themselves or both.

However, the more important step is creating the market structures and the incentives for actual
participation of end users, whether directly or through aggregators. This is one of the objectives
of the Reforming the Energy Vision (REV) effort in New York. The New York effort recognizes
that demand response benefits exist at both the overall grid level (associated with generation and
transmission limitations) and at the distribution level (associated with specific distribution
conditions and limitations). Market structures are needed that include the distribution system in
the delivery of pricing signals and incentives to the customer. There is growing evidence that
customers want choices in how they buy electric service that offer savings by taking more
control of their energy usage.

An example to review is Japan's demand response following the shutdown of their nuclear fleet
after the Fukushima accident. They initiated a rebate of $1,000 for customers to install a Home
Energy Management system that can coordinate with the electric utility and control local PV,
electric vehicle charging, energy storage and smart appliances. Japan started the establishment of
an infrastructure they can build on for years as customers have equipment that can be plugged
into these controls. The important part of this rebate approach is that it also established standards
that had to be met for the controllers so they could be integrated with grid operations and
management. Once enough controllers are deployed, market forces can take over and most, if not
all, controllers will meet these standards.

UNITED STATES SENATE
COMMITTEE ON ENERGY AND NATURAL RESOURCES

Hearing Questions for the Record
The Honorable Al Franken

Advancing an Integrated Electric Grid
Tuesday, March 17, 2015

Questions for Dr. Howard

> 1. *In the coming decades, we are going to need to deploy a lot more renewable energy. But transmission capacity is a big barrier to integrating more renewables into our grid. I understand that we currently have a lot of unused capacity in our transmission system because operators do not have access to real-time information for the temperature of their lines. Would dynamic line ratings help us take advantage of the full capacity of our transmission system? And what are some of the other advantages of this technology, in terms of reducing transmission congestion and keeping electricity costs low?*

1A: Would dynamic line ratings help us take advantage of the full capacity of our transmission system?

1A Answer: At present, while we can know the instantaneous overhead line rating, this can only support 1 to 2 hour ahead ratings. The state of Dynamic Rating Technology today enables the knowledge of the instantaneous overhead line rating which supports 1 to 2 hour ahead ratings. This information can provide transmission operators flexibility to ensure reliability when operating the grid when dealing with unforeseen situations. The real-time information on ratings does not, however, allow operators to schedule specific resources to deliver energy across congested flowgates as these decisions are made days to many hours ahead. As a result, present technology does not yield sufficient additional capacity from the existing system to enable reduced investment in installing new transmission assets or upgrading existing transmission assets. In order for the full capacity benefits of dynamic ratings on the order that could potentially defer some investment to be achieved, predictive ratings which are greater than a day ahead are needed. Research in this area is currently underway.

1B: What are some of the other advantages of this technology, in terms of reducing transmission congestion and keeping electricity costs low?

1B Answer: If predictive dynamic ratings are developed with more than a day ahead capability, the enabled additional capacity would result in less congestion; thereby resulting in potentially reducing electricity costs.

> 2. *A recent pilot study showed that dynamic line ratings could help us integrate more renewable*

energy into our grid. Do you think this could be a useful tool for reducing carbon emissions in the electricity sector?

In the U.S. Department of Energy-funded EPRI/NYPA study it was found that there was a correlation between the local wind farm's generation capacity and the lines dynamic rating due to the increased local wind speed cooling the transmission line. It should be noted that the sections of line monitored were close to the wind farm and in open terrain, so that both the wind farm and monitored line sections experienced the same ambient (wind, temperature) environmental conditions.

If the critical sections of the transmission line servicing the wind farm were not spatially close, or were in a different terrain (for example in a wooded valley), the local environmental conditions may be different and the correlation may not exist. Consequently, each transmission system design is different and needs to be analyzed on a case-by-case basis to determine if this correlation would theoretically exist and then followed by a measurement study to confirm this. An industry accepted practice to implement such a study to verify and confirm this correlation does not exist and would require research and verification.

UNITED STATES SENATE
COMMITTEE ON ENERGY AND NATURAL RESOURCES

Hearing Questions for the Record
The Honorable Marie K. Hirono

Advancing an Integrated Electric Grid
Tuesday, March 17, 2015

Question for Dr. Howard

1. *In Hawaii, more than 90 percent of our energy is produced from imported oil. We pay some of the highest prices for electricity in the country, about three times the national average.*

    In Hawaii, we decided to set a goal of no longer being the most oil-dependent state and instead get 40 percent of our electricity to come from renewables by 2030. This is the most ambitious goal in the country.

    In the six years since we established the Hawaii Clean Energy Initiative, Hawaii now produces 18 percent of our energy from renewables.

    The report from EPRI you provided as part of your testimony notes that photovoltaic solar sources represent less than 2 percent of total U.S. generation capacity. In comparison, the report notes that solar made up over 20 percent of the electric generation in Germany in 2012, coming from 1.3 million homes and businesses.

    What regulations and technologies have other countries developed that we could apply in Hawaii and in the rest of the United States as we seek to use more intermittent renewable energy sources?

When discussing generation resources, especially the incorporation of variable renewables, it's important to note the difference between generation capacity and generated energy. In Germany, solar installed capacity as of October 29, 2014, was 38.124 GW and total installed capacity in Germany was 177.14 GW, equaling 22 percent of installed capacity from solar[1]. Regardless, German grid operators rely substantially on resources from neighboring countries such as France, Switzerland, Poland, and the Czech Republic to help balance supply and demand (using electricity imports and exports). This has been effective for managing renewable variability and uncertainty of generation supply. In addition, Germany, and other situations like Ireland and Texas with very high penetration of renewables, maintain virtually 100% reserves of conventional generation when the renewable energy sources are not available.

---

[1] Bruno Burger, "Electricity production from solar and wind in Germany in 2014." (Presentation from Fraunhofer Institute for Solar Energy Systems ISE. Freiburg, Germany. December 29, 2014.)

The situation in Hawaii is more problematic because the islands lack interconnections with neighboring systems and have aggressive goals for renewable energy expansion. Success in achieving those goals will likely require significant advancements in grid infrastructure and control architectures that have yet to be demonstrated on a significant scale. Fortunately, there are technologies under development that will help Hawaii achieve its renewable energy targets. EPRI is currently doing research in these three areas (in close cooperation with Hawaiian utilities):

- **Smart Inverters** – The interface between photovoltaics and the grid, advanced inverters are sensitive to the conditions around them, and can better control their output to reduce potential negative impacts, as well as aid system operators in the event of a grid disturbance. The transition to smart inverters in Hawaii has already begun with local efforts under the Reliability Standards Working Group (RSWG) including the local utility, equipment suppliers, solar developers, and other stakeholders.

- **Distributed Energy Resource Management System (DERMS)** – Especially in Hawaii, where the overwhelming majority of solar development is at the residential-level, grid operators need a way to aggregate multiple resources so their output may be forecasted as well as quickly and effectively managed when necessary. DERMS is a tool that allows multiple local resources to be represented as a larger, virtual resource that is visible and controllable by grid operators.

- **Utility-Dispatched Energy Storage** – Energy storage is often talked about when referring to renewables integration. However, it is important to note that the control of that storage is just as necessary to consider as the storage itself. Especially when utilized as a supply/demand balancing resource, integration of storage with the entire resource portfolio available to the utility is critical to effective utilization.

- **Demand Response** – Consumers can play a role in providing the flexibility that is required with high penetrations of renewables. In Ireland, there is substantial research on how thermal storage in customer heating systems can be used to balance variations in wind generation. Electric vehicle charging is another resource that has substantial potential and is the subject of important research efforts in Maui. Effective use of demand response resources for these purposes requires substantial efforts on standards, communications protocols, and integration of these resources into grid operations.

UNITED STATES SENATE
COMMITTEE ON ENERGY AND NATURAL RESOURCES

Hearing Questions for the Record
The Honorable Joe Manchin III

Advancing an Integrated Electric Grid
Tuesday, March 17, 2015

<u>Questions for Dr. Howard</u>

1. *Dr. Howard, we have heard a lot today about increases in customer-sited generation and other changes to the grid. You specifically advocate for an Integrated Grid that is more flexible than the current system.*

   Can you walk me through how an Integrated Grid would contribute to overall reliability and how we can accommodate distributed energy resources while ensuring continued grid reliability?

Maintaining the reliability of the electric service will continue to be the most important objective of the electric utilities. Distributed resources must be integrated based on their benefits to both consumers and the grid. They cannot have a negative impact on reliability.

While much of the emphasis on distributed resources is reducing the environmental impacts of electricity generation by integrating more renewable generation, distributed resources can also have reliability benefits. This is one objective of the New York Reforming the Energy Vision (REV) effort – integrate distributed resources for economic and reliability benefit. Reliability benefits can be achieved when local resources help limit the impacts of limitations on the transmission and distribution system and when they can provide backup power following outages (microgrids). These are very important applications and an increased research focus at EPRI and elsewhere. The challenge to make this reliability improvement objective a reality is effective integration of the distributed resources with grid planning and operations. This is the focus of the integrated grid efforts – make the distributed resources actually operate as an integral part of grid planning, operations and management.

2. *In your testimony, you lay out four items for policymakers to enable an Integrated Grid, including "assessment and deployment of advanced distribution and reliability technologies." What do you believe would be effective policies for this Committee to consider related to these reliability technologies?*

The Electric Power Research Institute is an independent scientific research organization for the public benefit, and as such we seek to inform policymakers by only providing information on scientific and technical topics resulting from our research. We do not provide an opinion on specific policy recommendations.

UNITED STATES SENATE
COMMITTEE ON ENERGY AND NATURAL RESOURCES

Hearing Questions for the Record
The Honorable Lisa Murkowski

Advancing an Integrated Electric Grid
Tuesday, March 17, 2015

<u>Questions for Dr. Howard</u>

1. *Please describe what ancillary services are and how they contribute to grid reliability. How can distributed energy sources contribute to these needed grid reliability services, and under what time frame?*

The FERC definition of Ancillary Services includes Scheduling and Dispatch, Reactive Supply and Voltage Control, Frequency Regulation and Frequency Response Services, Energy Imbalance Services, Spinning Operating Reserves and Supplemental Operating Reserves. These are all services that are needed to support the transmission of energy from resources to loads while maintaining reliable operation of the transmission system.

With the increasing application of renewable resources that have variable and intermittent characteristics, there is a growing trend to define additional ancillary services that are being lumped into a general category of flexibility services. This can include balancing services for renewable variations, ramping services to provide support when fast changes occur in generation and fast frequency regulation services that may be necessary based on these fast changes in generation.

While many of these services are often associated with traditional "spinning" generation, Distributed Energy Resources (DER) can also contribute to many of these services. For certain services and certain DER types, the technical capability may already be present, and the system may already be able to leverage this capability; for example, distributed energy storage provides frequency regulation in PJM. If not already available, technical capabilities can be required through interconnection standard for some services. For example, the ability to provide voltage control from inverter-based distributed generation is well understood but not commonly utilized. Similarly, communication and control standards can allow the system to leverage DER to provide certain services. Finally, there may be some situations where some types of DER may not be able to provide certain services. For example, black-start capability is likely not possible from rooftop solar PV.

2. *You recommend that "the existing and emerging technologies must adapt to a future state to make the most of current power system investments" which will include "enabling central generation to perform flexibly as well as variable generation to contribute toward system capacity and ancillary services." What technologies have emerged that will make the adaptation you call for more likely? How widely are these technologies being deployed today? Are they being deployed at the necessary rate? If not, why not?*

The future state mentioned is a state where the power system is more flexible, connected and resilient. Emerging technologies include various forms of energy storage, more intelligent control of distributed generation (smart inverters), a variety of new and existing transmission and distribution technologies, such as High Voltage Direct Current (HVDC) or Flexible AC Transmission System devices (FACTs) and demand side technologies that enable customer side participation in the grid (such as smart electric vehicle chargers and grid interactive heat pumps). They also include grid operation tools that can interface with both existing and emerging resources.

Existing resources will also adapt to the future state. This includes, as mentioned, increasing flexibility from central station generation. EPRI has a number of research initiatives identifying opportunities to obtain increased flexibility from central generation, including gas, coal and nuclear generation. Currently, many generation resources are being retrofitted to provide increased ramping ability and minimum generation levels, and new resources are being installed that offer significantly quicker start times, ramp rates and increased operating range. Wind and solar generation technologies may be able to contribute to certain Ancillary Services, through controlling power output when available. This technology, currently being used in a number of wind plants, is known as Active Power Control and allows for wind (and potentially solar in the future) to provide some of the required Ancillary Services that manage system frequency. Wind and solar can also provide reactive power control, and most transmission connected wind in the U.S. currently has this capability.

3. *What do you believe the biggest game changer will be for the future of the grid?*

The electric grid has operated very well for the last 120 years, largely in the same way since it was built in a hub and spoke approach. This design is turning out to be limiting with the more complex demand for power, growing penetration of renewable generation (especially at the distribution level), the increasing role of demand response as a resource, the potential for energy storage and new approaches to improve reliability including microgrids. With the growing importance of a wide variety of resources distributed at the edge of the grid, the need for new approaches for integration of these resources with the overall planning and operation of the grid is critical. Achieving these new grid planning and management approaches (including support of new market structures) will be a game changer. It will allow us to realize the full, dynamic capability of our system while making the best, most efficient use of the resources available to us today. In EPRI's Integrated Grid discussion, the line is used: "making local energy optimization an integral part of global energy optimization."

4. *What technologies are you most excited about and why? Also, what is the timeframe for development and at what cost?*

While EPRI is agnostic to specific brands and models of technology, we are most excited about the suites of technologies that will enable inexpensive energy storage, extreme energy efficiency and consumer interaction and real-time control with the grid. Consumers and markets will largely determine timeframe for development and cost.

> 5. *You testified that, "tremendous advancements that are now occurring in how electricity is generated, delivered and now personally managed" will lead to "changes in regulations and [the] management of the power system." Please provide examples of three "tremendous advancements" and share, if there is one, EPRI's projection of when each such advancement will be deployed on the grid at a scale that will materially affect grid operations. Also, with respect to each such advancement, please outline the benefits and, if any, burdens it presents for electric customers and for the grid collectively.*

Below are examples of important related advancements:

1. Continued advancements in wind generation technology – turbines that are now as large as 6 MW, advanced controls that can provide reactive power, voltage control and characteristics of inertia to support grid requirements. There are still issues that need to be solved – more accurate forecasting, variability that can require ramping resources (market mechanisms for getting this capability from other resources are now being evaluated in Ireland and California) and intermittency but we are learning how to integrate significant quantities of wind generation to the grid – Ireland, Hawaii, Texas and California are all examples of this.

2. Solar PV Generation and Smart Inverters. Solar generation has also made major advancements, especially in the cost of this generation. Continued cost reductions will make this a very important part of our generation mix in the future. Besides the cost, smart inverters are improving our ability to integrate the PV generation with the power system. These inverters can be controlled to better match the needs of individual distribution systems for voltage control, reactive power and performance during disturbances. As with wind, there are still limits on how much can be effectively integrated due to variations and uncertainty (the recent solar eclipse in Europe is a good example where Germany had to incorporate more than 8 GW of additional reserve generation). Forecasting is even more complicated for solar due to local variations that are difficult to forecast.

3. Energy storage advancements could be the answer for many issues still existing for both solar and wind. Costs are being driven down by consumer electronics applications and automotive. Utilities are exploring many different storage technologies for large applications.

4. Perhaps most important, consumers are becoming part of the equation for meeting peak demand requirements and managing variability and intermittency of these renewable generation sources. Many loads in buildings and the home can be managed in a flexible way – water heating, air conditioning and heating, pool pumps, smart appliances. Japan's promotion of Home Energy Management Systems (HEMS) and Building Energy

Management Systems (BEMS) after Fukushima resulted in shutting down all nuclear generation is an important example of customer resources. The New York plan for widespread integration of consumer resources with the grid is another example. New York correctly recognized that there are still tremendous challenges in both market structures and technologies in order to achieve this integration.

There is still a lot of work to do in these areas but the combination of these ongoing developments will result in a more optimized power system when combined with clean traditional generation that runs in an optimum way and supports the reliability and resiliency that is required for today's society.

6. *You testified that "the entire power system is changing at a fast pace, driven by technology and customer expectations." How prevalent are the changes you cite? How is technology driving the change? Is there good data on customer expectations? What role has tax policy played in the pace of change?*

Basic changes in power system technology are occurring throughout the system – more sensors, more communications, more automation, and better controls. These technologies are resulting in better visibility of issues, better management of the infrastructure and improved planning and operations. One example is the Electric Power Board (EPB) in Chattanooga. Partially through the ARRA grants, they implemented complete advanced metering for all customers, a fiber communication infrastructure that goes to every customer on the system and an entirely new automation infrastructure that restores the great majority of customers automatically whenever there is a fault on the system. This has improved reliability and has put the EPB in a position to take advantage of new technologies (customer resources, energy storage, and distributed generation) through their communications and control infrastructure.

Technologies including advanced wind turbines, solar panels, smart inverters, energy storage, home energy management systems, smart thermostats, electric vehicle chargers, microgrids, advanced sensors and many others will continue to change the way we manage and operate the grid. Tax policies may accelerate certain technologies by making them more cost attractive; however, technologies that can improve performance and economics of planning, operating and managing the power system will continue to find applications based on their merits regardless of tax policies.

7. *You note that "Technologies like smart thermostats are resulting in unique ways customers can manage their energy." What is the percentage of thermostats that are "smart"? How quickly are smart thermostats penetrating the market? What is the projected smart thermostat market share by 2020, 2025 and 2030?*

Smart thermostats, defined as having two-way communications capability and graphical user interface, currently represent a small but rapidly growing segment of the thermostat market. Estimates vary, but the best available sources estimate a current installed base of approximately 2.5 million smart thermostats in the U.S., which represents approximately 2% of the residential market share. Through 2020, smart thermostats are projected to grow at an annual rate of 40%, reaching about 11% of U.S. residential market share by 2020. If growth were to continue at that pace, smart thermostats would reach 50% market share by 2025.

Longer term projections beyond 10 years are subject to even greater uncertainty. It may be equally plausible for smart thermostats to become the de facto standard technology for households with broadband access (which would be nearly ubiquitous by 2030, since 85% of U.S. households currently have broadband access).

8. *With regard to Distributed Energy Resources (DER), you testified that: "Customers, energy suppliers, and developers are increasingly adopting these DER technologies with the aim to supplement or supplant grid-provided electricity." Please provide statistics showing the rate of market penetration for DER and EPRI projections, if these exist, for DER market share or penetration for each of the following years, 2020, 2025, and 2030.*

EPRI uses statistics on deployment of DER technologies that are available from a range of industry associations and third-party consultancies (e.g., the Solar Energy Industries Association [SEIA], the American Wind Energy Association [AWEA], the Energy Storage Association [ESA], Bloomberg New Energy Finance, and GTM Research, among others). EPRI also leverages data from the Department of Energy's Energy Information Administration (EIA) that portrays historical trends as well as annual energy outlooks for some DER technologies out to 25 years.

In general, DER encompasses a number of different technologies – including solar PV, wind, energy storage, demand response, fuel cells, and diesel/natural gas generators, among many others. While the outlook for these individual technologies varies (and are sometimes linked), forecasts indicate that nearly all will experience considerable growth over the next 15 years. One technology, solar photovoltaics, offers representative insight into the growth of "edge-of-grid" distributed energy resources. Per the graph below, sourced from SEIA, cumulative installed PV capacity has risen from roughly 2.5GW to nearly 20GW in the five years spanning 2010 and 2014. Moreover, expectations are that cumulative PV installations will roughly double in the next two years. As a result, the percentage of total U.S. generation that comes from solar PV will have increased from 0.1% to 1.6%. The longer term outlook to 2020 and beyond is also a continued increase in PV installations given falling price points, emerging financing arrangements that are increasing access to a broader swath of the consumer population, technical advances, and the segment's progressive commercial maturity.

As a point of reference, local generation today amounts to roughly 10% of U.S. energy, according to EIA.

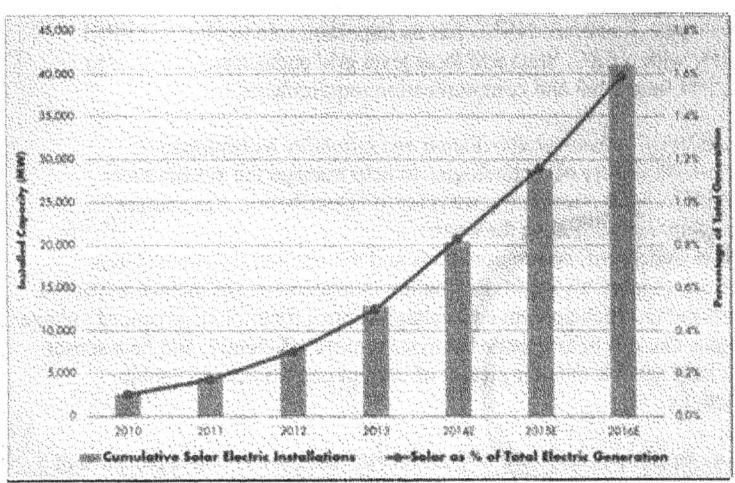

Source: Solar Electric Industries Association, 2014

9. You stated that "EPRI views the following four items as important considerations for policymakers to enable an Integrated Grid:
    1. Interconnection rules and communication technologies and standards;
    2. Assessment and deployment of advanced distribution and reliability technologies;
    3. Strategies for integrating DER with grid planning and operation; and
    4. Enabling policy and regulation."

a. Do you believe that the FERC should establish interconnection rules?
EPRI's role is to research and advise on the technical value of interconnection standards. We do not offer opinions on the institutional source for such rules.

b. What are examples of "communications technologies and standards"?
Communication systems involve many layers and many associated standards. In the area of an Integrated Grid, example standards include:
- IEC 61850-90-7 (standard function definitions and data models for distributed resources)
- DNP3 AN2013-001 (application layer communication protocol for connectivity to distributed resources)
- IEC CIM 61968-970 and MultiSpeak 5.0 (Enterprise integration standards for managing aggregations/groups of resources)
- CEA 2045 and other standards to make it possible for equipment and appliance manufacturers to build standard products that will be compatible with a wide variety of options for communications and integration

c. Who should set standards for communication technologies? What is NERC's role?
As interconnection rules progress to include requirements for smart functionality, this smart functionality will depend on communication interfaces and protocols that are

386

required to support these functions. NERC has an important role to play and EPRI collaborates closely with NERC. State and local level grid codes can also serve to effectively document functional and communication requirements.

d. What are examples of "advanced distribution and reliability technologies"?
This can include a wide variety of technologies to help manage the system more efficiently and reliably. Most commonly, we refer to "Distribution Automation" technologies that allow reconfiguring the system automatically to improve reliability. This includes smart switchgear, communications and automation functions. Other technologies that can support advanced distribution management can include advanced metering (supports outage management and voltage control functions), advanced voltage control technologies that can help operate the system more efficiently, and new sensor technologies that help provide a better picture of current conditions and enable optimization functions.

e. What is the role for the private sector with respect to "assessment and deployment of advanced distribution and reliability technologies"?
The private sector continues to develop new switching devices, control solutions, sensing technologies and control strategies to support Distribution Automation and advanced distribution management. These innovations focus on overcoming the key challenges for deployment which include:
- Minimizing the up-front investment
- Minimizing the ongoing O&M costs
- Maximizing the reliability improvement
- Enabling reliable and cost effective communications to a wide geographic area

What is the role of the federal government?
One of the key challenges that the federal government can impact is the need for reliable, secure and cost effective communications solutions to support advanced distribution solutions. Presently, utilities are deploying a variety of public and private solutions using licensed and unlicensed frequencies in a variety of architectures. Deploying advanced distribution solutions is particularly difficult for utilities covering large rural territories, where communications solutions often become a barrier.

f. What are examples of "strategies for integrating DER with grid planning and operation"?
Strategies to integrate DER with grid planning and operation can be implemented at both the bulk system level and at the local distribution level as well as at an industry-wide level. Developing an understanding of how Transmission System Operators (TSO's) and Distribution System Operators (DSO's) can develop strategies to coordinate with each other effectively is one key area in which EPRI is supporting the industry to enable DER contributions to system needs.

An example of a strategy to include DER in planning at the distribution system level is the adaption of planning processes to include a determination of the capability of an existing local network to integrate DER using the hosting capacity method, mindful of the interaction with the bulk system [Chapter 5, IG Benefit-Cost Assessment].

At the bulk system level, one example is to ensure that planning processes, such as generation and network reliability evaluation, are adapted to consider the capabilities that DER could provide to enhance reliability within the limits of local networks. EPRI's Integrated Grid Benefit-Cost Framework report highlights many of the strategies needed to realize an Integrated Grid [Chapter 7, IG Benefit-Cost Assessment].

Finally, new strategies are required to ensure transparency and effective communication regarding the approach taken to plan and operate the power system with DER for customers, regulators, utilities, reliability organizations and system operators. The implementation of the Integrated Grid framework itself is one strategy to support this goal.

EPRI is currently conducting a range of projects to develop the tools needed to enable planners and operators to effectively transition to planning and operating practices that reflect an Integrated Grid.

g. What is the role of the States with respect to such strategies? What is the federal role, if any?
The Electric Power Research Institute is an independent research organization for the public benefit, and as such we seek to inform policymakers by only providing information on scientific and technical topics resulting from our research. We do not provide an opinion on specific policy recommendations.

10. What is the effect of increased penetration of DER on consumers who do not install DER?
From a technical standpoint, utilities utilize screening practices to identify how much DER can be accommodated without degrading grid performance. In cases where adverse issues are found to be caused by DER, grid upgrades are performed in order to maintain power quality and reliability for all customers.

UNITED STATES SENATE
COMMITTEE ON ENERGY AND NATURAL RESOURCES

Hearing Questions for the Record
The Honorable James Risch

Advancing an Integrated Electric Grid
Tuesday, March 17, 2015

Question for Dr. Howard

1. *We hear plenty about how modernization via new Smart Grid technologies increase the attack surface of the grid, making it easier for adversaries to find ways in to critical operational networks and systems. Yet modernization also includes the introduction of new sensors and greatly increased intelligence in the grid, presenting us with the opportunity to enhance security situational awareness. Do you see the nation taking advantage of this opportunity, and if not, what new courses of action would you recommend?*

Yes, the nation is taking advantage of this opportunity, but challenges still exist in addressing cyber security for the electric sector. Because of the constantly changing threat environment, utilities are implementing technologies to address new cyber security risks and vulnerabilities. Two examples are listed below.

In the American Recovery and Reinvestment Act (ARRA) of 2009, cyber security was a critical component used to select the various grant winners. One of the technology areas was wide area monitoring, protection, and control (WAMPAC). WAMPAC systems constitute a suite of different system solutions aimed at meeting various wide-area application requirements. The objective is to analyze operations across the transmission system and communicate information about disruptions or changes in power flow. This enables a more expansive view of the bulk transmissions system while providing dynamic operation details. This information is used by operators to better assess and visualize system operation states and is important to the reliability and security of power systems.

Another direction is the development of Integrated Threat Analysis Centers (ITACs). With ITACs, utilities are beginning to evaluate cyber security threats to their communication networks in a way that integrates this information with other traditional information about equipment health status and power system status. Utilities have begun to integrate this information into a more comprehensive and consistent picture, for use by power system operators and communication system operators, to provide a system-wide view and to improve coordination of responses by the operational and information technology staff.

UNITED STATES SENATE
COMMITTEE ON ENERGY AND NATURAL RESOURCES

Hearing Questions for the Record
The Honorable Jeff Flake

Advancing an Integrated Electric Grid
Tuesday, March 17, 2015

<u>Question for Dr. Howard</u>

1. *In the recent EPRI report, The Integrated Grid: Realizing the Full Value of Central and Distributed Energy Resources, that you referenced, the German example of subsidizing distributed generation (DG) without considering DG's effect on the integrated grid serves as a cautionary tale. As you are well aware, here in the U.S. there have been federal subsidies for several DG technologies for over 20 years, and I am concerned that like Germany the U.S. has not planned for the impacts of widespread DG on the integrated grid. What policy considerations are there, particularly at the Federal level, for ensuring that the U.S. grids can operate in a robust manner while integrating widespread DG?*

While we believe the Integrated Grid Benefit-Cost Framework can provide a guide to necessary considerations for structuring and approaching policy and policy decisions – and while we strive to inform policymakers – The Electric Power Research Institute does not provide an opinion on specific policy.

EPRI is an independent research organization for the public benefit, and as such we seek to inform policymakers by providing information on scientific and technical topics resulting from our research.

Combined Questions from Senate Committee on Energy and Natural Resources

Senator Lisa Murkowski

Question 1: *Please describe what ancillary services are and why they are needed to maintain grid reliability. How can distributed energy sources contribute to these needed grid reliability services, and under what time frame?*

The U.S. Federal Energy Regulatory Commission (FERC) defines ancillary services as "those necessary to support the transmission of electric power (from seller to purchaser) to maintain reliable operations of the interconnected transmission system." Ancillary services include a range of supporting activities, such as load following, reactive power-voltage regulation, system protective services, loss compensation service, system control, load dispatch services, and energy imbalance services. Ancillary services support the primary services of generation, transmission, and distribution of electricity and as such, contribute to overall reliability of the power system.

Distributed energy sources present an opportunity to generate and store electricity through a more decentralized and diversified system that can enhance energy security and resilience. Distributed energy sources are options for individuals and communities to obtain clean, reliable energy. Distributed energy can include a greater mix of generation types that can potentially provide greater energy reliability. With distributed energy, load following can be done by a wider variety of storage types, especially in the case of highly intermittent generation sources. Distributed energy also includes demand response, which helps stabilize the grid on hot days and lower power costs. Advanced microgrids that are fully integrated within the distribution system may allow distributed resources to provide ancillary services at a more locally aggregated level.

Providing an exact timeframe for the integration of distributed energy resources is challenging, as the concept encompasses many different activities, each progressing on its own timeline. In general, distributed energy has been a growing trend since the 1970s when federal laws began to require utilities to purchase excess power from home generation. Adoption has accelerated recently as more homes and businesses have added solar generation capabilities. The trend is expected to continue as more consumers begin to transition to power generators through increased use of renewables and storage technologies, such as electric vehicles.

Rural communities that are far from central power generation have much to benefit from these changes, and should also be supported as early adopters of new technologies. For example, Argonne scientists have visited researchers at the University of Alaska to discuss how storage can be a key enabler for isolated communities.

Question 2: *In your written testimony, you highlight the three goals of the Grid Modernization Laboratory Consortium – 10 percent reduction in the societal costs of power outages; a 33 percent reduction in the cost of utilities' reserve margins, while maintaining reliability; and a 50 percent reduction in the cost of integrating distributed* *energy sources with the grid. What is the timeline to reach these goals and how much will it cost to get there?*

As will be described in the forthcoming Grid Modernization Multi-Year Program Plan, the GMLC intends for the three goals of the Grid Modernization program to be met by 2025. In the first five years, the national laboratories and their partners will develop and verify a suite of new grid capabilities and features that are foundational to a new grid infrastructure. As these capabilities and tools are developed in the lab-led R&D program, regional partnerships with private-sector stakeholders will be needed in 2020-2025 to translate research outcomes into practice and validate the expected value of new technologies and applications, resulting in a modernized grid.

The President's FY16 budget request for Grid Modernization is $356 million, which indicates the sustained level of investment in R&D needed over the next decade for success. While the up-front investment in this grid modernization effort is substantial, the GMLC expects the investment to be repaid through savings from a modern, more reliable, and secure grid. Current estimates indicate a future grid could result in savings of up to $7 billion a year, in addition to providing the United States with a needed upgrade to this critical piece of our country's infrastructure.

Electric Power Research Institute (EPRI) estimates a figure of $300B to $500B as the needed investment in the grid over the next 20 years, consistent with the goals of supporting grid modernization while maintaining low energy prices to the consumer, The grid modernization investment through the President's budget represents a modest cost within this envelope that will deliver through R&D the best possible outcomes.

Question 3: *In your written testimony, you state, "an underlying challenge is the grid's increasing use of microprocessors. While they provide many benefits, microprocessors make a system more complex and can slow it down." Please describe the research that needs to be done to ensure that adding more microprocessors does not hurt the reliability of the grid.*

The electric grid is tremendously complicated. The National Academy of Engineering refers to the grid as the "supreme engineering achievement of the 20th century." That said, it is a "just-in-time" system, with electricity being generated essentially as it is consumed. Electrical generation facilities and grid operators need to continuously match generation with demand. The more sensors and field controllers available, the greater level of control grid operators can exert over the distribution of electricity.

The architecture of the power grid has changed with the integration of more microprocessors. Previously, microprocessors were limited primarily to supervisory control and data acquisition (SCADA) systems and the main operations control systems. Today, microprocessors are multiplying across the system – at substations, on poles, and as part of control devices, sensing systems, and communication systems. As a result, decision-making is becoming increasingly distributed and decentralized. For example, distributed automation devices push decision-making closer to where actions occur – without operations center involvement. Some utilities are developing completely new control systems to deal with this proliferation of microprocessors. The increased complexity needs to be balanced against the gains in system visibility, control, and efficiency. Any proposed research must address the cost/benefit balance across these variables.

The increase in microprocessors can also increase the complexity of protecting the grid from potential cybersecurity threats. Traditional methods of cyber protection, such as patching security vulnerabilities, are not as effective for industrial control systems due to their long life cycles and the relative difficulty of updating the systems. These systems require new research regarding proactive cybersecurity defenses that are implemented from the initial design phase and address the multi-year lifecycle of deployed systems. In addition, much of the software development and hardware equipment manufacturing is conducted outside the U.S., making it difficult to fully protect the supply chain for computer equipment and software. Proposed research in this area should analyze risks associated with supply chain issues.

Analyzing this balance between complexity, speed, and risk is a central consideration of the grid resiliency research taking place at Argonne and other national labs. Analysis of the impact on performance and resiliency is a key consideration of any changes to the grid and is considered on both individual device and grid-wide levels.

Question 4: *What do you believe the biggest game changer will be for the future of the grid?*

One of the most profound changes coming to the future grid is a concept we refer to as "Grid Democratization." It is a true game-changer that has the potential to turn the traditional model of the grid on its head.

The U.S. grid was designed for centralized generation and one-way distribution where large power companies and cooperatives made decisions that affected millions of electricity consumers. Under the new model, much of that power will begin shifting toward the regional, local and even individual level as small-scale renewables become more common, while microgrids and local generation spread across the country. Energy storage will allow small groups or individuals to purchase or generate electricity at opportune times and store it for use or release back to the grid during peak need – and at peak price. New sensors and device controllers will empower consumers to make decisions on how and when they can most economically use electricity. When energy storage becomes efficient and reliable, energy becomes a currency; rather than energy being priced in dollars, dollars will be priced in energy. In short, a democratized grid gives individuals the tools and information to take control of their use of electricity.

An analogous way to think of this concept is the radical change our computing infrastructure has undergone over the past several decades as we have moved to distributed computing resources. Gone are the days when most of the computing power resided on large mainframes and users were relatively passive recipients. Massive advancements in technology have distributed that computing power to the individual, giving them much more control. Centralized computing and large servers farms are still critical to the system – as will be the case with large-scale electrical generation on the future grid – but much of that power has been shifted to the individual.

This democratization will fundamentally shift the role of the consumer from passive to active generator, buyer, and seller, sometimes referred to as a "prosumer." Some utilities already have hour-to-hour pricing and alternative power purchasing mechanisms, which allow homeowners to buy and store electricity when it is cheapest. This allows retail customers to change their behavior and alter their use and their cost of power, and it also enables demand response on the retail grid where the customer can further impact power use under utility control and signals.

Key technological mileposts on the path to grid democratization include:
- Distributed energy resources and renewables
- Grid storage
- Standardized two-way communication between plug-in vehicles, buildings, and distributed energy resources to enable local, community, and regional demand management.
- Financial transaction platforms

Question 5: *What technologies are you most excited about and why? Also, what is the timeframe for development and at what cost?*

The grid of the future will require development of hundreds of new devices and technologies, but there are a few that stand out as particularly exciting, and necessary, developments.

The effort of Argonne and other national labs to create the National Power Grid Simulator (NPGS) will be a key breakthrough that is necessary to develop the future of the grid. NPGS is a foundational step to allow us to thoroughly understand the status of our current grid and help develop a road map to a truly reliable and resilient grid of the future.

Once constructed, NPGS will allow us to see in real time how the grid functions, identify exposures and illuminate discussions on where additional resilience should be built into the system. NPGS will also allow us to create a national user facility where all types of grid stakeholders can assemble to conduct the research necessary to facilitate our drive for a new type of grid.

While the idea of the NPGS dates back to the 1990s, recent years have seen an increasing pace of development that would aid in its establishment. NPGS will require creation of a computer model with extreme fidelity and massive data sets about existing infrastructure. The data sets are nearly complete, but the development of the modeling and simulation will require approximately five years and an appropriate level of funding. The required talent to build the NPGS exits across several national labs. A model owned and operated by the national labs in conjunction with academia and private industry could take us a long way toward understanding the grid and events that have the potential to impact its functioning.

Determining the precise cost of NPGS is challenging, as the concept is still evolving. However, we would estimate the cost of development of the facility and required technology at approximately $20 million per year for five years. Once constructed, operation of NPGS would likely require an operating budget of around $20 million per year.

The other technology worth mentioning is energy storage. Radical advancements in energy storage have the potential to improve many key elements of our daily life, none more so than our electric grid. Great leaps forward in energy storage will enable increasing use of renewables and will help increase resiliency, so the grid is less vulnerable to major external events.

On a small scale, energy storage advancements will enable our vision of a democratized grid where homeowners can take control of their own power needs, buying or generating electricity, then storing it for later use or for resale to the grid when demand peaks. Energy storage is also a game changer for transportation, which also has profound implications for the grid. The path forward for energy storage is discussed in other questions, but it is a priority that occupies large portions of Argonne and our sister labs, as well as academic institutions and private industry.

Energy storage encompasses a broad swath of technologies, both developed and developing, with different cost and adoption rates. Some grid energy storage applications are already installed in California, New York, Alaska, Illinois, Washington, and Hawaii, with more adoption planned across the country. This storage uses existing technology, but research still needs to be done to refine these batteries and reduce costs to facilitate wider use. Batteries for the grid have a current cost of between $600 and $1,000 per kW-hour; we need to trim that to $100-125 per kW-hour to make storage competitive with peaker plants.

Over the long term, we need investment in breakthrough technologies to allow industry to progress to advanced Li-ion and beyond. Development time is three to seven years for advanced Li-ion and 10-20 years for beyond Li-ion. Success will require investment from government research and private industry for scale-up and production.

This multifaceted research effort balances the short-term needs to advance current technology and reduce costs with the long-term requirement to invest in breakthrough technologies that will lead the U.S. to the next generation of storage technology. Our goal, is that storing electrons in a battery – and retrieving them – will be as simple and as low cost as depositing and withdrawing money from a bank. This is not a short term goal, of course, but it will be one of the transformational achievements of this century.

Question 6: *Lithium ion batteries are the industry standard in energy storage – from cell phones to some grid scale storage. The Joint Center for Energy Storage Research, (JCESR) located at Argonne National Lab, is focused on transformative battery research. Can you describe more fully the goals of this research? What do you think the batteries of the future will look like? What part of your research is focused on keeping costs of energy storage systems low?*

The overarching goal of JCESR is to perform the basic research required to discover breakthrough materials and electrochemical systems to enable the commercial development of energy storage systems that will perform beyond the physical limitations of Li-ion batteries. Li-ion batteries were introduced to the market in the 1990's, based on research of two decades earlier, and their performance has been growing steadily at the rate of a few percent each year. On current rates of improvement, this technology will not be impactful in the mass market of vehicles, and on grid storage, for another two decades. From a fundamental perspective, it is believable that the ultimate electrical storage technology could have an energy density comparable to that of gasoline, which is about a factor of 100. As an aside, the best solar PV technologies exceed 30% efficiency, and LED lighting approaches 80%. So we believe that electrical storage is currently under-researched, partly because the whole periodic table of the elements is in play. JCESR aims to fill the front end of the nation's innovation pipeline with the knowledge required to ultimately move us beyond the performance limitations of existing battery systems. In this way, the federal government appropriately balances risk by funding a portfolio of research that will enable next-generation advanced Li-ion systems through DOE's Energy Frontier Research Centers (EFRCs) and Office of Energy Efficiency and Renewable Research (EERE) while also supporting research to deliver on the ambitious objective of discovering new materials for technologies that do not yet exist. We expect that the eventual market for electrical storage will be complex: lightweight but expensive solutions for consumer electronics; lightweight but cost-competitive solutions for vehicle transportation; low cost solutions for the grid. Each of these technologies will have to meet different specifications for energy density, rate of discharge, weight, and cost. It seems likely that multiple technologies will be developed for different applications. The pull from grid applications is the most recent, and it may be there that the most immediate gains from new technology development are to be had.

The specific goal of JCESR is to make the materials-level discoveries required for the development of battery systems with five times the energy density at one-fifth the cost of those benchmarked in 2012. The ultimate technological goal is to deliver early-stage prototypic battery cells. Note that the cost component is a core part of the goals of the program.

JCESR performs "techno-economic (TE) modeling," a process originated through EERE research on Li-ion batteries. In TE modeling, battery systems and manufacturing plants are developed using computer models, with industrial entities as partners in the model creation. TE modeling is used to project materials innovation into a deep understanding of pack-level manufacturing costs for a variety of batteries, including Li-S, magnesium-ion, and flow batteries.

TE modeling ensures that we incorporate industrial perspective on engineering and cost reality while conducting translation of materials-level discoveries into real systems. In this way, TE modeling ensures our materials research stays on target for both the performance and cost required by the original equipment manufacturers and end users. Some of our exploratory research is designed to answer fundamental questions of phenomenological behavior at the atomic level. But even that research is aimed at understanding materials behavior so we can identify and develop materials that enable higher performance at lower cost. It is an unusual experiment to be incorporating TE modelling at an early stage of research, but we believe it is important in order to focus quickly on deliverable solutions to the market.

An important goal of the JCESR approach is to dramatically increase the energy density and decrease the cost of energy storage systems while carefully coupling our research with the supply chain. JCESR aims to enable product development by industry for both grid and automotive applications. The ultimate goal is to discover and enable the disruptive technology that can be rapidly translated and commercialized by U.S. industry. JCESR partners include industry precisely for that purpose.

The batteries of the future must deliver the required performance at a cost that is competitive with gas peaker plants for the grid and internal combustion engines for transportation. There will be multiple "winning" technologies. Winners may include advanced, next-generation Li-ion batteries that contain a host of newly discovered materials and engineering innovations; flow batteries using novel concepts for the grid; lightweight batteries that are beyond Li-ion; and possibly even advanced lead acid batteries. The makeup of the battery will be immaterial to the end user, who will simply see the battery of the future that meets the performance and price goals necessary for their energy storage needs.

Question 7: *In your testimony, you discuss the importance of continuing to collect information about the grid through things like phasor measurement units and smart meter devices. Can you provide further details on how this increased information will lead to better grid reliability? What do you think the biggest challenges to managing this data will be?*

Data collection is a critical component of creating a smarter, more resilient grid. The collection of real-time data allows continuous monitoring of the grid, which enables operators to spot issues before they become major problems, and even anticipate issues before they occur. Phasor Measurement Units (PMUs) provide a good illustration of how this type of data acquisition can aid in creating a future grid.

PMUs increase data acquisition rates from once every five seconds to 30 times a second, and therefore offer a much more complete picture of what is happening on the grid. Currently, many important events cannot be observed at the slower time scale.

Analysis of PMU data helped identify many problems that were previously unknown, such as oscillations associated with improperly configured generators, generators interacting across the grid, and various transient events. Understanding and correcting these problems will help improve grid reliability.

PMUs are also being incorporated into high-speed control systems, where controls frequently need to operate in fractions of a second. This type of system is being explored for voltage and frequency regulation, with the potential for reducing the reserve margins on the grid by running it as a more finely tuned machine, enabled by the high-speed data provided by PMUs.

In order to realize the potential, however, it is necessary to connect these PMUs together into a national sensor network. The National Power Grid Simulator (NPGS) is a key element of realizing the potential of the PMU network. The NPGS would provide the wide area situational awareness needed to avoid an event such as the 2003 Blackout. PMUs offer a much more complete picture of what is happening on the grid. In Argonne's current work with ComEd we are finding transient features that have never been observed before. We are using ComEd's data logs to correlate these features with normal and abnormal behavior, then studying the behavior prior to events to see if there are ways to predict impacts and develop the ability to stop or mitigate the consequences.

Question 8: *How many registered electric vehicles are in the United States right now? How many were registered in 2010? What are your projections for the number of registered electric vehicles in 2020 and 2025? And on what grounds do you base your projections?*

Sales of electric vehicles have increased dramatically over the past five years as technology advances have made the vehicles more attractive to consumers. Approximately 300,000 Plug-In Electric Vehicles (PEVs) have been sold in the United States as of January 2015.

The PEV category encompasses battery electric vehicles (BEVs), such as the Nissan Leaf, that are fueled by electricity only; plug-in hybrid electric vehicles (PHEVs), such as the Plug-in version of the Prius, that are fueled by both electricity and gasoline; and extended range electric vehicles (EREVs), such as the BMW i3, which use an engine to drive a generator that recharges the battery once it is expended. Nonetheless, this remains a small market, though one which is growing.

Industry numbers indicate that in 2010, there were an insignificant number of these vehicles on American roads. However, analysis by Argonne indicates that EVs are on track for growth that significantly outpaces hybrid vehicles. For instance, the Prius hybrid, which does not use wall-plug electricity, was released in 1999, and the first U.S. PEVs were released at the end of 2010. By comparing sales of each for the first 31 months since their respective release, PEVs are outselling hybrids by a factor of more than two to one.

A sampling of industry projections indicate that American manufacturers are committed to plug-in vehicles. By 2020, Ford projects that from 10-25 percent of its fleet will be 'conventional' hybrids, PHEVs, or BEVs. GM states it will produce some 500,000 PEVs by the year 2017, including the Chevy Volt and the upcoming Bolt EV. Independent analysis also supports substantial growth in PEV sales. Navigant Research projects worldwide sales will reach 6.6 million units by 2020 and constitute 2.4% of global auto sales by 2023. Navigant expects U.S. consumers to purchase 2.7 million units by 2023.

Of course the sales of electric vehicles are influenced by factors well beyond their efficiency and price, particularly the cost of fossil fuels. The markets are variable internationally, but one might note that low energy prices in the USA provide a basis to support manufacturing to develop an export market to less favored countries who have significant problems in pollution and fossil energy costs.

Question 9: *In its FY2016 budget request, the Department of Energy has asked for $10 million for Transformer Resilience and Advanced Components, a program designed to advance the understanding of electromagnetic pulses (EMP) and geomagnetic disturbances (GMD) on transformers and other grid components. What new research is Argonne currently doing in this area? What research has Argonne done in the past in this area? How are other DOE National Laboratories participating in this effort?*

Serious geomagnetic disturbances (GMD) are predictable and have a risk to our infrastructure at the 1% level, which is not to be neglected. Electromagnetic pulses (EMP) are inherently unpredictable. It would be wise to have a grid resilient to both.

Argonne has conducted a number of projects to evaluate the effects of EMP and GMD on transformers and other grid components. For example, Argonne analyzed the impacts of a GMD event on the electric grid for DOE and presented results to federal, state, and local stakeholders at the Northeast Regional Energy Assurance Exercise in 2011. As part of this analysis, Argonne identified substations in the region most at risk from a major solar storm, as well as possible methods to replace failed equipment. In addition, Argonne is working with the Department of Homeland Security (DHS) to assess the potential impacts from commercially available EMP generators on critical infrastructure while evaluating various approaches to protecting these assets from EMP threats. In a previous initiative with DHS, Argonne examined the potential impacts of a solar storm on the energy sector and the use of an EMP generator on critical nodes in the natural gas pipeline network.

Looking more broadly across the national lab community, Los Alamos National Laboratory and Lawrence Livermore National Laboratory have developed models of EMP effects that can be used with an appropriate model of the grid. Idaho National Laboratory has run tests on the impact of the magneto-hydrodynamic effect, which is the major risk to transformers from EMP and GMD. This is but a small sampling of the work going on across the national laboratory complex.

Question 10: *Your testimony mentioned the need to protect grid infrastructure assets such as large transformers at substations. Please describe the Department of Homeland Security status on developing modular transformers. How far away are we from developing and implementing standardized transformers in key high-voltage classes? Are any such efforts currently underway in the Administration?*

While Argonne has worked closely with DHS on many research efforts related to grid protection, we have not been involved in the entire scope of DHS work in this area. I believe DHS officials would be best positioned to report on the entirety of their work on grid infrastructure. We would be happy to help facilitate such a briefing from DHS and contribute to this effort as we can.

Question 11: *Could you please elaborate on the role the national labs play in enabling innovation, specifically as it relates to energy storage solutions for microgrids and distributed energy solutions?*

One of our nation's most precious assets is what we refer to as our innovation ecosystem, which brings together national labs, academia and private industry in an environment that gives the nation a powerful competitive edge, providing solutions to major national challenges and fueling economic growth. This system is the envy of the world, which is why so many other nations are trying to copy our model.

Argonne is proud to be a part of this ecosystem, as we recognize the whole of the system is much more powerful than the sum of its parts in addressing national scientific priorities. One of those priorities is clearly energy storage and the role it will play in the future grid. By engaging our national labs, universities and private industry through this innovation ecosystem, we are poised for major breakthroughs that will continue to assure leading status of the U.S. in grid technology.

JCESR is a perfect illustration of how the innovation ecosystem can drive breakthroughs in critical areas like energy storage that will enable the future grid. JCESR brings together national leaders in science and engineering from five national labs, five universities and four private sector partners. By pooling the knowledge and resources of these partners, JCESR is poised to pave the way for energy storage beyond Li-ion.

While JCESR focuses on the next generation of energy storage, Li-ion is still evolving in important ways that will serve as a bridge to the future grid. Argonne is actively involved in that work, along with other national labs. Argonne discovered and developed a novel lithium ion cathode material, called nickel-manganese-cobalt (NMC), that has since been licensed to and commercialized by BASF, LG Chemical, and General Motors. NMC is one component in the battery cells LG manufactures for both GM and Ford. These same cells are incorporated into grid-level battery systems manufactured by AES, which has multiple commercial projects across the U.S. that will result in more than 600 MW-hours of storage on the grid.

Another example of innovation in grid research can be found in the Regional Resiliency Assessment Program (RRAP), led by DHS with support from Argonne. Through these regional assessments across the nation, Argonne is evaluating the risks of extended and cascading power outages due to natural and manmade hazards. Assessments conducted through the RRAP identify the most critical facilities in a community that require more resilient power. Distributed energy sources are considered as a resilient power option for a community or infrastructure system to obtain clean, reliable energy in the face of power outages. While RRAPs primarily focus on identifying the risks of power outages and the locations most likely to experience these outages due to a particular disaster, these analyses increasingly inform broader policy decisions and provide technical assistance to advance resilient power across the country.

Lastly, Argonne is working with an industry and university consortium to develop a first-of-its-kind cluster of networked microgrids in Chicago. Our microgrid research focuses on the development of advanced microgrid controllers and intentional sectionalization of the distribution system into virtual microgrids to achieve a self-healing resilient grid. We have also begun collaborating with the University of Alaska to better understand the scope of implementing microgrids within Alaskan communities.

Question 12: *You stated that it might take up to 20 years for current energy storage technologies to meet the desired improved performance parameters. Can you provide a better sense of what these performance parameters are? Can you provide more details on the types of technologies that are being investigated and how these differ from current-day technologies?*

Current storage technologies, such as Li-ion, have been improving steadily in efficiency and cost performance at the rate of a few percent a year. We need a factor of about 5 in both capacity and cost to enable a 300 mile range for an electric vehicle at a competitive price, or to provide on grid storage competitively with peaker plants. The current technologies are on track to get there in a few decades at the current rates of progress. This is not negligible – it has taken a cellphone from the 1990's brick model to a lightweight device (though that device is now mostly battery). The creation of energy storage solutions that meet the needs of the future grid is a substantial challenge that will require the type of cooperation between labs, industry, and academia demonstrated by JCESR. Achieving this goal will require a long-term effort and the best work of all participants.

The required performance parameters vary somewhat by application, but generally include higher energy density, higher power density, long life, and lower cost. In some applications, such as frequency regulation in the second-to-second time frame on the grid, higher power density might be the goal. Other applications, such as short-term storage for renewable energy production, might focus on higher energy density. In all applications, however, the predominant performance parameter is a combination of reliability and cost.

For the grid, an example of the type of technology investigated by JCESR is non-aqueous redox flow batteries. Aqueous systems are limited to 1.4 V, at which water is electrolyzed. By discovering materials that function in a non-aqueous solution or suspension, the combination of a higher operating voltage combined with materials that can store more energy in a smaller volume will result in lower cost and higher delivery of both power and long-term energy. For this application, the goal is cost reduction of the inactive components – such as tanks, pumps, and valves – that balance the potentially high cost of non-aqueous solvents.

Although the grid can accommodate larger systems, our TE models show that smaller batteries for the grid can often cost less, even when the active materials cost more than materials that use water. Although volume is often not a primary driver for the grid, as it is for transportation, it is a secondary driver because lower volume correlates to lower cost. Research shows that if we can achieve higher concentration of active energy storage materials in a flowable system at a higher voltage, then we can create a smaller system.

Another important example is the approach of the national laboratories to delivering higher-energy, more stable, lower cost materials for next generation Li-ion battery systems. Upon discovery of such materials, industry will be able to capitalize on existing and growing manufacturing infrastructure to move quickly to lower the cost of Li-ion systems that will compete in cost and performance with diesel generators and gas peaker plants.

The timeframe for these types of technologies is in line with past major scientific undertakings. Development of breakthrough technologies, from concept to full commercial production, typically takes 10 to 20 years. Silicon-based integrated circuits, cellular technology, personal computers, solar energy, and aviation technology all took one to two decades to move from laboratory concept to commercial production and adoption.

In energy storage, the history is similar. Li-ion was first identified at Exxon in the late 1970s, and the first commercial Li-ion battery was introduced by Sony in 1991. Full-scale commercial activity for automotive and grid applications has only occurred within the last decade. The full journey from discovery to commercialization to adoption is required to realize the true economic potential and societal impact of new, materials-based technologies.

JCESR is working to shorten this process through coordination and cooperation with industry; through cooperative research; and through carefully negotiated intellectual-property agreements. This type of public-private partnership model can be developed to expedite translation of basic research to adoption. This model can be employed while maintaining the important tenet of fairness of opportunity that is a cornerstone of the U.S. free-market system. JCESR is one such model of an effective public-private partnership that increases the interactions between national laboratories and industry without exclusive arrangements for intellectual property. We are actively exploring the opportunity to develop other consortia based on the same model to expedite innovation and commercialization.

Question 13: *In your testimony you mentioned the work done at the Joint Center for Energy Storage Research (JCESR), which is the DOE energy storage research hub housed at Argonne.*
   a. *Can you discuss the lessons learned from this consortium of national labs, academia and industry?*
   b. *Would you regard this consortium as a successful model for promoting technological innovation and useful collaborative efforts among stakeholders?*
   c. *How can the federal government best support such public-private partnerships, and what is your opinion on its current level of engagement?*

a. The creation, launch, and operation of JCESR has taught us several lessons that will be important moving forward with these types of public-private partnerships. First and foremost, early engagement of the partners – both traditional research partners like national labs and academia, as well as industry – is key to developing trusting relationships and the creation of important and compelling work scope. This early engagement enables the work to commence quickly once the funding is in place. A second lesson is that research with a clear technological goal is not for everyone, and that this is acceptable, and, in fact, encouraged. There must be a balance in the portfolio of research across the nation, from basic exploration to discovery that is mission-oriented. A third key lesson is that larger projects are required for larger problems. One positive side effect of larger funding modalities like JCESR or the EFRCs is that the entity can shift its research when discoveries dictate a change. These shifts are difficult in projects funded at smaller levels. Another important lesson is that a deep connection to industry is vital to keep our basic research focused in the right direction.

Finally, we have discovered that a team of complementary skills is a must. Physicists and chemists need to work with engineers; experimentalists with theorists and computational scientists; academic researchers with commercially-focused scientists and engineers. This mix is vital to create the productive friction between basic discovery and the urgency associated with the profit-motive of commercial needs.

b. We have found through our work leading JCESR that this model is extremely successful at fostering the type of collaboration necessary to address these large challenges. The combination of research philosophies from the labs and from industry creates a productive friction that results in more rapid innovation. Specifically, the combination of scientific and engineering disciplines creates pressure that results in new ideas.

Much of industry does not have the capability to perform long-horizon discovery research using tools like synchrotrons and supercomputers. Likewise, most basic scientists have long-developed careers focused on deep understanding of materials behavior down to the atomic level. The productive mix of basic scientists focused on scientific understanding with engineers focused on technology delivery results in discovery on a more specific timeline than if the scientists did not work with the engineers. This interaction also enables the engineers to witness and sometimes drive discovery by the scientists that would not be achieved by the engineers alone. It is not a perfect system; sometimes the academic-oriented scientists are frustrated with the immediacy of industry's goals, and sometimes the industrial partners are frustrated with basic discovery's deliberate methodology, but the mix truly drives productivity when aiming research toward a specific use.

c. Based on our ongoing success with JCESR, it seems the federal government clearly has a vital role in effectively fostering these types of collaborations. The federal government can support such public-private partnerships by allowing for exploration of new ways to develop legal structures, specifically structures that encourage private investment and participation that better capitalizes on the existing public investment in research. The federal government could also help by rewarding these types of partnerships with funding from various federal sources.

A pair of recent studies from Brookings and the American Academy of Arts & Sciences may be helpful in consideration of this question. The reports, which discuss ways to support science research and the national labs, are available here:

http://www.brookings.edu/research/reports/2014/09/10-national-labs-andes-muro-stepp

https://www.amacad.org/content/Research/researchproject.aspx?d=1276

Question 14: *It is well known that the national labs play an important role in the development of innovative advanced reactors that can potentially offer clean, safe and secure future energy solutions. Can you please comment on the potential for these advanced nuclear reactor technologies increase the resilience and reliability of the grid of the future and the role of the national labs in realizing such reactors? Please identify barriers to the advancement and deployment of these technologies.*

Nuclear energy is a critical component of the nation's electricity generation capacity and will need to grow in capability over the coming decades, given that it provides unique benefits in resiliency and reliability that other sources of electricity cannot match. In terms of grid reliability, nuclear is the only source that provides consistent, reliable, and carbon-free base-load energy, which can offset variable-supply sources, such as renewables.

Nuclear energy has supplied electricity to the U.S. national grid since the late 1950s and at this time supplies about 20% of the overall national electricity supply, accounting for over 65% of the nation's carbon-free electricity. Nuclear energy has zero emitted greenhouse gases during operations. Nuclear is also one of the nation's most reliable base-load energy sources, as demonstrated by its availability during the polar vortex (cold wave) of 2013-2014, when other energy sources struggled to deliver needed electricity. The diversity of energy supply is critically linked to the welfare of U.S. citizens and national security of the country.

Nuclear technologies should be advanced to continue to provide a reliable source of electricity, especially as the grid becomes more dependent on variable energy sources like wind and solar. Advanced nuclear technologies are necessary to supply the grid with reliable, competitive base-load energy in a deregulated market.

Challenges to incorporation of advanced nuclear technologies into the grid of the future include low natural gas prices and regional distortions from tax credits and similar market-shifting mechanisms for other energy sources, which hinder nuclear from competing on a level playing field based on its technical merits. These factors, if left unchecked, could create a barrier to advanced reactors and result in the shutdown of more existing nuclear power plants across the country. This issue is exacerbated by the rules in use by the electricity Regional Transmission Organizations (RTOs) and is more pronounced in unregulated energy markets in the U.S. Preservation of current nuclear power plants serves as an important bridge to future advanced reactors that can provide transformational changes to the effective use of nuclear fuel resources, the minimization of nuclear waste, and other beneficial attributes of an advanced energy system.
In the near term, nuclear plants would benefit from serving purposes other than supplying electricity to the grid to enhance their benefit to the nation. For example, nuclear facilities that use a fraction of the heat or electricity produced for other revenue-generating purposes, such as water desalination, hydrogen production, or process heating, are currently being investigated. The larger question of energy storage during non-peak hours for release during peak demands is also being considered as a way of making nuclear energy more competitive. At the same time such storage could be used to reduce variability on the electricity grid from renewable energy sources. In the long term, advanced nuclear systems that provide the capability to greatly increase fuel utilization and reduce nuclear waste generation are currently being studied and technologies are being developed.

National laboratories – including Argonne, Idaho National Laboratory (INL), Oak Ridge National Laboratory (ORNL), and others – have a crucial role in helping create the necessary components to make current-generation nuclear systems competitive in this environment. INL is leading an effort to create hybrid nuclear systems that combine power production with other missions for current-generation nuclear power plants. INL and Argonne are leading national efforts to develop next-generation advanced nuclear systems, such as a very high temperature reactor system that uses high thermal efficiency for conversion of heat to electricity and fast reactors that allow for transformational use of resources and minimization of nuclear waste. Current funding from the DOE Office of Nuclear Energy supports these efforts, but additional resources would be needed to enable rapid progress on the development and deployment of advanced nuclear plants. An additional challenge is the licensing pathway for advanced nuclear plants that could delay deployment, a challenge for the Nuclear Regulatory Commission (NRC) to develop the necessary regulations on advanced concepts within the current regulatory regime.

Question 15: *Recognizing that projections are difficult to make, please provide your best estimates for the timeframe for wide-scale market penetration of the most promising grid technologies.*

The most promising grid technologies at this time are arguably PMUs and those devices generally referred to as Smart Grid technologies

PMUs are currently on the grid in fair numbers, with more than 1,200 devices deployed, but their complete integration into utility operations is still a few years away. National standards are being refined, and companies are developing commercial packages and seeking demonstration projects at utilities to test these systems. Most of the current deployment of PMUs has been on the transmission grid, however there is growing use on the distribution grid, especially in places like San Diego and Hawaii where there is a large deployment of rooftop solar. With so much generation and storage moving to the distribution grid, the increased information speed and the situational awareness that PMUs provide will be crucial. Planners at utilities now recognize that the hottest day of the year may no longer be the most challenging day of the year for grid operations.

Currently, the transmission grid is reasonably well covered and penetration continues to increase, now at utility expense. This is due, in part, to PMUs being installed at an incremental cost during system control upgrades, which cover the cost of the connection system. Use of PMUs on the distribution grid remains exploratory, but could lead to the deployment of tens of thousands of PMUs in the next 10 years.

Smart Grid technology has expanded dramatically in the past few years by establishing two-way communication between the utility and the consumer. The ARRA helped deploy millions of smart meters, and many companies are building meters and selling applications. Current estimates indicate more than 50 million smart meters are deployed, or roughly 32 percent of all meters. Achieving full market penetration could still take years, but the technology has clearly made dramatic inroads.

Smart meters enable many associated technologies, including smart thermostats, appliance controls, and other technologies that enhance safety, security, and increase the democratization of the grid for a wider range of customers. These technologies continue to make inroads with consumers, and widespread adoption of many of these devices is ongoing.

Roof-top solar continues to grow rapidly. Smaller-scale storage is growing, particularly in California, with laws that require storage for utilities. The market for storage on the customer's side of the meter is growing rapidly, as well. Electric vehicles already impact the way the grid operates, and when vehicle to grid (V2G) is available, consumers will be able to use the EV's battery to sell power back to the grid or supplement their local power system. In rural areas, especially off-grid, as well as selected suburban areas, there is potential for rapid development of microgrids incorporating local small-scale power generation and storage.

Senator Joe Manchin III

Question 1: *Dr. Littlewood, some states and regions, including right here in DC, have mandated smart meters on everyone's home. Other utilities and state PUCs, including in my state, have sought a more selective deployment of these meters.*
In addition, there have been experiments where companies such as Google have allowed for home energy management systems that essentially skip over the utility meter and provide consumption information directly to the consumer through a chip in a thermostat or appliance.

Is it your view that we should widely deploy these smart meters or can we take a more selective approach? In other words, is the only solution an expensive smart meter on every home or are there benefits to a more selective application of these, as is being done in my state and others?

I certainly wouldn't presume to make broad generalizations about what is best for all states, areas or individuals. With such a massive and diverse piece of infrastructure as the national grid, which encompasses thousands of electricity providers and about 150 million customers, each situation will have unique requirements and will likely demand a somewhat tailored solution for achieving the goal of a modern, resilient grid.

I think what we can all agree on, in the broadest of terms, is that more information to empower consumers and more communication within the grid are positive developments that will be keys to ensuring our grid in the future.
Interestingly, the example in your question illustrates how the solution likely involves elements of all approaches. Google/NEST is a fascinating system that puts powerful information in the hands of homeowners. However, my understanding is that it currently requires a smart meter to serve as the primary link to the utility. NEST can monitor and control individual devices and systems, but a smart meter is required to monitor the entire home for energy usage and relay that information between the consumer and utility. In some places, smart meters are also used for providing energy pricing information to consumers for use in decision making. These meters can potentially provide other valuable information between utility and consumer.

As the future grid evolves, I suspect you will see more and more of this type of hybrid system, with small sensors and controllers embedded in many devices and appliances to monitor energy usage, coupled with larger information channels to the utilities. In fact, Argonne researchers are right now investigating ways to create tiny sensors that can be embedded into many types of devices. Armed with this type of information, consumers will be able to fully understand their energy usage and take full advantage of the democratized grid.

Senator Mazie K. Hirono

Question 1: *Hawaii is home to the Maui Smart Grid Project, which, with federal support, is evaluating new smart grid technologies to enable a cleaner, more efficient energy system on that island. Part of the project examines energy storage technologies as a way to increase reliability and lower the cost of electric service.*

Could you please elaborate on the research that Argonne is doing to develop energy storage systems, and explain when people in Hawaii might commonly see those systems in their communities?

The people of Hawaii and the rest of the country likely already have these types of energy storage technologies incorporated in their communities. Integration of storage technology is ongoing across the country in a variety of areas, but is being done so seamlessly that it often goes unnoticed.

While our DOE-funded Li-ion research is focused on enabling electric-drive transportation, materials discovered at Argonne are being used in battery technology for electronics and stationary storage, as well as automotive applications. The same LG-Chemical cells that are in the Chevy Volt – which contain cathode material invented at Argonne and licensed to BASF, LG and General Motors – are used in AES Energy Storage systems for the grid. While AES technology is not yet part of the Maui project, AES has been awarded a large contract by Southern California Edison to supply hundreds of megawatt-hours of storage to California. Similarly, Argonne has a history of working with the A123 company, which supplies Li-ion batteries into the Maui Smart Grid Project.

In the longer term, batteries that move beyond the performance of Li-ion will likely be delivered commercially in 10-15 years. The basic research and early-stage prototypes are being discovered now, and that discovery process itself will take at least five years. Commercialization that follows will take 5-10 years. This drive to prototyping and commercialization is a vital part of the nation's effort to enable breakthrough technology that moves the world beyond the limitations of Li-ion.

Such a portfolio of materials research balances risk for the investment. The Joint Center for Energy Storage Research (JCESR) research is not necessarily a higher risk in terms of discovering new materials, as new materials discovery and development is difficult regardless of the technology at the end use. However, with next-generation Li-ion, there is a small but growing manufacturing base in the United States that can be tapped into, so the delivery of new technology is relatively seamless from a manufacturing perspective. JCESR aims to deliver technology that reaches past the physically-limited performance of Li-ion, so it will likely deliver novel discoveries that require new manufacturing techniques. Hence, by performing research within a portfolio like this, risk is balanced and mitigated.

Question 2: *Which three goals would you recommend Congress or the Administration set to help spur modernization of the grid over the next ten years?*

For this recommendation, I would defer to the Grid Modernization Laboratory Consortium (GMLC), of which Argonne is a proud member. The GMLC has brought together the greatest minds in grid planning from our national laboratory system to create a cohesive and effective vision and plan for creating the future grid. Their efforts can be used to inform policy decisions by the Congress and Administration.

The GMLC has set an appropriately ambitious goal over the next 10 years of reducing the societal costs of power outages by 10 percent, cutting by 33 percent the cost of reserve margins while maintaining reliability and cutting in half the costs of wind, solar, and other distributed generation integration. I believe if the national labs work with academia and industry to drive the types of innovations necessary to achieve these goals, we will be well on our way to achieving the reliable and resilient grid we are all seeking.

Question 3: *Camp Smith in Hawaii is the headquarters to the U.S. Pacific Command, Special Operations Command Pacific and Marine Forces Pacific. Camp Smith is building new on-site generation sources and has hosted microgrid demonstration projects with the intent of becoming a microgrid to become the first military installation in the U.S. to have the ability to maintain all critical operations in the event of a power outage. This is part of the Department of Defense's Smart Power Infrastructure Demonstration for Energy Reliability and Security effort, also called SPIDERS.*

I am glad that Secretary Moniz established the Grid Modernization Laboratory Consortium in November, but meeting our nation's energy challenges will require as much collaboration as possible.

How much are you coordinating and cooperating with the Department of Defense on grid modernization, and do you discuss how to apply the technology and knowledge learned to civilian uses?

Clearly, finding a way to move our current grid into a more reliable, safe and resilient future will require cooperation across government, academia and private industry. The SPIDERS effort is a fascinating undertaking that illustrates that necessity.

DOD and DOE signed an MOU in 2010 to collaborate on energy challenges of interest and concern. That MOU really gained steam recently when it was named as one of the DOE's "Big Ideas" under the direction of Energy Secretary Ernest Moniz. DOD identified microgrid technology, and SPIDERS in particular, as a top priority for the MOU, which has led to great progress on the effort. While Sandia National Laboratories are leading the SPIDERS effort, all members of the GMLC are supporting, as appropriate, and following the outcome.

Microgrids are a promising technology that will play an important role in the grid of the future. Argonne has been researching microgrid technology for some time now, and no doubt new information and approaches will come out of the SPIDERS effort that will inform our efforts moving forward. Like all of the members of the GMLC, Argonne is committed to fostering this type of collaboration to ensure the energy future of our country.

Question 4: *As a country, we need to develop protocols for dealing with potential cybersecurity attacks on our grid. How should we do this? Who is best positioned to* *lead development of the protocols — private enterprise, the government, or both working together?*

Given the importance to the nation of our power grid and the increasingly sophisticated adversaries we face today, cybersecurity is a great concern to all who are working to create a more secure and resilient grid. This concern cannot be addressed by any one entity, but is instead a broad-based threat that requires an "all hands on deck" approach, incorporating the strengths of government, academia and private enterprise. In this area, Argonne is particularly interested in the Industrial Control Systems (ICS) that monitor and control the nation's electrical grid, as well as the traditional information technology systems that support operations. The nation's electrical grid is exposed to cyberattacks on a regular basis and must be protected from all threats. The communication to and from machines that operate the electric grid travels over both private and public networks. This improves operating efficiency but increases vulnerability. Focus on understanding the threats, vulnerabilities, and consequences is necessary to defend against external attack.

Protocol and standards development require a concerted partnership between private enterprise and the government. The willingness and ability to share cybersecurity information can be enhanced through economic incentives, liability protection for private industry, and making information available publicly. Greater information sharing will, in turn, increase capabilities to evaluate mitigation strategies, address emerging challenges, such as technology supply chain risk, and better understand and respond to complex and ever-changing cybersecurity threats.

The North American Electric Reliability Corporation (NERC) and the Federal Energy Regulatory Commission (FERC) have recognized the need to implement cybersecurity controls in the energy sector, particularly the electric subsector. For example, NERC Critical Infrastructure Protection Standards outline basic cybersecurity and physical security controls. The Nuclear Energy Institute and the Nuclear Regulatory Commission developed a set of standards for cybersecurity in the nuclear subsector that plant operators have adopted and implemented. While substantial opportunities remain to develop cybersecurity protocols for all critical infrastructure sectors, these programs demonstrate that public-private partnerships are well-positioned to lead effective standards development and promulgation.

The standards development process requires input and feedback from public sector stakeholders, industry groups and associations, equipment manufacturers, software developers, operators, system integrators, and other interested parties. The recommended approach is to follow proven standards development processes that include initial, open discussions to elicit concerns and potential solutions; development of a draft standard; an open comment period for stakeholder review; and follow-on forums to discuss and refine the standard moving forward. The process used by the National Institute for Standards and Technology in developing the Cybersecurity Framework offers a sound example of an inclusive approach that built on subject matter expertise from government, private sector, and academia.

Senator Ron Wyden

Question 1: *As many of you have testified, energy storage is the Swiss Army knife of the electric grid. Energy storage has the potential to solve many of the problems discussed in this hearing. Storage helps integrate renewables, it can help keep the lights on during extreme weather events, and it makes the whole grid operate more smoothly. It's no secret that I think energy storage is a good deal for the American public; the tool just needs sharpening to live up to its potential to reshape the grid. In particular, the recent progress to bring down costs and increase efficiency must continue. The barriers to storage aren't just cost, though, utilities also aren't used to working with storage yet, and regulations for storage vary from state to state. I continue to favor providing tax incentives for energy storage to encourage its deployment, but what else should be done to give energy storage the boost it needs? Please don't restrict your answer to the jurisdiction of the Senate Energy Committee, please also comment on useful policy solutions within the jurisdiction of the Senate Finance Committee.*

This is a critical question that will require input from all levels of government. As a national laboratory, Argonne is focused on providing the type of groundbreaking research that will enable the types of breakthroughs required to fulfill the promise of energy storage to transform our grid. While we do not advocate for particular policy initiatives, we have discovered through our work much information that could be helpful for policy makers as they consider this important challenge for the nation.

Insight into potential avenues to address this issue can be found in the policies and operation of the grids in California, Alaska, and Hawaii. California has developed policy specifically aimed to deliver storage to the grid, and that policy has already driven a 250MW procurement of battery systems by Southern California Edison. Much will be learned in the coming years about the efficacy of this policy and the performance and integration of the variety of energy storage systems being deployed in California. In Alaska, the grid was built essentially as a series of connected and disconnected micro-grids, or islanded grids, which could provide insight into the deployment of microgrids in other parts of the country. Renewable sources of wind and solar are being integrated into these islands with battery systems, which is another interesting development. In Hawaii, the penetration of renewable energy sources has already begun to transform the normal operation of the grid, requiring battery systems to enable the individual consumers to best capitalize on the democratization of energy production.

A careful study of Alaska and Hawaii, in particular, could help us understand both the creation and operation of microgrids and the disruption of "normal" grid operation. While both states have unique circumstances that differ from other parts of the United States, such as islands and remote locations, grid developments in Alaska and Hawaii could be studied to inform decisions on the future of the rest of the nation's grid. This study can then be used to make a reasoned case for appropriate policy.

The grid's vast nature and diverse construction are two final issues that will require study to illuminate policy decisions on storage. Nuanced decisions will be required in both technology and policy to accommodate broad regional differences. No single group has yet been assembled to provide a thorough answer to the questions this diversity poses. Triggered by the Joint Center for Energy Storage Research (JCESR) outreach activities to the broader community, Argonne is working with the Electric Power

Research Institute (EPRI), the Electricity Storage Association, and others to develop and execute a series of workshops across the nation to gather the organizations, people, and information required to identify the needs of each area of the country, so appropriate technology and policy can be developed.

Clearly, the policy discussions that enable transformative technology on such a critical piece of infrastructure as the grid will need to be conducted carefully and thoroughly. Argonne is happy to be amongst the many entities providing the science and engineering that will help illuminate those discussions.

Question 2: *What opportunities are there for the federal government to play a role in further unlocking the potential of demand response in America using the smart grid? Is it simply setting communications standards for appliances? Is it providing funding for pilot projects? I'm particularly interested to hear your thoughts on how to help grow demand response opportunities for homes and small businesses.*

As the United States moves toward a cleaner grid with significantly larger shares of variable resources such as wind and solar, we will see a dramatic increase in the need for grid flexibility to balance the intermittent supply. A range of technologies can provide this flexibility, including next-generation, fully-automated demand response. Demand response (DR) is enabled through the proliferation of intelligent consumer devices and appliances, along with the deployment of smart/advanced meters, estimated to have reached 50 million units or roughly 32 percent of all meters. Smart meter penetration in the residential sector is slightly higher than in the commercial sector. FERC estimates that retail and wholesale market demand resources currently provide the potential for approximately 55 GW of peak load reduction.

The adoption of common protocols such as OpenADR, Smart Energy Profile 2.0 and Green Button Progress has allowed progress in uniform standards for communicating DR pricing, signals, and usage data. However, with the plethora of new enabling technologies offered by companies entering the growing home automation and smart-home market, FERC's recent demand response assessment sees the proliferation of incompatible technologies as a near-term challenge that needs to be addressed. If we want to unlock the full potential of demand response, we also must address the slow adoption rate of time-based electricity rates, particularly by residential customers. Estimates put the number of households on dynamic rates at less than one percent. Fully and effectively engaging the consumer remains the subject of ongoing research. DOE's Grid Modernization Lab Consortium (GMLC), of which Argonne is a participant, recognizes the potential of demand response in the overall grid modernization effort. The GMLC promotes research in advanced distribution feeders and advanced grid planning and analytics platforms that would enable more demand-side options for consumers while contributing to leaner reserve margins in grid operations. The GMLC also advocates for improved energy management systems to enable self-awareness of building status, incorporate multiple sensor inputs and enhance building capability to participate in demand response, provide ancillary services, and facilitate widespread building-to-grid integration. This could be underpinned by establishing a federated test bed that would facilitate testing of various demand response scenarios and building energy management systems.

Utilities and state and local governments are playing a role in driving increased use of demand response, as well, by providing funding and incentives for pilot projects.

In the end, maximizing the potential of demand response in the future grid will require the efforts of interested parties at all levels, from the federal government to device manufacturers, utilities and state and local governments, ending with the most important stakeholder of all: consumers.

**U.S. Senate Committee on Energy and Natural Resources
March 17, 2015 Hearing: Electric Grid Innovation**

**Questions for the Record Submitted to Dr. Jeff Taft
from Senator Lisa Murkowski**

Question 1: Please describe what ancillary services are and why they are needed to maintain grid reliability. How can distributed energy sources contribute to these needed grid reliability services, and under what time frame?

Answer 1: FERC defines ancillary services as "those services necessary to support the transmission of electric power from seller to purchaser given the obligations of control areas and transmitting utilities within those control areas to maintain reliable operations of the interconnected transmission system." It identifies six broad categories of ancillary services that the electric system must supply to accomplish this, ranging from scheduling and dispatch of generation to system protection. Among these are a set of specific services, traditionally supplied by power plants that are required to meet requirements for stable and reliable operation of the grid. These are the most common use of the term "ancillary services." NERC and regional entities establish the minimum amount of each service that is required. From FERC's *Energy Primer*, these are:

- **Regulation** matches generation with very short-term changes in load by moving generator output up and down every few seconds to maintain system frequency at 60 hertz. Failure to do so can result in collapse of an electric grid.

- **Operating reserves** are needed to immediately replace a generator that trips off line unexpectedly. It is provided by units that can act quickly, rather than coal, or nuclear plants that are slow to respond. There are three types:
    - **Spinning reserves** can be provided by a generator that is operating with some unloaded (spare) capacity and capable of increasing its output within 10 minutes.
    - **Non-spinning reserves** come from generating units that can be brought online in 10 minutes.
    - **Supplemental reserves** come from generating units that can be made available in 30 minutes.

- **Reactive power** is the portion of power that establishes and maintains electric and magnetic fields in magnetic equipment, including rotating machinery and transformers. It is necessary for transporting AC power over transmission lines, and is consumed as power flows. The relative production of real and reactive power by generators is adjusted to maintain the proper supply of reactive power to prevent the grid from collapsing.

- **Black start** is provided by generators – such as hydroelectric facilities and diesel generators –that can start delivering power without any outside assistance from the electric grid to power pumps, fans, or other equipment. These are the first to be started up in the event of a complete collapse of the grid.

Basically what this all means that our power grids are operated in finely tuned dynamic balance at all times, with adjustments being made as often as every four seconds to maintain proper power flows, and to keep parameters such as voltage and system frequency within narrow limits.

U.S. Senate Committee on Energy and Natural Resources
March 17, 2015 Hearing: Electric Grid Innovation

To do this requires a number of capabilities that are in a sense auxiliary to the main business of generating power and delivering it to users. These auxiliary functions require various technical means and processes, and are the ancillary services. They may be supplied by utilities, or may be supplied by third parties, and in some regions are acquired for the grid through organized central power and energy markets.

Since many ancillary services involve adjusting power flows, it is possible for Distributed Energy Resources (DER: demand response, distributed storage and generation) to perform some of these services when coordinated in sufficiently aggregated quantities. This can include adding energy to the grid from distributed generation, reducing loads via demand response, or taking incremental amounts of energy from the grid and placing them into storage temporarily. It is feasible now for DER to perform some ancillary services and with increasing penetration of DER in the grid, and in fact this is gradually being allowed by wholesale market operators, Coupled with the development and deployment of advanced controls, power electronics, and storage, it is feasible for DER to play a significant role in providing ancillary services in the five to ten-year time frame.

Question 2: In your testimony, you highlight four technologies that are key to modernizing the grid – sensing and data analysis, high voltage power electronics, fast and flexible bulk energy storage, and advanced planning and control methods and tools. Which of these do you believe is the most readily deployable? Which one of these faces the biggest challenge to scalable, commercial deployment?

Answer 2: Each of the technologies in the list faces challenges in deployment. Advanced sensing and high voltage power electronics require deployment in very many grid locations and so represent deployment cost issues regardless of the cost of the components themselves. In addition, sensing requires adequate communications networking, which is not in place everywhere. Data analysis and advanced planning and control methods require new tool development and either high performance computing or distributed computing, depending on the implementations, and so face both development and industry adoption cycle time challenges. Storage and power electronics must continue the downward drive on costs and in the case of storage, open questions on how to value storage-based functions and services must be resolved.

Of the technologies in the list, storage is the most readily deployable, whereas advanced sensing and power electronics for distribution grids face the biggest challenges to scalable deployment due to deployment logistics and economics.

Questions 3: In your testimony, you discussed the importance of continuing to collect information about the grid through things like phasor measurement units and smart meter devices. Can you provide further details on how this additional information will lead to better grid reliability? What will be the biggest challenges to managing this data?

Answer 3: Grid sensing, which includes the operating state of the grid, but also includes monitoring the health and accumulated stress conditions on grid components, as well as factors affecting the grid and its parts, such environmental conditions and makes it possible to react automatically to grid events and conditions at speeds that keep up with the ever-faster dynamics of the grid. This is how, for example, a grid can automatically reroute power in the event of a line fault or change the loading on a transmission line or transformer in hot conditions so as to preserve or extend component life and therefore avoid or minimize the impact of an outage. Such sensing makes it possible to see conditions building up in the grid that, if left unattended, will result in wide or local area blackouts, and to take action in advance to mitigate the problem or to act quickly and automatically to restore service when it is lost. Monitoring of grid component health enables predictive maintenance and proactive management of asset life, both of which are important to maintaining overall grid reliability.

The biggest challenge to collecting this data is not the actual sensing, data transport, or data storage issues; it is the ability to extract useful information from the data flood automatically and reliably and then to connect that information to decision and control processes. Without the information extraction and the connection to decision and control, the latent value of the raw data cannot be realized. The value of the data is equal to the value of the decisions that use it (decision here includes the step by step control outputs that use such data as well). The chain of data measurement, analytics, and decision/control must be complete, otherwise sensing and data collection is just a cost with no benefit.

Question 4: The Pacific Northwest Smart Grid Initiative encompassed many states, utilities and the Pacific NW National Lab. In your testimony, you highlight the Department of Energy's cross-cutting initiative on grid modernization. What lessons learned from the Smart Grid Initiative could be applied to this effort moving forward?

Answer 4: The Pacific Northwest Smart Grid Initiative showed that it is feasible to have many non-utility assets partner with the grid in overall operation of an energy system. This means that homes, small businesses, commercial buildings, distributed generation, storage, microgrids, and other resources can be coordinated in a manner that is both automatic and beneficial to all concerned. The net result is greater flexibility in incorporation of diverse energy resources (such as the high penetration of wind in the Pacific NW), more efficient grid operation, and better opportunities for "prosumers" to participate in energy ecology. The project demonstrated how this can be accomplished in regions that do not have deregulated wholesale markets. Beyond previous demonstrations of transactive energy approaches, it made significant advances in the functionality of the control and coordination approach by incorporating a look-ahead signal. At the same time, the project pointed to the need for more rigorous approaches to both grid architecture and distributed control and coordination, so as to ensure smooth and effective operation of the full energy system, comprised of the utility assets and the consumer and third party stakeholder assets.

Question 5: In your written testimony, you dive more deeply into what the grid will look like in 2030. What do you believe the biggest game changer will be for the future of the grid?

Answer 5: The biggest game changer will be the conversion of the grid from a one-way electric energy delivery channel to an N-way energy network that supports broad access and innovation in energy production, delivery, and usage. The technologies listed in my testimony all support such an evolution, but additional changes will be needed in industry structure, power and energy

U.S. Senate Committee on Energy and Natural Resources
March 17, 2015 Hearing: Electric Grid Innovation

markets, grid control and coordination, building control and operation, and it will need the incorporation of fast flexible storage at multiple scales throughout the grid.

Question 6: What technologies are you most excited about and why? Also, what is the timeframe for development and at what cost?

Answer 6: I am excited about the combination of fast flexible storage, high voltage power electronics and advanced controls for grid functions. Regardless of how our grids evolve (and they will evolve differently in different regions and under differing organizing principles and emerging trends) this combination represents a new type of grid component that will become a standard part of new grid designs going forward. Its ability to support multiple uses simultaneously and to decouple source and load volatilities represent an important advance in power grid technology under any future grid scenario.

Question 7: Can you provide further details regarding the effect of information and communications technologies on the future of grid innovation? What are the challenges and opportunities in more fully integrating the grid with communications technology?

Answer 7: Information and Communication Technologies (ICT) have been a part of the grid one way or another for decades. However, during the era of what some refer to as Grid 2.0 (the last two decades where much effort has gone into what was known as the "Smart Grid", peaking with the ARRA work and still continuing to some extent) the convergence of ICT and the grid greatly accelerated. The result of that work was to improve the overall "intelligence" of the grid. Now as we move into what some are calling Grid 3.0 and most call grid modernization, the value of ICT is even greater that it was before. New requirements and emerging trends for the grid resulting in faster behavior have only added to the need for automated sensing, measurement, data transport, and data processing of various kinds. The transformation from centralized control to distributed control and coordination cannot take place without ubiquitous communication technology in particular. To fully integrate the extended energy system that includes the grid along with many assets not owned by the utilities and to enable prosumers to interact with the grid and each other, to foster energy innovation via the grid, ICT is a core technology and its continued convergence with the grid is not just important, it is assumed.

Grid communications is not without challenges. For wireless communications, spectrum allocation is often an issue. Some types of wireless communication that have been fine for smart meters have proven problematic for more demanding applications of distribution automation. Communication networks can also be channels for cyber-attack and so network security issues must be handled rigorously. The inclusion of the internet as part of a utility communication channel, while not typical, has been considered and would only expand the security issues for utilities.

Communications technology is a core enabler for grid modernization since utility assets are geographically dispersed. The introduction of DER and prosumer interaction with the grid can only happen with the use of communication technology. In addition, the combination of advanced transmission level sensing (phasor measurement units) with wide area networking offers an opportunity to significantly improve wide area grid protection and control by enabling

U.S. Senate Committee on Energy and Natural Resources
March 17, 2015 Hearing: Electric Grid Innovation

closed loop control for grid stabilization and for backup protection, as well as enabling adaptive protection that adjusts to grid conditions automatically instead of being fixed at installation time. Communications is the nervous system for the grid and increasing grid intelligence is not possible without it.

Question 8: During the hearing, you referenced the California ISO's goal of 1.3 GW of energy storage by 2020. What are the biggest challenges California faces in meetings this goal within five years?

Answer 8: The California Public Utilities Commission has set annual procurement targets for the three Investor Owned Utilities that add up to a total of 1325 MW by 2020, and these are broken down into three categories: transmission connected, distribution system connected, and behind the customer meter. The transmission connected portion is 700 MW, and the CAISO received over 2000 MW of storage projects (almost entirely battery) in its interconnection queue in the April 2014 request window. There will be another request window in April 2015, so more requests may be submitted. Last year CAISO clarified certain rules regarding storage interconnection to enable storage projects to enter the queue, which is officially a "generator" interconnection queue. The upshot is that the tariff allows treating storage as a generator that can have negative output, which greatly clarifies and simplifies interconnection for these projects (i.e., instead of viewing them as a combined generator plus load, which would then require them to participate in two interconnection processes). California also has a market participant construct called "Non Generator Resources" (NGR) that was specifically designed for storage - the resource can have both negative and positive output, and the economic dispatch algorithm keeps track of its state of charge so as to optimize its charging and discharging cycles to meet grid needs. Storage facilities using NGR can provide energy and ancillary services in the wholesale market.

There are several large challenges facing California in meeting this goal. From the standpoint of the CAISO, the challenges are how to pay for the new infrastructure and how to obtain maximum benefit to customers and the grid from the multiple services that storage can provide. This second point is important because most market models presume that an asset has a single function, something that is not the case for storage. Consequently, the issue of how market rules should be designed to include storage is of significant import.

From the standpoint of the vendors of storage components and storage based-services, the challenge is to understand the economics of storage as a component of the grid, and since storage has certain unique properties compared to other approaches for supply ancillary services, resolving how to value the multiple ways storage can be of use and determining who the "buyer" of storage services is are open questions. There are also questions about control of storage, integration of storage to the grid, as well as best location of storage units and whether therefore location should be factored in to the value of storage.

U.S. Senate Committee on Energy and Natural Resources
March 17, 2015 Hearing: The Electric Grid

Questions for the Record Submitted to Dr. Jeff Taft
from Senator Ron Wyden

Question 1: As many of you have testified, energy storage is the Swiss Army knife of the electric grid. Energy storage has the potential to solve many of the problems discussed in this hearing. Storage helps integrate renewables, it can help keep the lights on during extreme weather events, and it makes the whole grid operate more smoothly. It's no secret that I think energy storage is a good deal for the American public; the tool just needs sharpening to live up to its potential to reshape the grid. In particular, the recent progress to bring down costs and increase efficiency must continue. The barriers to storage aren't just cost, though, utilities also aren't used to working with storage yet, and regulations for storage vary from state to state.

I continue to favor providing tax incentives for energy storage to encourage its deployment, but what else should be done to give energy storage the boost it needs? Please don't restrict your answer to the jurisdiction of the Senate Energy Committee, please also comment on useful policy solutions within the jurisdiction of the Senate Finance Committee.

Answer 1: As an employee of a non-profit organization working at a Federal facility, I can only provide facts and technical insights. Certainly it is necessary to continue to drive costs of storage downward, but in addition, advanced controls to make storage behave in useful ways are also needed. Storage is unique among power grid components in that it can receive energy, release energy, and retain energy. For those classes of storage that are fast and flexible, many useful functions can be provided *simultaneously*, provided the necessary control systems are available. These functions include the well-known operations of smoothing out variations in variable energy resource like wind and solar, as well as providing electrical backup to support grid resilience. But they also include providing stabilization of the grid to help "ride through" small disturbances as well as ancillary services like support for system frequency regulation. Additional issues for storage include not just engineering integration of storage to grids, but for those part of the country that have organized centralized power and energy markets, they also include how to integrate storage into wholesale markets and how to value and regulate storage, especially storage that may be owned and operated by merchant or third party groups. At the distribution level, similar issues are emerging as regards the use of storage as a Distributed Energy Resource (DER) and a question there is whether storage will be treated as something that is dispatched from the system operator level and therefore subject to Federal regulation or whether there should be a distribution level means to manage storage so that it would be regulated at the State level.

Two major challenges facing deployment of storage are 1) the diverse ways in which storage is treated in various regulatory environments and markets, and 2) the absence of monetization for some of the many services can provide. Efforts to harmonize the regulatory treatment of storage across different jurisdictions would be helpful in creating a more predicable market and regulatory environment, facilitating deployment of storage

U.S. Senate Committee on Energy and Natural Resources
March 17, 2015 Hearing: The Electric Grid

where it adds value. Efforts to establish regulatory and market mechanisms to allow utilities and customers to receive financial benefits for the various use cases and functions storage performs would facilitate deployment where prudent. Enabling establishment of long-term agreements which recognize those benefits would provide financial stability for capital investment in storage deployment. Additional efforts to address industry acceptance, including codes and standards and multiple demonstrations (important due to the diversity of energy storage technologies), are also important to facilitate storage deployment.

Question 2: What opportunities are there for the federal government to play a role in further unlocking the potential of demand response in America using the smart grid? Is it simply setting communications standards for appliances? Is it providing funding for pilot projects? I'm particularly interested to hear your thoughts on how to help grow demand response opportunities for homes and small businesses.

Answer 2: Communication and interface standards are always important and much work has been done in this area for Demand Response. More remains, especially as regards commercial buildings, but the larger issues that require attention relate to how demand response is integrated at the functional level with grid operations, how buildings and homes can operate in a more nearly peer-to-peer manner to exchange energy and energy services, and how distribution grids in particular can become broad access networks that support innovation in energy services such as demand response, not just to aid grid operations, but more broadly to enable general energy transactions among interested stakeholders, including the owners of homes, businesses, and commercial facilities. In this regard, the distribution utilities will pay a key role that likely will differ somewhat from the one they play now. Instead of being one-way electricity delivery channels, they may become more general electricity network operators, facilitating the use of their networks for many kinds of energy transactions. Doing this will require changes in distribution grid infrastructure, and also changes in the roles and responsibilities of distribution utilities. The implications of these changes have impact at the bulk system level as well, so that infrastructure and especially control system changes will need a new architecture to guide investment.

Some states are working on how to create markets for DER, including demand response, so that it is possible for entities of any size to participate. Such an approach has potential for broadening the participation of small business and homes in demand response by enabling them to do more than just respond to variable electricity rates, but to become suppliers of valued services to the grid and to their neighbors.

One of the roles that DOE can provide in this area is the fostering of an overarching architectural vision for the grid, to be developed in concert with stakeholders of all kinds. The unique convening power of the Federal government and ability to marshal resources including the National Laboratories, programs such as ARPA-E, and organizations such as NIST, working in partnership with industry, the states, and regional stakeholders is the only means of tackling such a huge complex transformation of our energy systems. In

U.S. Senate Committee on Energy and Natural Resources
March 17, 2015 Hearing: The Electric Grid

addition, DOE can help by supporting development of new methods of control and coordination that go beyond interface standards to help ready the grid and buildings, appliances, electric vehicle chargers, rooftop solar PV, etc., to be partners in the overall operation of a true energy network, to the benefit of all stakeholders.

A general, advanced grid-customer interface and control/coordination strategy is required that maintains free will on the part of customers, yet encourages them to allow their homes and business to collaborate with the grid by offering a degree of flexibility in their usage patterns in exchange for an incentive or a lower energy bill. Such a user interface must support the variety of transactions that utilities might use to engage customers, yet provide a degree of generality and uniformity that helps utilities, customers, and technology vendors by simplifying deployment. It also must sort out which are the most advantageous opportunities for customer loads to participate in providing to the grid, since many if not all of the multiple operational objectives that storage could address could, at least in part, also addressed by flexible, grid responsive loads. Hence, a coordination scheme capable of engaging flexibility from demand, storage, and other DERs of all sorts should create a level playing field in which the most capable and least expensive assets are utilized. This is a big challenge that is just beginning to be addressed by Grid Modernization and some of the on-going efforts to construct transactive energy concepts consistent with the architectural vision.

The creation of a shared architectural vision, the development of advanced interface and control/coordination technology, combined with activities in the states around redefining roles and responsibilities of distribution providers, and the development of markets for DER, including demand response, will enable the country to find the best ways to enable the use of demand response appropriate to each region's unique goals and constraints.

U.S. Senate Committee on Energy and Natural Resources
March 17, 2015 Hearing: The Electric Grid

Questions for the Record Submitted to Dr. Jeff Taft
from Senator James Risch

Question 1: Grid resiliency is an important component to any grid modernization effort moving forward. Will outcomes from the smart grid demonstration projects such as the Pacific Northwest Smart Grid Demonstration project include actionable items for the industry to enhance grid resiliency as technology is deployed?

Answer 1: The Pacific Northwest Smart Grid Demonstration project outcomes will include significant lessons in how to enable non-utility assets such as distributed generation and demand response of buildings to provide services to the grid and effectively participate as partners in the management of the grid. So in addition to showing how a well instrumented flexible and interactive grid can enable consumer savings and reduction in peak demand on a grid, it also shows how they can assist in managing the output from renewable generation to help maintain grid reliability despite this new source of variability in the system. During the Pacific Northwest Smart Grid Demonstration project, Bonneville Power Authority tested the system against a half dozen resilience scenarios to determine how the system would respond. In all but one of the cases, the system detected and developed an appropriate response to the event. Going forward, the lessons from this project will be used to inform utilities about resilience measures and to improve the smart grid technology to address reliability and resilience events in a robust manner.

Question 2: In recent years, Bonneville Power Administration experienced periods of oversupply during high water flow and high wind events, leading to frustration over balancing the system load and dealing equitably with the energy imbalance. Much of the Pacific Northwest Smart Grid Demonstration project centered on this very issue -- how to improve renewable forecasting and integration. What lessons did we learn from that project that will improve the grid and allow for better management of these situations?

Answer 2: A primary lesson from the Pacific Northwest Smart Grid Demonstration is that it is possible to have customers and utilities cooperate in the management of the grid, thus providing new degrees of freedom to deal with events such as those you reference. This can be done with automation on both ends so that it does not constitute a burden on consumers to participate and in fact can provide tangible benefits to consumers as well as to the power system as a whole. By providing mechanisms to enable loads to adapt to varying energy supply conditions (as opposed to the original grid operational mode of generation always following load) the *overall* energy system can absorb variations that present difficulties under the more constrained 20th Century grid model. The project showed how loads could be engaged to adjust their operation up as well as down, which can help in oversupply situations that are of relatively short duration. The distributed storage resources in the project provide even greater flexibility and extend the duration of the response. The project enabled coordinated customer and utility actions on a 5 minute time frame, much faster than traditional markets, which can, when broadly implemented,

U.S. Senate Committee on Energy and Natural Resources
March 17, 2015 Hearing: The Electric Grid

enable reduction in grid reserve requirements, and more effective use of existing generation.

Question 3: In a system where regulatory forces and utility business models mean operational equipment is purchased to run for decades, (not 3-4 years as is typical in IT); modernization presents us with an unprecedented opportunity to deploy systems that are "secure by design." The DOE recently issue guidance for procuring secure energy systems, and its Grid Modernization Lab Consortium (GMLC) initiative includes a push in its multi-lab Security and Resilience group to design security into new grid components. Along these lines, do you see enough being done to take maximum advantage of the current push to modernize?

Answer 3: The issue of security (both cyber and physical) is top of mind in the utility industry and much progress has been made in improving security, due in part to the NERC Critical Infrastructure Protection (CIP) rules that have gone through multiple generations of increasing rigor. A problem that utilities have is that many legacy devices and systems do not have advanced cybersecurity capabilities and it is not practical to replace all of them in a short time frame, so there is a continually shifting mix of old and new elements in the grid that will continue for some time. At the same time, adversaries are innovating faster than our supply chain can provide secure solutions. Utilities (using NERC CIP as an operational standard) have difficulties procuring uniform security functionality from the supply chain. Much of the supply chain for the grid's control system infrastructure comes from multi-national corporations. Often, utilities are faced with challenges in getting the cost of cyber security investments to be covered by their rate base (controlled by state public utility commissions). This gives the adversaries an asymmetrical advantage. Grid Modernization is widening the necessary holistic view of the issue, and considering an all hazards perspective that includes nature as well as people. Threat Intelligence, Active Monitoring, Incident Response and Recovery are also being considered along with R&D topics related to securing components of the grid.

When considering the current push to modernize, it is important to consider how to influence the supply chain to better support these national interests and how to provide incentives for the utilities to adopt innovative cyber security approaches that are more nimble and able to adapt to change like the adversaries are. A side effect of the regulations from NERC CIP is that compliance has become the focus as opposed to more dynamic approaches that are increasingly needed.

U.S. Senate Committee on Energy and Natural Resources
March 17, 2015 Hearing: The Electric Grid

Question for the Record Submitted to Dr. Jeff Taft
from Senator Joe Manchin III

Question 1: Dr. Taft, your written testimony touches on our changing fuel mix, as well as retiring baseload generation. You also mention the specific technologies you see as critical to the future of the grid. Can you explain how and to what degree new grid technologies can address the reliability concerns we will likely face?

Answer 1: The 20th Century US Power grid was developed according to principles that included the use of centralized dispatchable generation, very little energy storage, load-following control, and design/operation for reliability achieved in part through the use of large operational margins. As the need to include additional economic constraints into grid operation developed, and as new functional requirements began to emerge, the methods of sensing and measurement and for grid management and control have changed. In fact, actual structure of the grid has begun to change, resulting in new grid characteristics and behaviors. The specific technologies I cited in my testimony are those that are most influential in addressing the changes now happening to the grid.

- *Improved sensing and measurement* is a key to knowing what is happening and is about to happen on the grid – without this it is very difficult to ensure proper operation. One cannot manage that which is not measured.
- Because grid data comes in ever increasing volumes, it is necessary to have *automated methods for extracting the useful information from floods of data* – hence the need for advanced grid analytics.
- Next, we must connect this information to decision and control processes. Existing tools for grid management do not address many of the upcoming changes and so more *advanced grid planning and control methods* are needed and this requires advances in high performance computing and control mathematics.
- Having advanced measurement, analytics, and control methods is still not sufficient; actual control mechanisms must be available to implement new grid operations. This is where *high voltage power electronics* comes into play – it provides new flexibility in controlling power flows and regulating voltages and other key grid parameters, and interfacing new devices to the grid to maintain reliability.
- Finally, *storage* is a crucial new tool for leveling out fluctuations that can be caused by non-dispatchable energy sources like wind and solar, and can also provide the energy buffers to assist the grid during stressful events ranging from component failures all the way to extreme weather events and attacks.

The technologies listed above are crucial to a grid that responds and adapts to changing conditions, including changed grid behaviors, component failures, and stress events and so will have a major role in ensuring continuing grid reliability.

U.S. Senate Committee on Energy and Natural Resources
March 17, 2015 Hearing: The Electric Grid

Questions for the Record Submitted to Dr. Jeffrey Taft
from Senator Mazie K. Hirono

Question 1: Camp Smith in Hawaii is the headquarters to the U.S. Pacific Command, Special Operations Command Pacific and Marine Forces Pacific. Camp Smith is building new on-site generation sources and has hosted microgrid demonstration projects with the intent of becoming the first military installation in the U.S. to have the ability to maintain all critical operations in the event of a power outage. This is part of the Department of Defense's Smart Power Infrastructure Demonstration for Energy Reliability and Security effort, also called SPIDERS.

I am glad that Secretary Moniz established the Grid Modernization Laboratory Consortium in November, but meeting our nation's energy challenges will require as much collaboration as possible.

How much are you coordinating and cooperating with the Department of Defense on grid modernization, and do you discuss how to bring the technology and knowledge learned to civilian use?

Answer 1: The development of microgrids is of great significance in the US utility industry and much of the advanced work has been done in DOD facilities. Camp Smith is the third phase of the SPIDERS Joint Capability Test Demonstration (JCTD) project; the first JCTD supported by the DOD, DOE, and DHS. The DOE Office of Electricity Delivery and Energy Reliability, under Assistant Secretary Hoffman, funded multiple DOE laboratories to participate in SPIDERS, including PNNL, under a joint DOE/DoD Memorandum of Understanding. The purpose of this memorandum is "…to identify a framework for cooperation and partnerships between the Department of Energy (DOE) and the Department of Defense…".

A major focus of SPIDERS is to transition the technologies and lessons learned to the utility industry. In addition to acting as the official Operational Test Agent for SPIDERS, PNNL is the Assistant Transition manager in charge facilitating the dissemination of SPIDERS knowledge where possible; this has included formal presentations at conferences, workshops, seminars, and discussions with utilities, such as the presentation given to engineers at Puget Sound Energy on SPIDERS.

As SPIDERS concludes in 2015 with the final Operational Demonstration (OD) at Camp Smith, additional opportunities to coordinate the results with industry will be pursued.

The SPIDERS program has been very valuable, and as PNNL has staff members who have been directly involved in this effort and who are able to bring the lessons learned from SPIDERS to the Grid Modernization Laboratory Consortium program. More generally, we also collaborate with DOD on cybersecurity issues.

U.S. Senate Committee on Energy and Natural Resources
March 17, 2015 Hearing: The Electric Grid

Question 2: Which three goals would you recommend Congress or the Administration set to help spur modernization of the grid over the next ten years?

Answer 2:

1. Foster the development of a stakeholder-driven overarching architectural vision for the grid of the 21st Century

2. Advance certain core technologies that will be needed regardless of how the grid evolves; these are:

   o Sensing and data analysis – electronic sensing and automated information extraction that will require new data collection and analysis tools

   o High voltage power electronics – adjustable electronics for controlling grid power flows to replace today's on/off electromechanical switches

   o Fast flexible bulk electric energy storage – can act as the buffer that evens out various power fluctuations and mismatches that can occur with diverse energy sources

   o Advanced planning and control methods and tools – new approaches using advanced control methods suitable for the modern grid that will require next generation high performance computing coupled with new control mathematics.

3. Support the evolution of distribution grids into broad access networks to support advanced functionality and energy innovation that can involve stakeholders of any size.